A Short History of
Circuits and Systems

A Short History of Circuits and Systems

Editors

Franco Maloberti

President IEEE CAS Society and University of Pavia, Italy

Anthony C. Davies

King's College London, UK

Published, sold and distributed by:
River Publishers
Niels Jernes Vej 10
9220 Aalborg Ø
Denmark

River Publishers
Lange Geer 44
2611 PW Delft
The Netherlands

Tel.: +45369953197
www.riverpublishers.com

ISBN: 978-87-93379-70-1 (Paperback)
        978-87-93379-71-8 (Hardback)
        978-87-93379-69-5 (Ebook)

Library of Congress Cataloging-in-Publication Data: 20th May 2016

Editors: Franco Maloberti and Anthony C. Davies
Title: A Short History of Circuits and Systems

Contents

Introduction

The IEEE Circuits and Systems Society comprises about ten thousand members distributed around the world. It covers many disciplines in the circuits and systems area and addresses global goals by conceiving and pioneering solutions to fundamental and applied scientific and engineering problems.

The history of electricity dates back thousands of years. It all began in ancient Greece when, about 600 years before Christ, Thales, a mathematician from Miletus, discovered that a rubbed piece of amber was able to lift small chips of wood. That discovery reminded Thales of the legend of Magnus (from whose name we have the word magnetism) whose iron-studded soles were stuck to a special rocky surface. After centuries the electrostatic and magnetic effects were studied and experimentally verified to lay the foundations of electrical circuits. The contributions of the initial steps came from disparate disciplines, Luigi Galvani (1737-1798) was a professor of anatomy in Bologna, Italy. One day he observed that electricity caused muscular contraction of the legs of a frog. He concluded that there is some sort of 'animal electricity'.

It was just necessary to wait a few years to see astonishing evolutions that lead to the discovery of the pile by Alessandro Volta (Chapter 2), the observation of Øersted on the generation of a magnetic field with an electrical current, the explanations of Ampère, the studies of Joule, Coulomb, Faraday and many others whose discoveries are described in Chapter 3.

Other milestones in our history are the active devices whose use permits amplification and switching and, accordingly more reliable processing of electrical variables. The history of electronic devices is a mix of scientific achievement and economical implementations as described in Chapter 4.

The above created solid foundations of what we classify now circuits and systems. Applications sprung from the resonant circuit and its consequences in radio engineering. They were facilitated by use of filters, a scientific area that interested so many scientists and produced exceptional research outcomes. The circuits and systems area evolved toward many other facets, including nonlinear circuits, chaos, neural networks, fuzzy logic, signal processing, image processing,

1

and data conversion. More recently the use of sensors and actuators made possible a substantial further expansion of the circuits and systems domain into measurement and control of non-electrical quantities. Moving into the digital domain allowed the building of extremely complex systems whose design was made possible by very effective CAD tools. All this is recalled in Chapter 5.

Chapters 6 and 7 retrace the steps in the foundation and development of the CAS Society. Chapter 8 honours many of the great scientists and engineers who made the circuit and systems field so important for electrical engineering.

We are grateful for the substantial efforts made by the many authors of parts of this book, whom we asked to prepare their material within a very short time scale, and who produced such a substantial outcome. We do hope that our readers will find it a fascinating and interesting insight into the historical roots of our Society, and that it encourage further study and deeper understanding of the history of this subject.

Franco Maloberti
Anthony C. Davies

Franco Maloberti received the Laurea Degree (Summa cum Laude) from the University of Parma, Italy, and the Dr. Honoris Causa degree from Inaoe, Mexico in 1996. He was a Visiting Professor at ETH-PEL, Zurich in 1993 and at EPFL-LEG, Lausanne in 2004. He was the TI/J.Kilby Chair at the Texas A&M University and the Microelectronic Chair at University of Texas at Dallas. Currently he is Professor at the University of Pavia, Italy and Honorary Professor at the University of Macau, China. He has written more than 500 published papers, six books and holds 34 patents. He is the Chairman of the Academic Committee of the Microelectronics Key Lab. Macau, China. He is the President of the IEEE CASS, he was VP Region 8 of IEEE CAS, AE of IEEE-TCAS-II, President of the IEEE Sensor Council, IEEE CAS BoG member, VP Publications IEEE CAS, DL IEEE SSC Society and DL IEEE CAS Society. He received the 1999 IEEE CAS Society Meritorious Service Award, the 2000 CAS Society Golden Jubilee Medal, and the IEEE 2000 Millenium Medal, the 1996 IEE Fleming Premium, the ESSCIRC 2007 Best Paper Award and the IEEJ Workshop 2007 and 2010 Best Paper Award. He received the IEEE CAS Society 2013 Mac Van Valkenburg Award. He is an IEEE Life Fellow.

Anthony C. Davies born Rainham, Kent, England, 1936. B.Sc(Eng) Electrical Engineering, 1st Class Hons (1963) Southampton University, MPhil (1967) University of London, PhD (1970) City University London. General Electric Co. in Coventry (1961-63): main work on Filter Design and PCM systems design. Lecturer (1963), Reader (1970) and Professor and Director, Centre for Information Engineering (1982) at City University London. Professor, Department of Electronic Engineering, King's College London (1990-99). Visiting appointments: One year as Lecturer, University of British Columbia (1968-69); One year as Visiting Full Professor, Purdue University (1973-74); One year Royal Society Fellowship, British Aerospace, Army Weapons Division, Stevenage (1987-88). Consultant to Admiralty Underwater Weapons Establishment, Portland, Dorset. Retired 1999, awarded title 'Emeritus Professor' at King's College London, Visiting Professor (2002-present), Kingston University, Surrey, England. IEEE services: Region 8 History Activities Coordinator, IEEE History Committee; IEEE Region 8 Director and member of IEEE Board of Directors (2003-2004); Region 8 Vice President of CAS Society (Jan 1998 to Dec 2001); He has served as chair of UKRI Circuit Theory Chapter (later Circuits and Systems) and of the IEE Professional Group on Circuit Theory and Design. Memberships and Awards: Chartered Engineer, IET Fellow, BCS Member and IEEE Life Fellow. CAS Society Jubilee medal, IEEE Millennium medal in 2000.

The birth of Electrical Science

Alessandro Volta

Alessandro Volta in a 1827-8 design by Roberto Focosi, engraved by Luigi Rados.

Few scientists had alive the regard, the honours, and the awards as Alessandro Volta, and only a few also have left such rich documentation, such as publications and manuscripts, of their works. Napoleon's esteem for him was so great that it is told that in 1803, when Napoleon visited the Bibliothèque Nationale saw in one of the rooms a trophy of bay wreaths, with the words *Au grand Voltaire*. After reading this, Napoleon erased with the nails the last three letters, so that one could read '*Au grand Volta*'. Coming now to last century Albert Einstein believed that the invention of the battery was the 'foundation of all modern inventions'. In fact the practical applications of Volta inventions have been so rapid and sensational as to conquer even the literary field. A little more than forty years after Volta death, Jules Verne wrote: "*Il est un agent puissant, obéissant, rapide, facile, qui se plie à tous les usages et qui règne en maitre á mon bord. Tout se fait par lui. Il m'èclaire, il m'èchauffe, il est l'âme de mes appareils mècaniques. Cet agent, c'est l'électricité*".

This chapter presents, after a Volta bio note, how the pile of Volta was conceived and developed in the century of his discovery.

Volta's life

Volta was born in Como in 1745, from a family of Lombard aristocracy linked to the church. Pupil of the Jesuits, later entered in a seminary, he rejected the priesthood and the legal studies and expressed deep interest to science that, up from his teens, he studied on his own. While very young Volta started reading important scientific works, in particular in electrology. When he was only twenty four, he published his first dissertation *"De vi attractiva ignis electrici ac phaenomenis independentibus,"* and in 1771 a correspondence memory *"Novus ac simplicissimus electricorum tentaminum apparatus,"* so showing that he already achieved a solid background and uncommon observation and experimental abilities.

In 1771, aiming at strengthening the scientific teachings of the university finalized to providing the role of *"Central State School"* to the University of Pavia, the Empress Maria Theresa launches two successive plans: one for direction, discipline and economy and a second for the framework of science.

In 1774 he was appointed regent of the public schools in Como; the following year, through competition, became professor of experimental physics in the Gymnasium of Como.

He made several travel studies in Italy and abroad and enhanced his scientific knowledge getting in touch with the top scientists of the time. In 1777, he was in Switzerland, Alsace and Savoy. In 1780 he went to Florence to visit the Royal Museum of Natural History and Physics and to conduct naturalistic investigations. In September 1781 he visited Switzerland, Alsace, central western Germany, Netherlands, Belgium and, in late December, he traveled to Paris. He stayed there for four months working with Laplace and Lavoisier. In 1782 he arrived to London, where he stayed until June. Later, in 1784, he went to Vienna and Berlin.

The historical Volta classroom (University of Pavia, Italy.)

In 1778, the Austrian governor, confered to Volta the Chair of Experimental Physics at Pavia University. The following year he was appointed Rector of the University.

In 1785, in order to accommodate the large throng of people attending Volta's lectures, Emperor Joseph II, Emperor of Austria, ordered the construction of a

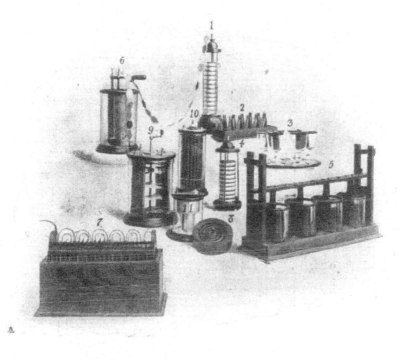

A series of the first century piles/batteries, 1st group: 1 Volta column pile; 2 and 3 the chain-of-tumblers versions of Volta pile; 4 Chevreuse column pile; 5 and 6 Wollaston piles; 7 Yung stuck troughs pile; 8 Hare spiral pile; 9 movement pile; Buchin-Tricoche pile, 11 Hulot pile. From 1899 Como e l'esposizione Voltiana, N. 4, June 10, 1899.

new classroom for the teaching of Physics. The architect, Leopoldo Pollack, was commissioned to do this work, which was completed in 1787. The original flat ceiling, damaged some time after 1828, was later replaced with the impressive shell-shaped ceiling that can be seen today.

The researches he did during the eighties, in meteorology, electricity, calorimetry, geology and chemistry of gases, allowed him to emerge as one of the most eminent scientists of Europe. In 1794, five years before the invention of the pile, the Royal Society awarded Volta the prestigious Copley Medal for his paper about the capacitor and electricity excited by contact of dissimilar metals.

In March 1800, Volta announced his invention of the pile to the Royal Society. In June of that year, Napoleon confirmed to the scientist from Como the role of professor of Experimental Physics at the University of Pavia.

In 1801 he traveled to Paris to illustrate the researches that led him to the invention of the pile. In the presence of First Consul Napoleon Bonaparte, he read the *"Memory on the identity of the electric fluid with the galvanic fluid,"* in a plenary session, at the *Institut National des Sciences et des Arts* that nominated Volta *Associé étranger* and awarded him gold medal.

Part of a letter dated March 20th, 1800, addressed to the President of the Royal Society, Joseph Banks, in which Volta announced the invention of the 'pile' or electromotor device, which was also included in the letter in the form of a sketch. The letter was later published in the Philosophical Transactions journal with the title: On the Electricity excited by the mere Contact of conducting Substances of different Kinds.

Despite the huge success he had the humbleness of a real scientist. On this point, it suffices to read what Volta wrote to his wife on November 7, 1801: *"Among many things that certainly please me, and which are too much flattering, I'm not so vain as to think I'm more than what I am, and to a life troubled by vainglory I prefer the tranquility and sweetness of domestic life"*. In the same year he was appointed as a member of the Council of Lyon.

In 1804, Volta presented his resignation from Pavia's University. Napoleon opposed this request with these words: *"I cannot agree with Volta's request. If the teaching activities tire him, then they have to be reduced. If he wants, he can teach a lesson a year, but Pavia's University would be hit at its core on the day I will allow such a distinguished name to be removed from the list of its professors. Indeed, a good general has to die on the battle field."* Moreover, the cross of the *Ordre National de la Légion d'Honneur* was awarded to Volta by Napoleon in 1805 and successively, in this connections, a few battle ships of the France Navy was named after Volta.

Other series of the first century of piles/batteries, (left) 1 Bunsen ; 2 Callan ; 3 Bunsen-Archerau; 4 Grove; 5 Maiche; 6 Arsonval; 7 Bunsen-Miergues; 8 Belloni; 9 Buff; 10, 11 and 12 Grenet; 13 Cloris-Baudet; 14 Kousmine; 15 Renault-Desvernay; 16 and 17 Trouv. (right) 1, 2 e 3 Daniell; 4 Callaud; 5 Daniell-D'Amico; 6 Daniell; 7 Olivetti d'Ivrea pour liquid pile; 8 American Daniell-Hill large surface; 9 Raoul communicating vases; 10 Meidinger flask; 11 Daniell-Trouvé with zinc and copper spirals. From 1899 Como e l'esposizione Voltiana, N. 7, July 1, 1899

Volta was also appointed Senator of the Italian kingdom in 1809, and was nominated Count in 1810.

With the collapse of Napoleon's empire, Imperial Austria regained control over Lombardy. In 1815, Emperor Franz I re-established the Order of the Iron Crown as an Austrian order and confirmed Volta as Professor of Philosophy at the University of Pavia, giving him the title of *Cavaliere dell'Ordine Imperiale Austriaco della Corona Ferrea*

In January 1819 he completed one last important task commissioned by the Austrian government: "*the reasoned report around the establishment of one or two institutions for the education of Engineers, Architects and Land surveyors*". Volta, secluded and shy man, the philosopher of nature in love with the sensible world, with great awareness, in a busy Lombardy who had been able to draw the best from the Enlightenment and the Austrian rigor, effectively dealt with the birth and development of engineering teaching. In a letter dated January 23rd 1819, Volta sent: "*To the Imperial Royal Government of Milan*" a rational report on the establishment of schools for engineers, architects and land surveyors: "*I hope, that I. R. Government will find herewith collected all the information required to establish this really important school. The proposed technical and practical study, which, in my opinion is quite frankly much better than those previously put forward; it can compete with the most well-founded Establishments of this type and complies with the current state of progress in physical and mathematical sciences. I strongly recommend it to the wise attention of the Government: and on this occasion I must repeat what I have had the honor to say on other occasions regarding the zeal, and knowledge of the*

Rector, Prof. Configliachi, to whom, after consulting other eminent Professors, I have principally entrusted the matter, given that it is a difficult but important issue."

In the same year, 1819, he retired to private life in his hometown Como, with his family. He dedicated the last years of his life to the religious practice and to pray, according to his education. He passed away on March 5, 1827 at age 82.

The discoveries and Inventions of Volta

A replica of Volta's pile. The original was destroyed in the fire of the 1899 Universal Exhibition; what shown here is a copy made in 1899.

Alessandro Volta taught at Pavia University from 1778, for 41 years. Mindful of the event, 100 years later, the University celebrated the anniversary of the Volta recruitment with outstanding events and initiatives. Among the initiatives, the most relevant for the scientific, cultural and social implications, were two.

The former was the delivery of an honorary degree to some of the world's most important scientists of the time. The sense of this initiative should be seen in terms of progress in electrology with a collection that, ten years before the Hertz's experiment, balanced action at distance (awardees W. E. Weber and L. F. Von Helmholtz) with contact action (awardees W. Thomson and J. C. Maxwell).

The latter was committed to an extensive presentation of the works, inventions, and instruments of Volta by the Dean of the Faculty of Mathematics and Natural Sciences and chair of Experimental Physics (Volta's former positions), Giovanni Cantoni. For the high reputation Cantoni enjoyed in the scientific world (he was called on to be a member of the most important academies and various international committee), it is worthwhile to present how the action and the role of Volta in the second half of the XIX century was perceived. Here, it follows a part of the speech in which he describes the sequence of steps that lead to the pile.

The first issue addressed was the methodological approach in which, in fact, Volta developed *"for the electric world"* the Galilean approach to research: *"after having found, by reasoning, a scientific truth he immediately proceeded to trial and re-trial by experimental evidence."* But this aspect was well described by Cantoni in the previous day's opening address: *"The concepts we are developing of the phenomena that nature puts in front of us, are always the result of the combination of what nature really says, and what our minds think to see: in shorter words, our concepts about phenomena always include a subjective part, in addition to the objective one."* In particular, he stated that *"Prior to Volta electricians most valuable, such as Franklin and Beccaria, talked of a special fluid or others, such as Euler and Aepinus, of ether, placing it as an agent of electrical phenomena, and treated it as something material, ... that now invading and now erupting from the bodies, regarded almost as a sponge for this simple matter, which triggered all special appearances observed in them. And the earth was now seen as an unfailing source, and now as an endless shelter of the electric substance; and often some of the bodies, under certain conditions, had to accommodate or provide it in very large amounts. ... "*

After this initial introduction, Cantoni described, with particular details, what he considered the most important and innovative Volta inventions: the electric condenser, the electrometer and finally the pile. The following are descriptions of these three inventions, mainly taken from the book of the celebrations.

The Electric Condenser

Cantoni considers that *"this wonderful discovery could not have been accomplished if his genius, really disposed to rigour on scientific investigations, had not firstly prepared a very refined method to reveal the existence of very weak electric charge, and, at the same time, an instrument able to realize exact measurements; in a word, if he had not first discovered the foundations of Electrometry which constitutes one of the most important titles of his scientific glory.*

Already in 1782, seeking to determine whether the atmosphere, even in clear and calm sky, had been in electric status, even when there was no clue also with the best electroscope of that time, he thought to benefit in a special way, of its electrophorus. With this one, the metal plate on which a thin layer of resin (a fourth of line), smooth and electrified at each point was put down, he carefully put over the shield (less wide of the plate), and he made it communicate, through a wire, with a sharp rod conductor of atmospheric electricity, while the plate communicates with the ground: after the first few minutes, took off his communication with the

The original early 19th century disc condenser kept in the Museo della Storia, University of Pavia.

atmospheric conductor, and, raising the shield, he had large electrical signs, though, the conductor itself alone did not give any sign of electricity. This thin drain electrophorus, or rather this thin concrete insulating, was called by Volta micro-electroscope, and even electricity condenser. This was one of the most useful among Volta inventions; as it levelled so many investigations of Electrology.

In the same memory of 1782, he also taught to build capacitors with solid semi-insulating (marble dry and very smooth, dry wood) sometimes naked sometimes covered by a thin paint, to which the shield shall directly apply, and that are good for cases in which the electric charges are so energetic, that, penetrating into the resin, would turn the condenser in electrophorus. And a more 'delicious' condenser was realized with another metal disc, flat and covered with thin layer of insulating paint (amber or lacquer), and a little more larger than the shield, useful only for cases in which the charges to be explored are very weak...".

In his note Cantoni indicates that: "*The most eloquent instrument used by Volta to declare the operation modality of the condenser, consists of a small equipment, which is stored in this physical cabinet, and responds much better than the so-called capacitor of Aepinus to prove the double electric induction, i.e. of the electric condensation*".

The electrometer and the first electrometry foundation

The Volta straw electrometer, built in the last quarter of the 18th century kept in the Museo per la Storia, University of Pavia.

Cantoni also said "*A few years later (1788) he studied with great foresight, and with rare luck, the topic of electrometry. The electroscope then in use, that of Cavallo, did not deserve the title of electrometer. In fact, the deviations of its tilting had any direct relation with the size of the electrical charges sent to the button of the electroscope, as Volta demonstrated by direct experience. He replaced the balls of elder pith, supported by metal wires, with two thin flakes, about two inches long and suspended to the hat by highly mobile rings, and surrogated the cylindrical or conical bell by a bottle with a square section by applying on one of its flat faces a little bow, made with strips of paper, concentric to the rings and subdivided in parts of fixed length ... Thus, the sensitivity was increased in relation to the greater extent of the two facing sur-*faces, *the small initial distance between them and the lower weight of the indices, and also the reading of the deviations, on the mentioned flat arc, was more sure, even when this deviation reached over 20 °. Now, what really mattered was to ascertain whether these deviations, at least within certain limits, are*

Alessandro Volta shows the pile to Napoleon. From a picture dated 1897 by Giuseppe Bertini (1825-1898), from the Volta Temple, Como.

proportional to electrical charges And he knew how to solve this sensitive issue, in sure manner, with various ingenious ways. ..."

Volta realized another way to assess the graduation by using cylindrical thin wire, which he previously studied, to demonstrate the high electrical capacitance. Given a charge to one of these cylinders, and verified the corresponding electrometric deviation, subdivided that charge into two or three parts equal to each other, by connecting the first cylinder with one or two others of equal capacitance, and then each of these, connected to the electrometer, produced a deviation equal to half or one third of the first one. He also experimented the subdivision of a given charge on two or three small Leiden Jars of equal capacitance, and had similar results.

He also considered to link the measurement of very weak charges to the voltage at higher and higher intensity, improving the thin straws electrometer, he prepared another one having larger and heavier straws, so that one degree of deviation of the latter corresponds to five degrees of the first.

Volta checked these performances through the artifices just indicated of the conductors and bottles of equal capacitance. Therefore, with the aid of these four instruments, he had the way of assessing the relative magnitude of very different charges (in a ratio of 1 to about two thousand).

Then adapting the electrometer with thin straws the condenser aforesaid, consisting of two metal discs of 2 inches in diameter, separated by an insulating thin layer of paint, the sensitivity of the electrometer was in a large area increased in a ratio of about of 1 to 120.

In a note of the cited volume Cantoni indicates that: *"The different electrometers with quadrant and with bottle, prepared by Volta and kept in this cabinet, are still well preserved ... This shows how much care and diligence in the package Volta put for his electrometers (maximum for straws ones), and how valuable were the electrometric measurements that he imagined."*

The Pile and the Voltian Electromotion

On this subject Cantoni says: *"This way he came to the direct and fundamental proof mentioned above, namely that of the electromotion due to the contact of two heterogeneous metals.*

But even here he had big difficulties that, with his peculiar experimental acumen, he was able to overcome. It is necessary to take care not only to the nature of the facing metals, but also to the disks which constitute the capacitor, which, in their turn, exert an electromotive action on the explored discs, which are brought in contact with the shield or with the electrometric disk; and we must ensure that the two contact surfaces are clean and flat, and that in detaching from one another, the one rises parallel to the other.

...

Then, Volta, after putting sequentially in contact metals of different nature, obviously taken in equal conditions, and determining the electrometric deviation given by each pair of them, could establish such a series, for which each couple acquires an opposite electrical state, depending if it touches a metal preceding it, or one metal following it in the series. Then the electromotion becomes more effective the further are the two metals in the series that are put in contact. Thus, for example, seeing that copper acquires a negative potential with zinc, while with silver becomes positive, it is argued that silver with zinc will constitute a pair of higher electromotive force than the two previously indicated. It is however a remarkable fact that metals with more difficult oxidation, such as platinum, silver and gold are at one end of the series, while tin, iron, lead and zinc, which are the most oxidized, are at the other end."

The Volta Circuit

"In order to utilize an electromotive force, caused by the conflict of two heterogeneous metals, to produce some physiological effects (e.g. contraction of the frog), or some chemical or physical actions, it must be set up a circuit, which, as Volta advised, requires at least three different bodies, one of which must be an imperfect conductor. So, if the circuit between the two aforementioned metals is closed using a third metal, a little different from them in electromotion and conductivity, the intent is not reached. Because this one, in its turn, applies over the first an electromotive action opposite to the one that exerts on the other; these two electromotions conceal each other, and outside will not be produced a significant effect; for the same reason any circuit formed with only two metal arches could not produce any effect, touching each other at both ends, although heterogeneous, since in each side it would be generated an electromotion equal

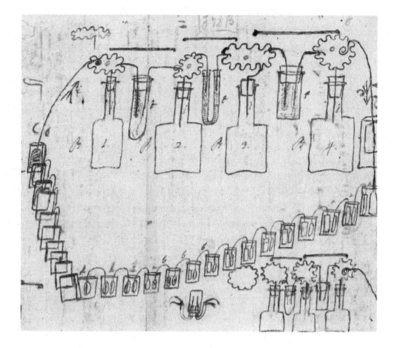

A sketch signed by Volta from a letter of August 1st, 1796, to Friedrich Albrecht Curl Gren, of Halle University, dealing with chemical effects occurring in the pile, here presented in the chain-of-tumblers version.

but with opposite direction. The same happens in a magnetic circuit, with a ring shape, where the internal inductive actions, still working and still eliding mutually, give no external magnetic action.

Volta, because of the opposition of Galvanic people, recognized he had to realize the circuit between the two electromotors with another material, whose conductivity and electromotricity are much smaller, as happens for liquids compared to metals. In this case, with two electromotors and a liquid conductor, an effective circuit is obtained, i.e. an electric movement from the metal electro-positive to the one electro-negative, which while acts as conductive arc toward the liquid, is affected by thermal or chemical action, as it is combined with some of the items of the liquid itself. Volta thought then to call conductors or electromotors of second class the liquids so used in the circuit, reserving to metals and solid conductors the name of first class electromotors.

With this distinction, in fact, the difficulty is not resolved, but only removed, at least from the theoretical side. This is due to the reason that this wet conductor, either acts as electromotor, as in the previous case of the circuit formed with three solids, two of which differ slightly in electromotive power compared to the third; or it acts as a simple wire, and in this case it is impossible to understand how it acquires so much ability to give form and efficiency to the *metals electromotion, in itself inactive. On this point, the opposition arose by Parrot and Dal-Negro, not entirely unfounded, to which Volta did not answer directly with hard reasons, but could respond with new and unexpected facts, as the water decomposition obtained by Nicholson through a Volta combination, and the oxidation of metals hard to be oxidable, such as copper and silver, where also achieved by Volta. The electromotive force of his pile, conveniently sorted, produced high-value effects, that is operated as a powerful chemical agent.”*

The Compound Electromotor

Finally Cantoni reaches the concept of pile: *“But these effects would not be achieved without another important invention, another discovery that may well be attributed to Volta, that is, an electromotor circuit could gain much more effectiveness, associated with the actions of multiple electromotive pairs, ordered in series, having gradually in sequence these three elements: an electro-positive metal, an electro-negative metal, a wet conductor, and again the two metals with the same order and the conductor wet, and so on.*

And here again he was able to give on this wonderful discovery a conclusive demonstration through his exceptional electrometer-condenser; because in the two aforementioned extremes of the series, the opposite electrometric voltage was precisely growing in direct proportion to the number of complete pairs, ranging from these two extremes of the series.

Indeed, this fact seemed consistent with his idea that the liquid, as interposed between one and another couple, was just doing the work of a simple conductor, instead of electromotor.

Then, the discovery of the pile of Volta, generally speaking, consists of two achievements, i.e. two basic facts: the electromotive force of different heterogeneous metals, forming a pair, in a liquid that touches them; and the gathering of electromotive forces in a single circuit composed of any number of pairs.”

All in one, the Cantoni highlights on the compound electromotor can be summarized by the two well-known Volta's laws:

- the contact between two different metals at the same temperature causes an electromotive force (*emf*) depending on the kind of the metals;

- in a chain of more than two different metals at the same temperature, the *emf* between the ends of the chain is the same that would arise if the two metals located at the ends would be placed in direct contact.

Materials fulfilling the second law are called first-class conductors (metals), while materials violating it are named second-class conductors.

Considering Volta's laws, the following remark can be put forward. Both kinds of conductors are essential to realize the pile in practice: in fact, in a chain composed of pairs of different metals, separated by means of a second-class conductor, the *emf* at the ends of the chain is the sum of the *emf*s originated by each pair (i.e. the pile as the series connection of basic elements): without second-class conductors interleaved, the effect of series connection would simply not exist.

Returning to Cantoni: *"However it should be noted that the brilliant chemical and thermal results obtained by the electromotive compounds depended also on nature of the metals; the Volta combinations on which especially he concentrated his studies were, luckily, the most effective. If he instead would concentrated to other combinations, that he also attempted but for a short period, as would those consisting of a single metal and two different liquids, or consisting of only solid materials (such as the dry cells studied by Zamboni), he would not have given much importance to his second discovery ... A dry cell, even with a great number of couples, in fact, gives considerable deviations to a voltage electrometer, but does not produce significant current for one of most precise galvanometers. Therefore, the power of these piles, appreciated by the amount of thermal or chemical action produced by the current, results, broadly speaking, negligible. Conversely even a single electromotive pair, consisting of two metals and one or two liquids one of which is chemically active (as with a Grove or Bunsen pair), can produce significant heat, magnetic or chemical effects into the components of the Volta circuit. ...*

Then Volta gave suitably greater importance to that form of compound electromotor, which he first named crown of cups, unlike the other more properly called pile, where the bimetallic discs are piled on each other, with the inter position of a moistened cloth with saline or acid solution, because in this case, due to the weight of stacked disks, the fluid is partially squeezed out of the cloth, and makes it more inconvenient to prepare and to dismount the device."

The Volta Legacy

Volta's more remarkable qualities were his perseverance in the physical experimentation, his acute observation of the natural phenomena, his patience in the repetition of experiments and his imagination in modifying the experiments to get the correct assertions and conclusions. Indeed Volta did not invent the pile 'by chance', but arrived to this discovery after applying a scientific method, by choosing accurately the chains of very poor objects (disks of different metals, disks of cardboard wet with salted water, little pieces of wood and string) resulting in the maximum difference of potential.

Providing for the first time a concrete and tangible primary source of electric power, the Volta pile opened up the way for previously unthinkable fruitful applications and conceptual horizons for electrical science. Indeed, in the years following the pile invention, the availability of the *DC* current made possible the detailed study of the magnetoelectric and electrodynamics interaction that generated the modern electricity. Not surprisingly, even today, illustrating the experiences of Charles-Augustin de Coulomb and André-Marie Ampère, treatises on electromagnetism usually start with: when the battery is connected. Through Volta, the first half of the new century, as part of its understanding of steady-state electricity, was destined to learn about: typology (Ohm), topology (Kirchhoff), energy (Joule) and finally dynamics (Ampère).

Volta has been clear and was conscious of the enormous practical application of his discoveries, but was also a good planner considering that in the document *"Technical and practical study ... according to the current state of progress of physical and mathematical sciences"* attached to the quoted letter for the promotion of the engineering studies, he anticipated that: *"by a pile in Como, by a wire and using the river water as a return, I could make a hot wire in Milan"*.

The importance of Alessandro Volta's work is recognized worldwide, but we think it is worth quoting the great esteem and appreciation expressed by leading scientists of the time, awarded with the honorary degree on that occasion which the Rector Corradi of the Pavia University called : *"Feast of science and of the University"* on the basis of the long correspondence, now fully assembled in the University of Pavia's Historical Archive, between these outstanding scientists and the Rector of the University of Pavia.

A telegram sent by Helmholtz on the very day of the Volta celebrations: *"The person you are honouring is a second Columbus in the sea of knowledge. He discovered areas full of miracles, May the current that he started carry us beyond."* And in a letter dated May 29th 1878, to Rector Corradi: *" ... an honour that I particularly appreciate because I received it the day on which your University is celebrating the memory of the greatest physicist of all times. ... this could not have been expressed in a better way and on a better occasion."*

A letter from Wilhelm Weber, dated August 6th 1878, to Rector Corradi: *"... The great honour that I have received on this occasion is all the greater for me because I received it for my work in the field of electricity on the basis of the recommendations of the Faculty who sees my work as an important continuation of Volta's great pioneering studies ... "*.

In two letters from Maxwell to the Rector of the University of Pavia: *"... Allow me to express to you and your learned colleagues my sense of the honour they have done me in connecting my name with your ancient and illustrious University, and with its great ornament, Alessandro Volta. ... the sentiment with which the whole civilized world regards the great electrician."*

In the anniversary other quoted comments were *"to honour one who with his invention of the pile, had amazed the world, and whose simple electric motor had provided"* (Von Humboldt), and *"[the pile is] the most valuable piece of equipment in the whole of science."*

The Volta scientific legacy is in its evolutionary process still in connection with the present times. As Nobel prize winner Carlo Rubbia highlighted in his speech on the second century anniversary of the pile discovery: *"Volta truly changed our way of seeing things in terms of electricity use, to the point that today, some 200 years later, we are confronted with a much more complex and new situation, since electricity has become the most noble of all the forms of energy possible."*

It is worth to remark that nature has always been a major source of inspiration for scientists who aimed at describing it in a mathematical way (Galileian approach) and also for scientists who, like Volta, were able to invent a new device by mimicking a natural process or phenomenon. This was just the case

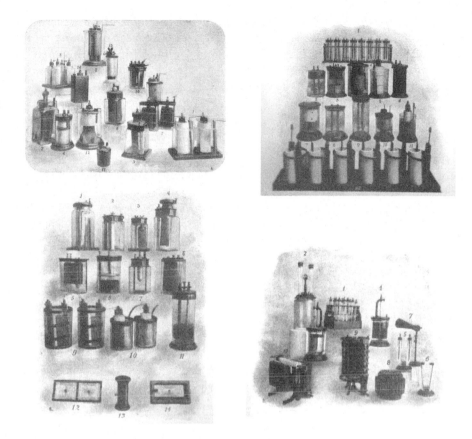

Four series of the first century of piles/batteries. On the top-left (1, 2, 3, 4, 5, 6, 7, 9, 10, 11) various versions of the Leclanché pile – From 1899 Como e l'esposizione Voltiana, N. 7, July 1, 1899

of the pile; in fact, in the letter to Sir Joseph Banks, Volta himself wrote: "*To this apparatus, much more similar, as I shall show, and even such as I have constructed it, in its form to the natural electric organ of the torpedo than to the Leyden flask, I would wish to give the name of the artificial electric organ.*" In conclusion to witness the personality of Volta, especially his modesty, here it follows part of a letter of the scientist to his brother, Arcidiacono Luigi, dated November 17, 1801, published on La Perseveranza the day after the Physics Conference, on Wednesday May 1, 1878: "*... leaving the jokes, I wonder how my old and new discoveries on the so-called Galvanism, which show that not more than pure and simple electricity moved by the mutual contact of different metals with each other, have generated so much enthusiasm ...*

... That's all that I did. As for the new apparatus, to which I have been directed by my discoveries, mentioned above, I thought it would produce clamor (and I told you, if you reminds, just after I built it, about two years ago, and

you had the opportunity to experiment the effects). But I would never have imagined that it would make so much clatter. Since one year and more all the newspapers of Germany, France, and England are full of them. Then here in Paris, you could say, there is real enthusiasm, because as for other things, the enthusiasm is added to what is fashionable ... "

References

V. Cantoni, A.P. Morando. *"1878: Volta Anniversary and Honorary Degree for Maxwell,"* in IEEE Antennas and Propagation Magazine, Vol. 53, No. 1, February 2011, pp. 205-210.

V. Cantoni, M. Mosconi, *"Volta Celebrations and Honorary Degree: The Exhibition at the University of Pavia's Historical Archive,"* IEEE Antennas and Propagation Magazine, Vol. 53, No. 6, December 2011, pp.225-230.

Virginio Cantoni, Adriano Paolo Morando: *"Alessandro Volta. Le onoranze del 1878 all'Universit di Pavia / The 1878 Celebrations at the University of Pavia,"* Silvana Editoriale, Cinisello Balsamo (Milan, 2011.

V. Cantoni, A. P. Morando, F. Zucca,*"Pavia 1878. Il mondo della fisica onora Volta,"* Milano, Cisalpino IEU, 2013.

A. Corradi, G. Cantoni, etc., (1878). *"The Monument dedicated to Alessandro Volta in Pavia, Memories and Documents,"* Pavia, Stabilimento Tipografico Bizzoni.

"Alessandro Volta in the second centennial of the pile 1799-1999 - Opening addresses and lectures given in Rome, Pavia and Como and of an Essay on the Galvani-Volta controversy," Como, Grafica Marelli, 2002.

V. Cantoni[1]*, P. Di Barba*[1,2]

1- Dept. of Elect., Computer and Biomedical Engineering, University of Pavia, Italy.
2 - Research Centre for the History of Electrical Technology - CIRSTE, University of Pavia, Italy.

About Volta and the "Voltaic Pile"

The word 'pile' is not found on an English dictionary but I found it by looking in an old English textbook, Ganot's Physics (chosen only because I have a copy, which belonged to my grandfather) which consistently refers to Volta's battery, and says '...known to this day as the Voltaic pile...' and also says '.... since its invention ...the general name of pile has been retained for all apparatus of the same kind ...' The book goes on to use the word 'battery' for the improved version developed by others. Ganot wrote his textbook in French, so the copy which I have is actually a translation into English. However it shows that the word 'pile' was used for a battery in the English language at that time, although it is certainly obsolete usage now.

The below shows my copy of the book, the 1878 translation from French, and the next is the Frontispiece. It was awarded as a prize to my grandfather (mother's father) in 1880 when he was at school in Taunton, Devon, England. He died 10 years before I was born. He was only eleven years old when he received the book, which perhaps suggests something interesting about the English education then compared to now.

Picture of the book "Ganot's Physics"

I remember as a small child that I spent much time looking through this book, and although it is sure that I did not understand much of it, the many illustrations were embedded in my mind. It is interesting that the frontispiece states that the book is for 'General Readers and Young Persons'. I am unsure of the upper limit of 'Young' at that time, but at least an England, this would probably be regarded as far too advanced a book for todays 'Young Persons'.

The author of the translation, E. Atkinson, is shown as being professor at the Staff College: Almost certainly this is the College associated with the Royal Military College at Sandhurst.

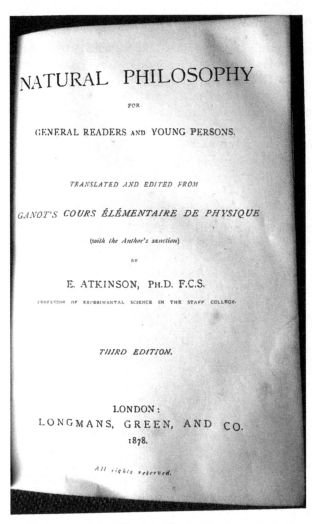

First page of the "Ganot's Physics" book translation.

As far as I have been able to discover, the school in Taunton which my grandfather attended still exists as Queen's College, originally named the West of England Wesleyan Proprietary Grammar School. It was founded in 1843 to give "a regular and liberal course of education" and arose from 'dissatisfaction' of Methodist and Wesleyan Christians (called 'non-conformists') at the education then available to them via the established 'Church of England' educational system. The title "Queen" refers to Queen Victoria, the name being adopted to recognise her Golden Jubilee.

Going back to Volta and the "Voltaic Pile" in the next two pages there are figures taken from the book which shows Voltas invention and the words which state that it is known as a Voltaic pile.

Anthony C Davies,
Life Fellow IEEE,
Emeritus Professor, King's College London, England

457. **Voltaic pile.**—Reasoning from this theory of contact, Volta was led in 1800 to the invention of the marvellous instrument which

Fig. 393.

immortalised him, and which is known to this day as the *Voltaic pile*. Wishing to multiply the points of contact, and to collect the electricities produced by each, Volta arranged, as represented in figure 393, a disc of zinc, a disc of copper, then a round piece of cloth moistened with acidulated water, then again a disc of zinc, a disc of copper, a piece of cloth, and so forth, care being taken always to preserve the same order. What was to be expected from such a combination? Arago says, ' I do not hesitate to assert, that this mass so inert in appearance, this pile of so many couples of

Page of the "Ganot's Physics" book showing the Voltaic pile.

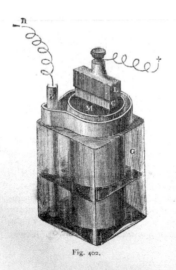

Fig. 402.

467. **Leclanche's battery.**—Each element of this battery (fig. 402) consists of a rod of carbon L placed in a porous pot, which is then tightly packed with a mixture of pyrolusite (peroxide of manganese) and coke M. The porous pot is contained in an outer vessel G in which is the electropositive metal zinc Z. The exciting liquid is a solution of sal-ammoniac. This battery, from its simplicity, its constancy, combined with considerable electromotive force, is coming into use for telegraphs, and for alarums in private houses.

Page of the "Ganot's Physics" book showing the Leclanché pile.

for these are the *ways* through which the respective electricities emerge. It is important not to confound the positive *plate* with the positive *pole* or *electrode*. The positive electrode is that connected with the negative plate, while the negative electrode is connected with the positive plate.

462. **Voltaic battery.**—When a series of voltaic elements or pairs are arranged in such a manner that the zinc of one element is connected with the copper of another, the zinc of this with the copper of another, and so on, such an arrangement is called a *voltaic battery* ; and by its means the effects produced by a single element are capable of being very greatly increased.

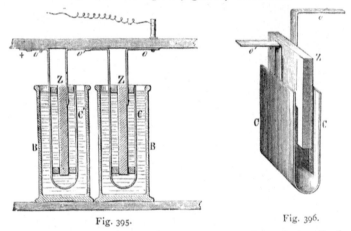

Fig. 395.   Fig. 396.

The earliest of these arrangements was the voltaic pile devised by Volta himself.

It will be readily seen that it is merely a series of simple voltaic couples, the moistened disc acting as the liquid, and that the terminal zinc is the negative and the terminal copper the positive pole. From the mode of its arrangement, and from its discoverer, the apparatus is known as the *voltaic pile* (457), a term applied to all apparatus of this kind for accumulating the effects of dynamical electricity.

The distribution of electricity in the pile varies according as it is in connection with the ground by one of its extremities, or as it is insulated by being placed on a non-conducting cake of resin or glass.

Page of the "Ganot's Physics" book with a detailed description of the Voltaic pile.

The Golden Epoch

The XIX Century was a surge in scientific development spreading across Europe and United States. The discovery of Volta triggered many experiments and discoveries that created the basis of the modern electrical engineering. One of the first relevant events was by accident. One day, in 1820 at the University of Copenhagen, Professor Hans Christian Øersted (1777-1851) was explaining to his students the functions of a battery. He used a wire connected between to the two poles to show the flow of electric current. By chance, on the table there was a small compass just under the wire. Øersted noticed that the needle of the compass was pointing to the east instead than north. Then, he disconnected the wire from the battery and observed that the compass needle returned pointing to the north. Øersted was astonished and puzzled. How electric current had that effect on the compass?

After the students left the classroom, he verified again the phenomenon to make sure there was not something wrong. The repeated verification he made proved that electric current causes magnetism. Øersted published the results and the news spread far and wide and reached André-Marie Ampère, professor in Paris.

The successive studies of Ampère on the current revealed that current produces heat and the experiments of James Prescott Joule demonstrated that *"the mechanical power is converted into the heat evolved by the passage of the currents of induction through its coils"*.

The Magic of the Electrical Current

Various effects originated by the electrical current were studied at the beginning of the nineteen century. Politically those were turbulent years but the turmoil around Europe did not prevent the scientific studies carried out by Ampère and Joule.

André-Marie Ampère

André-Marie Ampère [Wikipedia]).

The discovery of Hans Christian Øersted on the deflection of a magnetic needle placed near a wire carrying a current prompted the scientific curiosity of André-Marie Ampère. Within a short time he was able to explain that surprising effect by supposing that the current generates magnetic effect all round the wire. He found out that the attraction is magnetic, but no magnet is necessary for causing that effect. This observation was supported by many experiments that he designed to explain the relation between electric current flowing through various types of conductors and magnetism. For his experiments Ampère used detailed instruments like the one in the figure, called "Ampère's table". The experiments led to the formulation of the Ampère's Law of electromagnetism and gave rise to the best formal definition of the electric current of his time. The Ampère's studies, that were weekly reported before the Académie des Sciences, established a new science: electrodynamics. Then, he developed a mathematical theory that formally described the relationship between electricity and magnetism. Because of that theory and other successive studies and achievements André-Marie Ampère is considered the founder of the classical electromagnetism.

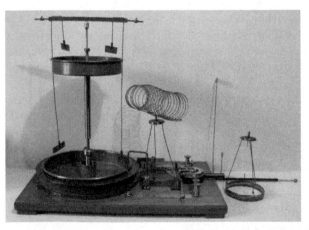

Ampère table. (Courtesy: "Le Zograscope" – Paris)

Ampère was born in Lyons, France, on January 1775. The library of his prosperous bourgeois family was full of books and this allowed the young André-Marie to implement the Jean Jacques Rousseau advice of avoiding formal schooling and pursuing instead of an *"education direct from nature."* With the help of many books he built an excellent self-education in several disciplines including mathematics and science. In 1803 he moved to Paris and just a year after he joined the École Polytechnique. He became professor of mathematics in the same school in 1809. That appointment was the first step of an outstanding academic career, including the prestigious chair of experimental physics at the Collège de France, despite the lack of a formal education.

Ampère was a genuine scientific genius committed to investigation, insight and innovation, but he was not a well-organized and disciplined person. Ampère frequently published papers and shortly after he discovered new aspects that he did not incorporate into the original manuscript; since he couldn't change the first paper, he used to publish a new version with contained the additions. The result was multiple versions of the same publication, like the ones in the reprint of the Ampères Mémoirs that includes publications substantially different from the originals.

Ampère anticipated the existence of the electron, discovered the fluorine, and he properly grouped chemical elements over half a century before Dmitri Mendeleev produced his periodic table. The SI unit of electric current, the ampere, is named in his honor.

James Prescott Joule

James Prescott Joule (Salford, 24 December 1818 Sale, 11 October 1889) was a sickly child of the owner of the Salford Brewery. From age eleven, he started his studies with a series of private tutors. In 1834 his father wanted that his two sons were educated by the famous chemist John Dalton, since he must have considered the physics and chemistry of liquids and gas relevant to the brewing industry [1]. Indeed, years later the young James managed the family brewery, but at the same time he was also passionate about research in electricity. It seems that his house was full of voltaic batteries and other electrical and mechanical apparatus. Many years later, in an autobiographical note he will write that *"Dalton possessed a rare power of engaging the affection*

James P. Joule [Wikipedia]).

of his pupils for scientific truth; and it was from his instruction that I first formed a desire to increase my knowledge by original research".

Likely motivated by the ambitions of replacing the brewery's steam engines with electric motor and improve the efficiency of these last, since 1938 he published his first scientific papers in the Sturgeon's Annals of Electricity. In 1939 he started further studies of chemistry with John Davies.

At the end of 1840 is paper on *"On the Production of Heat by Voltaic Electricity"* was rejected by the Royal Society of London. However, in 1841 those contents appeared in the Philosophical Magazine [2]. The paper reported that the heat manifested by the proper action of any voltaic current is proportional to the square of the intensity of that current, multiplied by the resistance to conduction. It was the Joule's effect!

The question about how much work could be extracted from a given source, led him later to investigate about the convertibility of energy.

In 1843, at a meeting of the British Association for the Advancement of Science held in Cork, Ireland, he presented a paper titled *"On the Calorific Effects of Magneto-Electricity, and on the Mechanical Value of Heat"*. This reported the experimental results proving that the heating effect he had quantified in 1841 was due to generation of heat in the conductor and not its transfer from another part of the equipment. In particular, it reports that *"The quantity of heat capable of increasing the temperature of a pound of water by one degree of Fahrenheit's scale is equal to, and may be converted into a mechanical force capable of raising 838 lb to the perpendicular height of one foot."*

This claim was a direct challenge to the caloric theory (i.e. heat could neither be created or destroyed) which had dominated thinking in the science of heat according to Antoine Lavoisier in 1783 and Sadi Carnot in 1824. The young Joule, working outside either academia or the engineering profession, had a difficult path with respect to the general philosophical convincement of the caloric theory.

In 1843 Joule's father built an expanded laboratory for the young James.

In 1844 his paper on the same subject, but addressing experimental proofs by using gas (air), was presented to the Royal Society [4] and denied for publication, and again published in the Philosophical Magazine the year later [5].

The concepts expressed about heat and mechanical work led later to the identification of the concept of the conservation of energy, i.e. the first law of thermodynamics!

In the next year his reputation and popularity rose. William Thompson, later known as Lord Kelvin, considered Joule's findings and in the next years the two scientists established a cooperation that led to the definition of the absolute temperature.

As a pupil of Dalton, he learned the atomic theory that must have led him to the intuitions about the kinetic theory of heat.

In 1850 was elected a Fellow of the Royal Society, of which he was made a member of the Council in 1857. In the same year received the honorary doctorate from Trinity College Dublin, and Oxford in 1860. In 1973 was elected president of the British Association for the Advancement of Science. In his honour, the unit of energy is named after Joule.

Today, in the era of Internet of Things, Joule docet ...!

References

[1] "Science in Victorian Manchester: Enterprise and Expertise," Robert Hugh Kargon, Manchester University Press, 1977.

[2] J.P. Joule, "On the Heat evolved by Metallic Conductors of Electricity, and in the Cells of a Battery during Electrolysis." Philosophical Magazine, 1841.

[3] J.P. Joule, "On the Calorific Effects of Magneto-Electricity, and on the Mechanical Value of Heat". Philosophical Magazine, 1843.

[4] J.P. Joule, "On the Changes of Temperature Produced by the Rarefaction and Condensation of Air". Proceedings of the Royal Society of London, 1844, p.171

[5] J.P. Joule, "On the Changes of Temperature Produced by the Rarefaction and Condensation of Air," Philosophical Magazine, 1845, pp. 369 – 383.

Domenico Zito
University College Cork

From the Charge to the Electric Field

The phenomena related to the electrical charge and the physical rules that govern them was initially developed by two giants of the of electrical sciences: Coulomb and Faraday.

Charles Augustin Coulomb

Charles A. Coulomb [Wikipedia]).

Charles Augustin Coulomb (Angoulême, 14 June 1736 Paris, 23 August 1806) received an education in philosophy, language and literature, mathematics, astronomy, chemistry and botany in Paris. He graduated as *Ingénieur du Roy* from the École royale du génie de Mézières (Royal Military Academy) in Paris in 1761.

Over the next twenty years he focused on mechanical and structural engineering for building fortifications. In 1773, after a long service in the island of Martinique in the West Indies, he moved back to France and presented his first work to the Académie des Sciences of Paris entitled "*Essai sur une application des règles de maximis et de minimis à quelques problèmes de Statique relatifs à l'Architecture,*" in which he structured advanced problems and solutions, defined the laws of static and dynamic friction, formalised the concept of shear stress and introduced the method to characterise the mechanical resistance of materials. In the next years spent in different locations across France, he carried out further studies on friction between two cords and the surfaces of two materials, reported in the work entitled "*Thorie des machines simples, en ayant réqard au frottement de leurs parties et à la roideur des cordages,*" for which he received a prize from the Académie des Sciences, of which was an elected member.

In 1781 he moved to Paris where a relatively new (i.e. had remained little more than an intellectual curiosity from millennia since ancient Greeks) exiting discipline was drawing the attention of the most influent scientists: the electricity!

In 1784 he published the memory entitled "*Recherches théoriques et expérimentales sur la force de torsion et sur l'élasticité des fils de metal*" [1] in which he developed a mechanical problem: the torsion force. In the reality this was just a background study: his intension was to build an instrument, i.e. the torsion balance, to measure the torsion and that he used later in his experimental investigations on the electric charge of the surfaces, electricity and magnetism.

On the basis of his experimental observations carried out through the torsion balance, from 1785 to 1791 Coulomb published seven memoirs in which he determined mathematically that the magnitude of the electric attraction or repulsion force between two point charges is directly proportional to the product of the charges and inversely proportional to the square of the distance between them [2], i.e.what we all know today as Coulomb's Law. In his honour, the unit of charge is named after Coulomb.

It is interesting to note how his background on mechanical friction forces between two surfaces and the outstanding abilities matured in formalizing mathematical problems led him to establish the fundamental basics of the electrical and magnetic theory and experiments.

It is also interesting to note how despite he posed the mathematical foundations of electric and magnetic theory, from his memoirs, [2], it appears that he was still using the concept of *fluid* (already in use among the philosophers of the ancient Greece in regard of heat, but to name a few) to explain the interactions.

Still reading through the seven memoirs, it is worth noting how he addressed a number of other basic aspects such as the absence of penetration of the electric fluid (i.e. the electric field, but we have to wait for Faraday before it is defined as such) inside the conductor materials and the redistribution of charges over a surface, and temperature effects.

He got famous for the seven memoirs; however these were only a part of his eclectic work in different research fields. From 1781 to 1806 he published other 25 memoirs, some of them in collaboration with other physicists and mathematicians such as Laplace.

In 1789 he retired from the Military Academy since the beginning of the French Revolution and left Paris. In 1795 moved back to Paris and was one of the first elected members of the *Institut de France*, which replaced the *Académie des Sciences*.

He took part to the definition of the new standard system of weights and measures during the revolutionary government. In 1802 was nominated as Inspector of the Public Education by Napoleon, establishing new lyceums across France.

Michael Faraday

Michael Faraday [Wikipedia]).

Michael Faraday (Newington, 22 September 1791 – Hampton Court 25 August, 1867) came from poor family. He received only the rudiments of education, learning how to read and write in a church Sunday school. In 1804, at the age of thirteen, he began to earn money by delivering newspapers for a book dealer and bookbinder. As apprentice for about eight years, he took the opportunity to read some books, being so a self-taught person. Many years later, in a letter to a friend he confessed that two books captured his attention: *the Encyclopaedia Britannica*, from which he learnt the first notions of electricity, and *Conversation on Chemistry* by Mrs. Marcet from which he gained his foundations [3].

In 1812 Faraday had the opportunity to attend the "Chemistry Lectures" by Sir Humphry Davy at the Royal Institution of Great Britain in London. This changed his life. Faraday recorded the lectures in his notes and sent a bound copy to Davy along with an application letter asking for a job. As a laboratory assistant he learned chemistry. Following an atomic theory by Ruggero G. Boscovich, according to which atoms were mathematical points surrounded by fields of attractive and repulsive forces, in those years Davy conceived the idea that chemical properties were determined not by specific elements alone, but also by the ways in which these elements were arranged together in molecules, and the chemical qualities of both elements and compounds were the results of the final patterns of force surrounding the point of atoms. Atoms and molecules could be placed under considerable strain, or tension, before the "*bonds*" that hold them together were broken. These strains were to be central to Faraday's ideas about electricity, as it will be explained hereinafter.

At the end of his apprenticeship in 1820 he mastered experimental and theoretical chemistry to the level that they could drive him in new researches and discoveries that have changed the world. Despite his important discoveries in chemistry, here we would like focusing on his research on electricity and magnetism carried out at the Royal Institution since 1821.

In 1820 after that Ørsted announced that the flow of an electric current through a wire produced a magnetic field around the wire, and Ampère showed that the magnetic force was circular, Faraday deduced that, if a magnetic pole could be isolated, it should move constantly in a circle around the wire. He also observed that a variable magnetic field produced an electric field [4]. On the basis of his in-depth understanding about the electrical phenomena and his background in chemistry (including the influence of the new atomic theory), he was the first to produce an electric current from a magnetic field, invented the first electric motor and dynamo, demonstrated the relation between electricity and chemical bonding, discovered the effect of magnetism on light, and diamagnetism [4], i.e. the Faraday's law as later named by Maxwell.

The first electric motor invented by Michael Faraday in 1821. [http://michaelfaradayelectricmotor. weebly.com/reaction.html]

In 1823 became a member of the Royal Society, in 1825 director of the laboratory of the Royal Institution and in 1833 John Fuller Professor of Chemistry.

Unlike all the other scientists of his time, he rejected the idea that electricity was a fluid as reported also by Coulomb. Instead, he thought of it as a vibration or force that was transmitted as the result of tensions created in the conductor. Since 1832 he wanted to prove that all forms of electricity had precisely the same properties and caused precisely the same effects. The effect was electrochemical decomposition, which led to the Faraday's two laws of electrochemistry. Faraday's work on electrochemistry provided him also a key insight about the static electrical induction. Every material has a specific induction capacitance. In honour of this the unit of capacitance is today named Farad.

By 1839 Faraday was able to bring forth a new and general theory: all electrical action was the result of forced strains in bodies. Later this evolved toward a new view of space and force. Space was not *nothing*, but a medium capable of supporting the strains of electric and magnetic forces. The energies were not localized in the particles, but in the space surrounding them. This was the birth of field theory that, as the same Maxwell admitted later, the basic ideas for his mathematical theory of electrical and magnetic fields came from Faraday, and that his contribution was to give the mathematical structure of the classical field equations.

Queen Victoria rewarded his devotion to science by granting him the use of a house at Hampton Court and a knighthood. Faraday accepted the cottage but rejected the knighthood to remain plain Mr. Faraday to the end. Among the many quotes attributed to him, it seems he said also: *"The secret is comprised in three words Work, Finish, Publish"*. As excellent communicator he was, maybe today he would have said: Publish or Perish!

References

[1] *"Recherches théoriques et expérimentales sur la force de torsion et sur l'électricit des fils de métal"*. Extrait des Mémoires de l'Académie royale des Sciences (1784), p. 65 Online available: http://cnum.cnam.fr/CGI/fpage. cgi-8CA121-1/117/100/416/0079/0316

[2] *"Mémoires sur l'électricité et le magnétisme"*. Extrait des Mémoires de l'Académie royale des Sciences (1785), p. 107) Online available: http:// cnum.cnam.fr/CGI /fpage.cgi 8CA121-1/117/100/416/0079/0316

[3] *"Faraday as a Discoverer,"* John Tyndall; The Project Gutenberg EBook. Online Available: http://www.gutenberg.org/files/1225/1225-h/1225-h. htm #1225

[4] *"Experimental researches in electricity,"* Michael Faraday. Reprinted from Philosophical Transactions, 1832-1852. Vol. 3. London, R. Taylor and W. Francis, 1855. Online Available: http://catalog.hathitrust.org/Record/ 001481343

Domenico Zito
University College Cork

The Laws of Circuits

The laws that regulate voltage and current in a circuit have been identified by two scientists: Ohm and Kirchhoff. Their contributions to the foundations of the circuit theory are described below along with some biographical notes.

Georg Simon Ohm

Georg Simon Ohm was a German physicist who developed the foundations of electrical circuit theory and was one of the founders of physical acoustics. In the period 1825 - 1827 he studied the relationship between the magnetic force exerted by a current and the nature of the conductor bearing the current. After a series of improvements of his experimental arrangements Ohm was able to observe that currents and voltages are the suitable variables for describing interconnections of conductors. However it is necessary to emphasize that this statement is only correct if G. R. Kirchhoff's physical interpretation from 1849 of Ohm's circuit variables is used. A crucial step of Kirchhoff was that

Georg Simon Ohm [Wikipedia]).

he interpreted Ohm's 'electroscopic force' as electrical potential, whereas the original interpretation by Ohm was similar to a charge density. However, it can be shown from a measurement point of view, that both physical quantities are related and therefore Ohm's law survived.

The fundamental laws of resistive circuits are presented in a detailed manner in Ohm's seminal monograph *Die galvanische Kette – mathematisch erklärt*, which was published in 1827 – English translation *The Galvanic Circuit Investigated Mathematically* from 1891. The most famous result is the relationship between current and voltage of a metallic conductor: Ohm's law, which was already presented in a paper of Ohm from 1826. However, also further physical laws of galvanic circuits – later called Kirchhoff's laws – are included in Ohm's monograph but these fundamental laws are formulated in a verbal manner.

Georg Simon Ohm was born in Erlangen, Germany, on March 16, 1789. He was the son of Johann Wolfgang Ohm, a mechanician at the Erlangen University, and Maria Elizabeth Beck, the daughter of a tailor. Ohm had a younger brother Martin, who later became a well-known mathematician, and a sister. His mother died when he was ten. Already in the early years Ohm and his brother Martin were taught by their father in the natural sciences and mathematics. With sixteen years he left the gymnasium and studied mathematics, physics and philosophy at the Erlangen University but for financial reason he had to stop his studies and became a private teacher in mathematics in Switzerland. In 1811 he returned to Erlangen where finished his doctorate with a thesis *"About light and colors"*.

Memorial for Ohm at the Technical University of Munich, Campus Theresienstrasse

After a short time as a lecturer in mathematics Ohm left the Erlangen University and in turn he was a teacher at a school in Bamberg. In 1817 Ohm followed an offer at the Jesuit Gymnasium of Cologne, Germany, where he became a teacher in physics and mathematics and supervised the cabinet of physical instruments and the chemical laboratory where he carried out many experiments. Around 1825 Ohm was interested in experiments in connection with the electrical conductivity. Since about 1825 Ohm was particularly interested in the electrical conductivity and therefore he carried out associated experiments. Ohm's studies of electrical conductivity in galvanic circuits were strongly influenced by J. Fourier's theory of heat.

Already in 1826 Ohm presented his first results but these experiments were performed using a wet cell. At this time these cells cannot deliver a constant current and therefore Ohm's results were faulty, such that most of his manuscripts, written for the *Annalen d. Physik u. Chemie* – a german journal – were criticized by some reviewers. Following an advice of J. C. Poggendorff, the editor of this journal, Ohm used a thermo-electrical source and finally he was able to obtain his famous relationship, which is known nowadays as Ohm's law. Although Ohm was successful with the derived formulas, doubted many colleagues in this time its physical interpretation. In particular, with respect to Ohm's interpretations of the physical variables came up objections, since they stood in contradictory with respect to the electrostatics. Eventually, G. Kirchhoff solved these difficulties in a paper from 1849. Ohm published about his fundamental physical principles in galvanic circuits in a comprehensive monograph from 1827.

Despite the rejection of many colleagues, Ohm received in 1833 a professorship at the *"Royal Bavarian Polytechnic"* in Nuremberg, Germany, and at least in its experimental part, Ohm's work was accepted by some prominent colleague, e. g. F. W. Gauss , W. Weber, G. T. Fechner, E. Lenz, and M. Jacobi. In particular, the British physicist C. Wheatstone was impressed of Ohm's book, translated it in English language and after its publication in 1841, Ohm received the distinguished Copley medal from the Royal Society of London.

Because of its growing recognition, Ohm received in 1849 a position as second curator of the mathematical and physical collection in Munich, Germany, and in 1852 he became a professor at the University of Munich. But only two years after the late recognition, Ohm died on July 6, 1854, in Munich in the age of 65 at the effects of a stroke.

Gustav Robert Kirchhoff

Gustav Robert Kirchhoff was a German physicist and one of the founders of mathematical physics who discovered the law of thermal radiation, developed together with R. W. Bunsen the optical spectroscopy and yielded fundamental contributions to the fundamentals of electrical circuit theory. Based on G. S. Ohm's fundamental results about current flows in conductive materials Kirchhoff presented in an appendix of his first scientific publication from 1845 two theorems about currents at nodes and voltages around a mesh in resistive networks which are called nowadays Kirchhoff's laws. The system of equations which can be derived from these laws along with Ohm's law is the mathematical basis for the theory of resistive circuits. In 1847 Kirchhoff associated a geometric object to an electrical circuit and became one of the founders of graph theory. By using the meshes of such a graph he was able to develop a new solution method for the system of linear equations of resistive circuits and therefore he founded the network topology. Subsequently Kirchhoff discussed Ohm's physical interpretations of the current flow in conductive materials in further publications from 1849 and 1857 and presented a new interpretation which was consistent with theory of electrostatics.

Gustav Robert Kirchhoff was born on March 12, 1824, in Königsberg (now Kaliningrad, Russia) as the son of Friedrich Kirchhoff, a lawyer, and Johanna Henriette Wittke. After he left the Kneiphöfische Gymnasium at age eighteen Kirchhoff studied mathematics and physics at the Albertus University of Königsberg. Under the scientific direction of F. E. Neumann, he worked since 1843 in the mathematical-physical seminar, which was founded by K. G. J. Jacobi. Already in 1845 Kirchhoff published still as a student his first scientific paper "*On electric conduction in a thin plate, and specifically in a circular one*" where the Kirchhoff's laws were included. As mentioned above a paper about the graph-theoretical solution of the circuit

Gustav Robert Kirchhoff [Wikipedia]).

equations was published in 1847. In the same year Kirchhoff graduated from Königsberg University and he married Clara Richelot, the daughter of his former mathematics professor, F. J. Richelot. Electricity remained a research area until the end of his life. In 1879 he was one of the founders of the worldwide first Association of Electrical Engineers in Berlin.

Then, Kirchhoff moved to the Friedrich-Wilhelms-University in Berlin where he got his habilitation (academic teaching license) and became an external and unpaid university lecturer from 1848 - 1850. In 1850, Kirchhoff and his family moved to Breslau, Prussia (now Wroclaw, Poland) and Kirchhoff became

the extraordinary professor of physics at University of Breslau. There he met
R. Bunsen who came in 1851 at this university. Bunsen left Breslau in 1852
and moved to the University of Heidelberg but Kirchhoff followed him in 1854.
Now, in collaboration with Bunsen a very productive phase of Kirchhoff was
initiated. Based on some previous work on spectroscopy Kirchhoff and Bunsen
founded the spectrum analysis in order to produce fingerprints of elements
since each element is characterized by its own pattern of colored lines. Very
soon spectrum analysis became an essential tool in chemical exploration. His
spectroscopic work led Kirchhoff to the investigation of the sun spectrum and
considerations for physics of the sun. Together Kirchhoff and Bunsen discovered
caesium and rubidium in 1861. Another milestone of physics were Kirchhoff's
researches about the thermal properties of radiation. Based on his experience
of spectroscopy he discovered the thermal law of radiation and introduced the
so-called black body in 1859. It is an idealized physical body that absorbs all
incident electromagnetic radiation, regardless of frequency or angle of incidence.
Forty years later the black body was studied by M. Planck and became a crucial
initial point of quantum theory.

*Celebrative stamp for the 150 years
of the Kirkoff's birthday (1974)*

Although Kirchhoff refused numerous honourable callings from other universities for a long time, he finally accepted a professorship at the Friedrich-Wilhelms University, Berlin and was appointed at the same time at the Prussia Academy of Sciences. In 1875 he moved to Berlin. Kirchhoff devoted the last years of his academic career mainly teaching and wrote the first four-volume textbook of theoretical physics. But also in the years in which Kirchhoff disease condition increasingly worsened, he published new research results. For example, Kirchhoff's scalar diffraction theory became an essential concept in optics.

Kirchhoff's disease eventually forcing him to abandon his lectures in 1886
and a year later he died peaceful on October 17, 1887 in Berlin.

Wolfgang Mathis
Leibniz University, Hannover

The Magnetic Inductance

The Øersted observation generated significant research on the field. In the first half of the XIX century Joseph Henry and Michael Faraday worked on the electromagnetic phenomenon of self-inductance and mutual inductance. Henry also developed practical devices based on electromagnets like the doorbell and the relay. Those were some of the first examples of practical use of the electrical science.

Joseph Henry

Joseph Henry (Albany, 1797 – Washington, 1878) came from a poor family. He attended a school which would later be named after Henry in his honour. He worked after school and at the age of thirteen became an apprentice watchmaker and silversmith. He loved theatre and came close to becoming an actor. The spark for science came at the age of sixteen through the book titled *"Popular Lectures on Experimental Philosophy, Astronomy, and Chemistry"* [1].

In 1819 he was given free tuition admittance at the Albany Academy, who offered him the equivalent of a college ed-

Joseph Henry [Wikipedia]).

ucation. Henry went beyond the coursework, reading avidly books in every area of science, as well as many other fields [2]. He supported himself by working as a schoolmaster then tutor and engineer road surveyor for the road construction between the Hudson River and Lake Erie. Henry excelled at his studies and intended to go into medicine, but from then onwards he was inspired by the career in engineering.

At the Albany Academy his interests were captured by the terrestrial magnetism, which led him to focus his research on experiments with magnetism. He was the first to coil insulated wire tightly around an iron core in order to make a more powerful electromagnet than those built to that date. Using this technique, he built a most powerful electromagnet at the time for Yale.

He continued to improve the devices with his research and, in 1831, created one of the first machines to use electromagnetism for motion that could be considered as a progenitor of the DC motor. It did not use rotating motion for power, but was merely an electromagnet perched on a pole, rocking back and forth. The rocking motion was caused by one of the two leads on both ends of the magnet rocker touching one of the two battery cells, causing a polarity change, and rocking the opposite direction until the other two leads hit the other battery.

This apparatus allowed Henry to recognize the property of magnetic inductance. Michael Faraday also recognized this property around the same time, and since he published his results first, he became the officially recognized discoverer of the phenomenon [2].

Henry left the Albany Academy for a Professorship at Princeton in 1832. In 1835 he used the magnetic induction to invent a relay, i.e. the electromechanical switch that would have later replaced the mechanical switch. Despite the tremendous future implications in the history of electronics and computers, and the all world, Henry privileged the scientific aspects of electricity and considered the relay as a laboratory trick to entertain students.

Samuel Morse later used the relay to carry Morse-code signals over long kilometers of wire, but generally the invention of Henry remained relatively unknown for several decades. We have to wait until the development of telegraph and phone communications, to see a massive use of electromechanical switches. Shannon will give later to this device the great impulse for the calculator machine revolution and future digital electronics.

Then he became the Secretary of the brand new Smithsonian Institution in 1846, i.e. a large museum and research institution.

In 1848 Henry worked in collaboration with his brother-in-law, the astronomer Stephen Alexander, to determine the relative temperatures for different parts of the solar disk. They used a thermopile to determine that sunspots were cooler than the surrounding regions.

In the spring of 1863 Henry became one of the founding members of the National Academy of Science, and served as president since beginning in 1867. In 1893 his name was given to the standard electrical unit of inductive resistance, the henry. Since a few decades, especially after the intuitions by Professor Robert Meyer at University of California Berkeley, the (nano) henry has become also popular among the members of the IC design community.

References

[1] Popular Lectures on Experimental Philosophy, Astronomy, and Chemistry, George Gregory, 1808. Online available: https://archive.org/details/lecturesonexper 00greggoog

[2] American Physical Society. Online available: https://www.aps.org/ programs/outreach /history/historicsites/henry.cfm

Domenico Zito
University College Cork

The two faces of an Equivalent Circuit

Every electrical engineer knows the methods of reducing the complexity of electrical network with the Thévenin's or the Norton's theorem. They are widely used in circuit theory to reduce any one-port network to a single voltage or current source and a single impedance. The use of the two theorems facilitate the study of networks and in some cases simplify the use of Kirchhoff's circuit laws.

Léon Charles Thévenin

Léon Charles Thévenin (Meaux, 30 March 1857 – Paris, 21 September 1926) studied in Paris. In 1876 received the diploma from the École Polytechnique and graduated as engineer from the École Supérieure de Télégraphie in 1879.

Since 1880 he worked for the Postes et Télégraphes (P&T), the public company of post and telegraphs of France. At the same time, he also used to hold courses on mathematics and was passionate about research on electricity.

On the basis of the Ohm's and Kirchhoff's laws, and apparently unaware of the principle of superposition by Helmholtz, in 1883 he published the formulae to derive the voltage-source equivalent circuit between two nodes of a linear electric circuit, i.e. the Thévenin's theorem as

Léon C. Thévenin [Wikipedia]).

we know it today. It was presented as a short - *Sur memory a nouveau théorème of électricité dynamique* - to the Academy of sciences of Paris. Indeed, in the paper published by Helmholtz in 1853, thirty years earlier and four years before Thévenin's birth, where he reported the principle of superposition [2], the derivation of the equivalent voltage source and its application was reported as well.

ÉLECTRICITÉ. — *Sur un nouveau théorème d'électricité dynamique.*   .
Note de **M. L. Thévenin.**

« *Théorème.* — Étant donné un système quelconque de conducteurs linéaires reliés (¹), et renfermant des forces électromotrices quelconques E_1, E_2, ..., E_n réparties d'une manière quelconque, on considère deux points A et A' appartenant au système et possédant actuellement des potentiels V et V'. Si l'on vient à réunir les points A et A' par un fil ABA' de résistance r, ne contenant pas de force électromotrice, les potentiels des points A et A' prennent des valeurs différentes de V et V', mais le courant i qui circule dans ce fil est donné par la formule $i = \frac{V - V'}{r + R}$, dans laquelle R

First lines the Thévenin's paper.

Reading through the original articles, the formulation by Thévenin (which includes also an elegant demonstration) mirrors an approach closer to electrical engineering with respect to the formulation reported by Helmholtz, which mirrors a more general approach perhaps closer to physics. Considering that it did not emerged in 30 years (and it could have been so for a longer time), it is worth acknowledging Thévenin's publication for its contribution in bringing this finding more in light.

By incise, Helmholtz, professor of physiology, was very eclectic and gave important contributions to the theoretical and experimental development of electrical engineering, as well as those less known to optics and acoustics, in large part related to his pioneering studies on the peripheral nervous system, in particular the sensory system, for which he could be considered among the precursors of the bio-engineering. In addition to his countless technical contributions to cross- and multi-disciplinary research, he contributed also to the development of a school of engineering as a whole. Hertz was one of his students! In the end, one could say that he did not need the attribution of the equivalent voltage-source theorem to get famous, indeed!

In 1888 Thévenin started teaching at the École Supérieure de Télégraphie, of which after was the director from 1896 to 1901. In this period he carried out studies on advanced implementations of electric networks. The requests for further funds to empower his research laboratory had effect only fifteen years later he left the institute. In 1901 he started a new job as director of printing machinery for post stamps.

Forgotten in the last years of his life, he expressed the wish to have only his family at his funeral, nothing on his tomb and only a rose in his home yard. Today his name echoes in all the technical schools and universities. Since I was a student I must admit that his name have given to me positive feelings, since the theorem makes students' and professors' life a bit easier!

Edward Lawry Norton

Edward L. Norton [Wikipedia]).

Edward Lawry Norton (Rockland, 28 July1898 – Chatham, 28 January 1983), after the military service as radio operator in the US Navy, in 1920 was admitted at MIT, where he graduated in electrical engineering in 1922.

After, he joined Western Electric Corporation in New York, which became Bell Laboratories in 1925. In that same year, he received the master degree in electrical engineering from the Columbia University.

Although his primary interest were in network theory, acoustical systems, electromagnetic apparatus, and data transmission a communications circuit theory and the transmission of data at high speeds over telephone lines, Edward L. Norton is

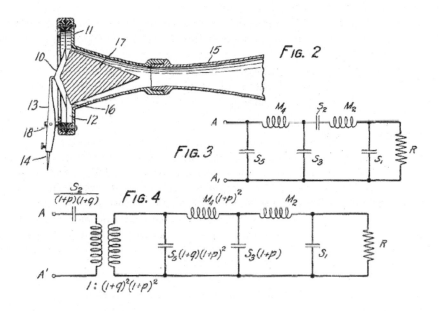

Norton's mechanical filter and its equivalent circuit. (from a Norton's patent)

universally recognised for development of the dual of Thévenin's equivalent circuit. Norton in the early 1920s was one of the first scientists who applied the Thévenin's equivalent circuit to simplify networks; then realised that in some cases it can be convenient to use an alternative method.

Norton wrote many technical memoranda but not so many publications because Norton preferred to stay in the background. Nevertheless his capabilities were recognised and highly valued [3]:

> *"Norton was something of a legendary figure in network theory work who turned out a prodigious number of designs armed only with a slide rule and his intuition. Many anecdotes survive. On one occasion T.C. Fry called in his network theory group, which included at that time Bode, Darlington and R.L. Dietzold among others, and told them: "You fellows had better not sign up for any graduate courses or other outside work this coming year because you are going to take over the network design that Ed Norton has been doing single-handed. [Taken from p. 210, A History of Engineering and Science in the Bell System: Transmission Technology (1925-1975)]"*

Norton also used the filter theory to model mechanical systems. He employed the same general approach used to describe electronic *"maximally flat"* filter. The figure above shows some of the diagrams of his patent describing a mechanical filter and its electrical equivalent circuit.

In 1926 Norton published the formulae to derive the current-source equivalent circuit, i.e. the dual circuit with respect to the Thevenin's equivalent circuit. In 1961 he retired after a productive career which included several patents, many technical reports and a few publications.

Today known as the Norton's theorem, the same result was also derived by Hans Ferdinand Mayer [4], but appeared that Norton had reported them first in an internal report [5].

It's true that Thévenin's or Norton's theorems refer to the two faces of an equivalent circuit, but to make life really a bit easier, it is wise to know when it is worth using the one or the others!

References

[1] L. Thévenin, *"Extension de la loi d'Ohm aux circuits électromoteurs complexes,"* Annales Télégraphiques (Troisieme série), vol. 10, pp. 222-224, 1883. Reprints: L. Thévenin (1883), *"Sur un nouveau théorme d'électricité dynamique,"* Comptes Rendus hebdomadaires des sances de l'Académie des Sciences, vol. 97, pp. 159161.

[2] Helmholtz, *"Über einige Gesetze der Vertheilung elektrischer Ströme in körperlichen Leitern mit Anwendung auf die thierisch-elektrischen Versuche,"* Annalen der Physik und Chemie, vol. 89, n. 6, 1853, pp. 211-233. [Online Available] http://gallica.bnf.fr/ark:/12148/bpt6k151746.image. f225.langFR

[3] Edward Lawry Norton Online Available: http://www.ece.rice.edu/ dhj/norton/

[4] H. F. Mayer. *"On the equivalent-circuit scheme of the amplifier tube".* Telegraph and Telephony, 15, 335-337, 1926

[5] D. H. Johnson, *"Origins of the equivalent circuit concept: the current-source equivalent,"* Proceedings of the IEEE, Vol. 91, Is. 5, May 2003, pp. 817-821.

Domenico Zito
University College, Cork

The path to Radio: from Maxwell to Marconi

Many people contributed to the invention of radio. Experimental work on electricity and magnetism made by Ørsted and Ampère, continued with Joseph Henry, and Michael Faraday. Maxwell predicted the existence of electromagnetic waves (EM) and developed the theory of electromagnetism. Heinrich Rudolf Hertz performed the very first transmission of EM waves in 1887 but he considered the results as being of minor practical value. In 1890, Nikola Tesla made research into high frequency electricity but it was around five years later that Guglielmo Marconi developed the first system for long distance radio communication. For his research related to that invention, Marconi received the 1909 Nobel prize (shared with Karl Ferdinand Brown) in recognition of the discovery of the phenomenon of radio waves propagation.

Maxwell

Everybody knows the famous *"four Maxwell's equations"* for the electric and magnetic fields. James Clerk Maxwell, born in Edinburgh in 1831, spent several years working in the math behind electricity and this is the reason we all enjoyed in our life the lovely *"Maxwell's equations"* of electro-magnetism, which are a true milestone in all advanced courses in electricity. Maxwell started his work in 1855, by publishing his first paper on experimental observations done several years before by Michael Faraday, and he ended in 1873, by publishing his Treatise on Electricity and Magnetism. His Treatise completed his work on math and understanding of electricity and related phenomena. In between 1855 and 1873, Maxwell got an incredible series of key understanding and

Monument dedicated to James Clerk Maxwell in Edinburgh

true inventions, such as the concept of displacement current (fundamental to his conceptual frame of electricity and magnetism) he introduced at his presentation to the Royal Society in London, in 1864.

Maxwell made the huge effort to link measurable circuit properties to only two (not more used) constants that express how readily electric and magnetic fields form in response to a voltage or a current [1]. Much less known is that James Clerk didn't write his *"famous four laws"*! Oliver Heaviside was the man that wrote, first time in the history of science, the now-so-called *"Maxwell's four laws of Electricity and Magnetism"*. In fact, the original theoretical frame defined by Maxwell in 1873 presented up to twenty equations creating relationships among several measurable properties of electrical circuits, while

Heaviside published only in 1885, the year after Nikola Tesla started working for the company of Thomas Edison, a revision of Maxwell theory of Electricity and Magnetism. This revision presents a condensed form of the original set of equations by reducing the equations counting from the original set of twenty to the well-known four equations. Interestingly, from the four Maxwell's equations written by Heaviside, it is easy to derive a simple wave equation that works for both the electric and the magnetic field, which is the basis of modern wireless communications: the equations of the electro-magnetic waves that describe both the light rays as well as the radio waves. This wave equation brings us to another pioneer of this story, Guglielmo Marconi and his invention of the radio. However, before moving to Marconi, we need to meet another pioneer of that golden age of science and technology: Nikola Tesla.

Tesla

Nikola Tesla was an amazing inventor, one of the true pioneers of the history of electricity. Eclectic and full of imagination, he has been a deserving competitor of Edison. As a worldwide known expert in the field of circuits and systems in electricity, he starts his carrier in Edison's enterprise. He has been indeed recruited by Edison once he was just arrived in New York. Tesla was born in Smiljan part of the Austrian Empire in 1856, and it is worth noting that this was the next year with respect to that of James Clerk Maxwell publishing his first paper on Faraday's observations.

Nikola Tesla [Wikipedia]).   Tesla arrived in New York in 1884, the year the people of France gave America the statue of Liberty, and he got immediately recruited by the Edison Electric Light Company. Thomas Alva Edison, at the age of thirty-two in 1884, had already established his generating station at 255-57 in Pearl Street to serve the entire Wall Street and the East River area of Manhattan.

It is well known that their relationship didn't stand for long (Tesla worked for Edison only few years) and they split, practically, to start the famous *"war of currents"*: Edison supporting the idea to produce and transmit Direct Current (DC) over long distances while Tesla on the side of producing and sending Alternating Current (AC). The diatribe was not trivial that time because, few years later, the true challenge became to produce electricity out of the Niagara Falls with the aim to supply electrical power to New York City. Well, Nikola Tesla moved out from the *"Edison court"* [2] and established his own company filling lots of patents on alternating current and the many ways to exploit it for illumination of towns. Less known is that George Westinghouse operated the first AC system in Buffalo, in 1887. Therefore, very interested in the business, he offered to Tesla about $60.000, including $5.000 in cash and, most important, $2.50 per horsepower of electricity sold. This royalty generated so high revenue rights to Tesla which created a serious problem to Westinghouse:

few years later, while Westinghouse was setting a very aggressive selling strategy by getting down prices in order to push out all the competitors, the royalty rights of Tesla rose up to $12 million! Clearly this *"embarrassing situation"* was a bankrupt situation for Westinghouse that had not enough money to pay back Tesla's royalties meanwhile obtaining the large majority of the market of electricity. Of course, this market has been based on the very true heritage of Tesla to the humanity: his polyphase engine that has been the core of the whole system to exploit electricity in the form of alternating currents. Therefore, the dilemma for Westinghouse and Tesla has been to bankrupt and leave the market in the hands of competitors (including Edison and his proposal to use DC) or to find the way to save differently the situation. In that crucial moment, Tesla recognized the big help he received by Westinghouse when no one was supporting him and decided to accept $216.600 to definitely sell his patents on AC to Westinghouse, renouncing to the rights on royalties originally foreseen by the previous contract. This part of the story shows, in the one hand, the big friendship established between the two men and, on the other hand, the big will of Nikola to donate his polyphase system to the world. Among the many different inventions that Tesla patented in his life, there is also something related to the radio. However, the true story of the invention of radio has been made by an Italian inventor and acknowledged pioneer in the commercial development of useful radio communications: Guglielmo Marconi.

Marconi

The history about Marconi and his invention move the first steps in the countryside of Bologna (Italy), in the years around 1890, where the young Marconi was not yet at the end of his adolescence. And, as already mentioned, his story is strictly bound, in some extent, to the wave equation that we can derive from the four Maxwell's laws about electricity and magnetism.

However, the early history of the wireless communications on very long distances started one century before the invention of Marconi, when Claude Chappe established, in 1793, the first system, called *télégraphe*, based on optical signs organized on the roof of signalling houses. These signs have been changed with the purpose to transmit signals over very long distances and then copied houses-by-houses by coherently setting the same signal on the roof of all signalling houses along the line. The system was so powerful and useful that, at the top of his exploitation, it connected Marseille, the main French harbour in Mediterranean Sea, with Paris, the capital of the nation. Of course, 102 years later the humanity was ready for a

The Charles Chappe telegraph system with big signs on the roof.

huge step forward on this track of wireless communication over long distances.

In summer 1895, Guglielmo Marconi, at that time only twenty-one years old, started making experiments outdoors trying to generate, send, and receive radio waves over considerable distances in the field. Fascinated by the work of Hertz, he studied in Bologna University thanks to the lectures of Augusto Righi, and he finally succeeded in optimizing the coherer, key component of any radio system at that age. As we have already written, Maxwell's equations are useful to derive a further wave equation that describes oscillations possible both on the electrical and on the magnetic fields. At that age, the concept of force fields was not yet established and the accepted idea was that the ether, a kind of soft medium embedding the entire universe, was capable of oscillations supporting the waves foreseen by the Maxwell's equations.

Fascinated by these ideas, the young Marconi worked hard to obtain a coherer that was far better than the ones on the hands of his competitors. Among them, e.g., Alexander Stepanovich Popov presented in the same year of the Marconi outdoor experiments, in 1895, a wireless system again based on a coherer for detecting radio waves at large distances. Also, Oliver Lodge provided a lecture

Celebrative stamp for the 100 years of Marconi's invention

in London in 1894, introducing the name of coherer and demonstrating radio communications within the space of a room. However, Guglielmo Marconi verified the transmission of radio waves over distances seriously working for real geographical communications, till the exploitation of the propagation phenomenon (the phenomenon of radio waves reflected by the ionosphere), done by him on 1901. The experiment that demonstrated the radio waves propagation shows how great and deep was the capability of Guglielmo about the science and technology of electricity and circuits and systems as well (he obtained the best ever coherer in that age), especially for applications to radio communication throughout the whole globe. The existence of the ionosphere layers was suggested earlier, but it was not until 1902 that Heaviside proposed this in a serious way, leading to the Heaviside Appleton layer being named in the way that it is.

References

[1] J. C. Rautio, *"The Long Road to Maxwell's Equations"*, IEEE Spectrum December 2014, 32-37.

[2] M. Cheney, *"Tesla - Man out of time"*, Simon & Schuster Inc. New York, (1981).

[3] G.R.M. Garrat, *"The Early History of Radio – from Faraday to Marconi"*, The Institute of Electrical Engineering, (1994).

[4] G. J. Holzmann, B. Pehrson, *"The Early History of Data Netrworking"*. AT&T Corporation, Los Alamos, (1995).

Sandro Carrara
EPFL, Lausanne

Magnetic Flux and Electromagnetic Waves

After Faraday and Maxwell, other two scientists have contributed to the underpinnings of the future of telecommunications: Weber and Hertz. Their contributions to electrodynamics and demonstration of the electromagnetic waves are described below along with a few notes about their life.

Wilhelm Weber

Wilhelm Weber developed with C. F. Gauss the first operational electromagnetic telegraph. Furthermore, in collaboration with R. Kohlrausch, he determined the ratio of electro-dynamical and electrostatic unit of charge which was essential for the reasoning of Maxwell's equations. With his particle and force oriented view of charges, Weber became a precursor of electron theory and metallic conductivity. His name is used as a unit of magnetic flux (the weber) as a tribute to him.

Wilhelm Weber [Wikipedia]).

Wilhelm Eduard Weber was born in Wittenberg, Germany, on Oktober 24, 1804. His father Michael Weber was a professor of theology at the University of Wittenberg. His mother Christiane Friederike Wilhelmine died, as Wilhelm was twelve years old and from the 12 brothers and sisters, only 4 brothers and 1 sister survived. In 1822 Weber became a student at the University of Halle, Germany, and studied philosophy. Weber's brother Ernst Heinrich guided him to natural sciences and together with him, Wilhelm Eduard published in 1825 his first monograph about wave experiments of fluids also applied to acoustic and light waves. Certainly, this work together with his brother Ernst Heinrich was a stimulus for Wilhelm Eduard Weber's studies in physics. Furthermore, Weber had a friendly relationship with E. F. F. Chladni, the well-known founder of the acoustics. In his doctorate thesis in 1826 as well as in his habilitation (academic teaching license) in 1827 he studied coupled oscillations in whistling tongues. Additionally, he published further results of this difficult problem mainly in Poggendorff's Annalen. Consequently, he became a professor at the Halle University in 1828. In the same year Weber met A. v. Humboldt and C. F. Gauss at the 7th Meeting of naturalists and physicians in Berlin, Germany. The two famous scientists were very impressed with the young physicist and probably these connection was helpful with respect to the occupation of an open professorship at the Göttingen University. In 1831, Weber moved to Göttingen. At that time Gauss began to study the magnetic field of the earth and in 1832 he published his first results. Gauss and Weber started a collaboration about the magnetics. At the suggestion of A. v. Humboldt to Weber and Gauss involved in international measuring campaign of the magnetic field. As a first result of their collaboration, Gauss and Weber developed in 1833 their electromagnetic telegraph. With the telegraph the magnetic observatory and the institute of physics in Göttingen were connected and made its constructors famous. It was the first of a series of improved devices for transmitting signals which became very successful in the United Kingdom and the United States. In 1836, the *"Magnetic Club"* (German: Magnetischer Verein) was founded in Göttingen and Gauss was the scientific director and editor of the *"results"*, in which the scientific results were published. Already in 1838 the *"Magnetic Club"* an international association had become, including the United Kingdom. In the same year, Weber lost his professorship in Göttingen for political reasons and he had to continue his work as a private person. Until 1842 six volumes of the *"Results"* were published where most of the articles came from Gauss and Weber. Weber's articles included details about improved or even new measurement equipment of the magnetic field, but also a dynamo machine and some results about electrics.

Weber's Memorial plaque, Lutherstadt Wittenberg, Germany (Wikimedia Commons)

Eventually, Weber accepted in 1843 a professorship at the University of Leipzig, Germany, but as the political situation changed he moved back to the University of Göttingen in 1849. Already in 1847 he developed an elegant theory of electrodynamic interactions based on the action-at-a-distance principle. However, as Maxwell's theory came up, which is based on local interaction, Weber's theory was not accepted anymore. However, Weber proceeded with the application his particle and force oriented view to the problem of conductivity. His ideas which were included in paper from 1875 *About the movement of electricity in bodies of molecular constitution* (in German: Über die Bewegungen der Elektrizität in Körpern von molekularer Konstitution), experienced in connection with H. A. Lorentz's theory of electrons and P. Drude's theory of metallic lead renewed attention. Moreover, in collaboration with R. Kohlrausch, Weber determined the ratio of electro-dynamical and electrostatic unit of charge by measurements in 1856 and his Electrodynamic Proportional Measures from 1864 containing a system of absolute measurements for electric currents, which forms the basis of those in usage. The former results were essential for J. C. Maxwell and his electromagnetic theory of light. 1881 Weber became emeritus.

Wilhelm Weber died in Göttingen on June 23, 1891 in the age of 86 years.

Heinrich Hertz

Heinrich Hertz discovered in 1887 the electromagnetic waves experimentally. Regardless of O. Heaviside and J. W. Gibbs, he formulated Maxwell's equations in the form known today. Therefore, Hertz became one of the founders of wireless communications although he never was interested in engineering aspects of his discovery. But Hertz was an eminent scientist who worked in many other areas. Especially, he opened the door to the research of atomistic models of matter and the special relativity theory.

Heinrich Rudolf Hertz was born in Hamburg, Germany, on February 1857. His father Gustav Hertz was a lawyer and judge. Hertz received his abitur at Hamburg Johanneum and in 1874 he began to study civil engineering at the Dresden Polytechnicum but he concentrated himself to mathematics and exact natural sciences. After his military service Hertz concluded his studies in 1877 at the Munich Polytechnicum. One year later he changed to the University of Berlin and studied with H. Helmholtz and G.R. Kirchhoff and tried to solve a problem that was given by Helmholtz, who formulated an award task for the Prussian Academy of Science. At this time Hertz

Heinrich Hertz [Wikipedia]).

was not successful and he obtained his doctor's degree with a thesis on "*About the induction in rotating balls* (in German: Uber die Induction in rotierenden Kugeln). In 1881 he changed to the area of elastic and plastic deformations of bodies and published a first paper.

After Hertz had gone to the university of Kiel in 1883, he extended his results of his 1881 paper and he received his habilitation (academic teaching license) with a thesis *About the contact of solid elastic bodies* (in German: Über die Berührung fester elastischer Körper). In 1886 Hertz became the successor

of F. Braun at Karlsruhe Polytechnicum where he began with his experimental research of electromagnetic waves. The stimulus to this work was again Helmholtz's award task of the Prussian Academy of Science. At first Hertz studied induction phenomena in isolators and the propagation speed electromagnetic effects along a wire where he found standing waves. He published his results in the Annalen der Physik in 1887 and 1888, respectively. Eventually, he discovered electromagnetic wave in the free space and its reflections, and finally Hertz studied these waves with dipoles and parabolic mirrors and published his results in 1888. By means of his studies Hertz showed

Clebrative stamp in honor of Heinrich Hertz (DB).

that light waves and electromagnetic waves are con-substantial and therefore he confirmed Maxwell's theory of electromagnetics. He finished this period of working with his 1889 paper *The forces of electric oscillations, treated according to Maxwell's theory* (in German: die Kräfte elektrischer Schwingungen, behandelt nach der Maxwellschen Theorie). Subsequently, Hertz became one of the Maxwellians (F. FitzGerald, O. Heaviside, O. J. Lodge, and W. Thomson). Additionally, Hertz studied other areas of physics; e. g. in 1887 he discovered the photoelectric effect. On applications of his discovery of electromagnetic waves in wireless communications Hertz was not interested. The civil engineer Huber asked Hertz in a letter in 1889, whether electromagnetic waves can be used for wireless telegraphy; Hertz gave him a negative answer; cf. *Ch. Süsskind: Hertz and the Technological Significance of Elekctromagnetic Waves. ISIS, vol. 56, no. 185, 342-345, 1965.*

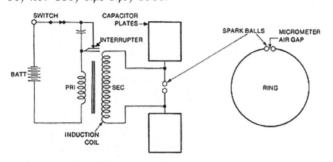

Schematic of the experiment made by Herts to verify the Maxwell's prediction: electromagnetic waves would be transmitted during a series of sparks. (source: http://people.seas.harvard.edu)

In 1889 Hertz followed R. Clausius as professor at the University of Bonn, Germany, and published the last two papers *About the basic equations of electrodynamics of resting bodies* (in German: Über die Grundgleichungen der Elektrodynamik fr ruhende Körper) and *About the basic equations of electrodynamics for moving bodies* (in German: Über die Grundgleichungen der Elektrodynamik fr bewegte Körper) (both published in Göttinger Nachrichten 1890). With his second paper, the problem was not solved by Hertz, but it opened the door to theory of relativity of A. Einstein. Furthermore, he discovered in 1892 the passage of cathode radiation through thin metal layers. This research area was proceeded by P. Lenard who received the Nobel prize for his discoveries in 1905. In 1894, Hertz's fundamental monograph about the new foundations of the principles of mechanics was published.

Heinrich Hertz died in Bonn on January 1, 1894 in the age of almost 37 years on a sepsis.

Wolfgang Mathis
Leibniz University, Hannover

Entrepreneurship in Electrical Engineering

Science and entrepreneurship are the two key components twisted together in the Siemens' life and his contributions to the development of electrical engineering. These are reported below along with a few biographical notes.

Werner von Siemens

Werner von Siemens was an entrepreneur and co-founder of the electrical engineering company Siemens & Halske – nowadays Siemens AG. He developed methods of electroplating, a machine for wrapping wires with gutta percha for the application in submarine cables, a pointer telegraph, the world's first electric elevator, and the so-called trolleybus. His most important invention was a viable dynamo-electric machine, where he used the concept of positive feedback. In a letter to Heinrich von Stephan, Siemens suggested in 1879 the first Electrotechnical Association in Berlin, Germany, and on the occasion of its naming, he coined the word "*Electrotechnik*" (English translation: electrical engineering) for the first time.

Werner von Siemens [Wikipedia]).

Werner von Siemens was born in Lenthe near Hannover, Germany, on December 13, 1816. His father Christian Ferdinand Siemens was a tenant farmers. Siemens attended the Lübeck Katherineum but he was not interested in classical languages and therefore he left this gymnasium without graduation. Siemens followed the advice of a teacher and in 1835 he moved to Berlin, Germany, in order to attend an artillery school where he was educated for three years. During this period of time he also worked as inventor. In 1842 Siemens invented a method for gilding by means of electrical current and in 1846 he developed an improved version of Wheatstone's electrical pointer telegraph, and as a result together with J. G. Halske he founded jointly a telegraph company ("*Telegraphen-Bauanstalt Siemens & Halske*") in 1847. In a short time their new company became very successful, especially with its international business with the United Kingdom and Russia. They developed railroad bells, Water counters, and a machine for gutta-percha wire insulation. After the dead of his parents in 1839/40 he had to earn money in order to support also his brothers and sisters. But there were collaborations with his brothers Carl and William. Under the leadership of Carl a telegraph network was built in Russia and William became successful with the manufacture and laying of sea cables in the United Kingdom.

But also in physical sciences Siemens was active. So in 1866 he discovered the principles for the dynamo-electric machine. The essential aspect of his

Stamps and banknote celebrating Werner von Siemens.

discovery was the application of positive feedback and the remaining magnetic remanence in the iron of the machine. Actually, the basic principle was already invented by Hungarian A. Jedlik (1851/53) and Dane S. Hjorth (1854) but their inventions were forgotten. Together with Siemens but probably independently C. Wheatstone and S. A. Varley have invented the principles of the dynamo-electric machine. Siemens immediately realized the importance of the dynamo and opened up new fields of application; e. g.: electric railway (1879), a street lighting with arc lamps (1879), an electric elevator (1880), a tram (1881) or the *"elektromote"* (a trolleybus) (1882).

Siemens also dealt with the organization of electrical engineering in Germany. He proposed the creation of chairs of electrical engineering at German universities and coined the word *"electrical engineering"* (in German: Elektrotechnik). Siemens also dealt with the foundation of the Imperial Physical Technical Institute (PTR) which was a precursor of today's National Metrology Institute of the Federal Republic of Germany (PTB) – comparable with the NIST in the United States and the NPL in Great Britain. In 1888, the German Emperor, Friedrich II, ennobles Siemens, and extended his surname to *"von Siemens"*.

In 1890 Siemens ended his active participation in the Siemens company, but he retained until his death a certain influence. Werner von Siemens died in Berlin on December 6, 1892 from pneumonia.

Wolfgang Mathis
Leibniz University, Hannover

Power Electric Circuits: Pacinotti and Ferraris

The pioneering age of power circuits

In April 1820 Oersted demonstrated that a current could make a compass needle move: this could be considered as the first electromagnetic actuator. Within a very short time, in September 1820, Ampère formulated his law linking the current to the magnetic field generated. Then, Faraday (1832), in an amazing sequence of carefully planned experiments, demonstrated that a time-varying magnetic field produced an electric field, and so an electromotive force *emf*.

Electromagnetic induction appears to have been taken up soon and used by machine designers. From then on, the fast development of electromagnetic energy converters, and relevant power circuits, followed: first motors and then generators, which are commonly called 'electrical machines'. These devices form the backbone of our industrial heritage, supplying electrical power for countless applications, without which modern society could not exist. From the present time, approximately 180 years on, we tend to overlook the past developments, so that the flow of ideas and applications seems to be a smooth and continuous process. In contrast, several discontinuities characterized the development of generators; the main discontinuity was the controversy between direct current (DC) and alternating current (AC) systems.

The subject is briefly revisited, with special emphasis on Pacinotti's and Ferraris' contribution. A *fil rouge* connecting their work in the industrial and cultural milieu of the time is proposed.

DC generators: the contribution of Antonio Pacinotti

Dynamo-electric machines were operating as early as 1832, when Pixii presented his prototype in Paris. It was basically an AC generator equipped with the Ampère two-sectors commutator for *emf* rectification. Many other generators were prototyped in the next twenty years, but all of them suffered from various limitations, like e.g. sparking effect at the commutator and power loss.

A substantial advance took place in the year 1863, when Antonio Pacinotti (1841-1912), an Italian physicist with the University of Pisa, published a paper in the journal *Il Nuovo Cimento*, where he presented the collector-ring machine. Although he had a degree in mathematics, in the paper Pacinotti did not present any mathematical model to explain the operation of his machine: the focus

Antonio Pacinotti [Wikipedia]).

was just on the prototype. In a sense, he followed the style of Volta, who never presented the mathematical model of the voltage-current law of the pile.

The collector proposed by Pacinotti was characterized by a copper wire which was spirally wound on an iron ring, so forming a multi-turn winding. The two ends of the wire were short-circuited and the ring, mounted on a wheel with vertical axis, was placed between the two poles of an electromagnet. Eventually, two sliding brushes were placed along the direction orthogonal to the pole axis, in such a way to externally contact turns located in diametrically opposed positions. After making the wheel rotate at a constant speed, a sinusoidal electromotive force (*emf*) was induced along each turn; in particular, unlike *emf*s were induced along two opposed turns. Therefore, the voltage between the two brushes exhibited always the same polarity: the originally sinusoidal *emf* was rectified by the sliding contacts, making a unidirectional *emf* available at the two terminals of the contacts. In other words, the dynamo was intrinsically an AC generator, delivering a full-wave rectified *emf* in the basic case of two-turn commutator. With a multi-turn commutator, an almost constant *emf*, approximating DC, resulted. Thanks to the generated power made possible by its configuration, Pacinotti's dynamo was mature for an industrial use. This way, the conversion of primary energy into electrical energy, and its transmission through a line to a remote end user, was technically possible. Moreover, exploiting the reversible operation of the commutator, in the end section of the energy transmission, dymamos operating as motors could have made mechanical power available. The idea was demonstrated at the Bologna Exposition (1869). Transferring energy at a distance was an antique dream that now became reality. Pacinotti realized immediately the applicative impact of his commutator system; regretfully, however, due to the political and economic conditions of Italy at that time, he was not able to patent and implement it as an industrial product.

Pacinotti's generator protoype (courtesy of Museum of Electrical Technology, University of Pavia).

In the year 1867 Werner von Siemens at the Berlin Academy of Sciences had presented a dynamo-electric machine with series excitation, which exploited the weak residual magnetism to activate the *emf* generation. It was a prototype of self-excited machine, characterized by a magnetic circuit exhibiting wide and short air-gaps: this way, the magnetic-circuit reluctance was reduced and, simultaneously, weight and size were decreased. At the Wien exhibition (1875), the golden medal was conferred on Pacinotti, thanks also to von Siemens who acknowledged him as the first inventor of the commutator based on the collector ring. At the Paris Exhibition in the year 1881 Pacinotti presented his original demonstrations obtaining the appreciation of many scientists, like e.g. von Helmholtz, while Edison asked to have an exact copy of the dynamo.

The fifteen most important years for the industrial development of the dynamo took place approximately in the period from 1871 through 1886, when the research and development activity aimed at increasing efficiency, reducing costs, and reducing the ripple of the output *emf*. The implementation of the drum-like armature was the crucial step: in contrast to the original Pacinotti's ring, the internal conductors of the armature circuit were placed on the external surface of the rotor and, therefore, subject to the induction field: this way, they became active conductors contributing to the electromechanical conversion. Moreover, by means of a multi-sector collector, the *emf* ripple at the armature terminals was substantially reduced. The industrial implementability of this configuration was eventually successful: at the Paris Exhibition in the year 1881, Edison presented a dynamo rating a delivered power equal to 120 CV (about 88 kW).

The first European power station: Santa Radegonda in Milan

The first thermal power station in Italy, and Europe, was the one of Santa Radegonda in Milan, so-called after the name of the former church which existed in the same place. It was located in front of the northern transept of the cathedral, where there is now the *Rinascente* building. The power station had been designed and built in the year 1883 by the Edison company under the direction of the chief engineer Lieb. It was equipped with four DC generators (Edison dynamos) rating a total power equal to 352 kW. Powered by steam motors, they supplied series-connected arc lamps lighting the cathedral square, the gallery, and *Teatro alla Scala*. The original idea of building the power station was due to Giuseppe Colombo, professor at the Technical University in Milan. Two years before, Colombo had visited the Paris exhibition, where he had realized the innovative impact of electric supply for lighting instead of gas supply. He succeeded to convince Edison of the technological feasibility of the project as well as entrepreneurs and bankers in Milan of the economical feasibility. In fact, lighting based on arc lamps and, subsequently, electric traction for tramway lines in the town were the first large-scale applications of DC power systems in Milan.

The technology of electric light had a deep influence also on the arts. In the year 1909 Giacomo Balla, famous painter affiliated to the Futurism movement, painted an oil picture entitled *Arc lamp*, subsequently bought by the MoMA in New York. In the scene, light rays originated by the arc lamp overwhelm even moonlight. Balla himself wrote: *"I painted the picture of the lamp during the Pointillism period (1900-1910); in fact, the glare of light is obtained by means of assembly of pure colours. The picture, besides being an artistic achievement,* has also a scientific value because I tried to represent the light by separating the colours which compose it".

The "tyranny" of the pile

Since, from the very beginning, piles had been used to supply electrical power, then it was obvious to everybody that direct current was required from a generator. In its very essence, the dynamo was an AC generator; therefore, the viable solution for rectifying the *emf* was to use a commutator. In a sense, this idea of realizing a DC generator by rectifying an AC *emf* set back the development of industrial electrical engineering by some 60 years, i.e. from Faraday's experiment in 1832 to the Lauffen experiment in 1891, that eventually determined the advent of AC for long-distance transmission.

Nevertheless, the direct current had to run its course, and so in the period from 1870 through 1880 there were extensive DC electrical devices and systems producing and using electrical power. The approach to the magnetic design was unscientific, empirical and pragmatic, and often quite wrong. It was not until 1873-1874 that a good understanding of the link between the magnetic flux and the field coil currents was achieved. In fact, during the period 1840-1873, the science upon which electrical engineering would be based had to establish itself. Initially, there was no agreed units, no reliable measuring instruments readily available, nor agreed ways of specifying magnetic materials. Entrepreneurs simply built machines that produced a voltage, initially using permanent magnets, and later coils.

The DC-AC controversy and the Galileo Ferraris contribution

GALILEO FERRARIS
nato a Livorno Vercellese (Piemonte) il 31 ottobre 1847
morto a Torino il 7 febbraio 1897.

Galileo Ferraris [Wikipedia]).

Galileo Ferraris was born in the year 1847, when von Helmholtz in Berlin published *On the conservation of force*. He died prematurely in the year 1897, when Charles "*Proteus*" Steinmetz, active at the General Electric laboratories in Schenectady, published *Theory and calculation of alternating current phenomena*. The scientific education of Ferraris was substantially influenced by the development of classical electromagnetism. In the year 1869 he obtained the degree in civil engineering at the Technical University in Turin after presenting the dissertation *On the teleo-dynamic transmissions of Hirn*, focused on a system able to transmit mechanical power at a distance. Three years later, in 1872, he presented *On the propagation of electricity in homogeneous solid media*. In 1878 he published two papers, entitled *On the proof of Helmholtz principles on the temper of sounds derived from telephone experiments*, and *On the strength of electric currents and extra currents in the telephone*, respectively. In 1873 the monograph *Theorems on the distribution of constant electric currents* was published. In 1881, he was the official representative of the Italian government at the Paris exhibition; in that occasion he presented two reports entitled *On the industrial applications of electric current*.

The first mature contribution of Galileo Ferraris was the scientific theory of the transformer, i.e. the first systematic study on Gaulard's secondary generator. In 1884 Ferraris was the organizer of the electrical section of the International Exhibition in Turin, where the Gaulard's generators were one of the main attractions. In three subsequent reports in the years 1885 and 1886, Ferraris proposed a full model of the device, considering the power loss in the transformer core under AC conditions at the frequency $\omega/2\pi$. Specifically, he introduced the time shift τ between magnetic flux density $B(t)$, which depends on the applied voltage $v(t)$, and field strength $H(t)$, which depends on the coil current $i(t)$. Ferraris proved that the power loss in the transformer core, due to eddy currents and hysteresis, depends just on the time shift τ. He referred the power loss to quantities measurable at the terminals of the transformers: in particular, he identified the expression of active power $P = VI\cos\Phi$, where V and I are the root-mean square values of voltage and current, respectively, Φ is the relevant phase shift, while $\cos\Phi$ is the so-called power factor. Eventually, Ferraris introduced a new model, characterized by an ideal two-phase transformer plus a short-circuited tertiary winding which carried the loss currents present in the core. This way, he proposed a model based on three equations; after manipulating them, the same results previously obtained by means of the model in terms of power factor were independently found. As a result, the classical theory of the transformer, incorporating the concept of efficiency, was clearly defined.

His investigations laid the groundwork for the industrial use of AC. In fact, going on with his studies on the transformer, Ferraris realized the need for a motor exhibiting spontaneous rotation: only this way, in fact, AC systems would have overperformed DC systems. Without such a motor, any implementation of alternators and transformers would have been of modest importance. In a technical note to Ganz company in Budapest, he expressed this idea saying that the *"transformer is nothing but a component of the whole power system"*. He understood that two or more AC magnetic fields could be superposed in such a way to obtain a rotating field, producing the same effects of a permanent magnet which rotates. The result was the famous Ferraris' theorem: given two equal coils with orthogonally directed axes, given two sinusoidal currents with same amplitude, same angle frequency ω and 90° out-of-phase, if each coil carries the relevant current, then a magnetic field rotating at the speed ω takes place at the centre point of the coils (two-phase field). The core of the achievement was just this: the mechanical speed of the magnetic field was coincident with the electrical frequency of currents injected in circuits being at rest. More generally, the result holds for three currents and three coils, if both electric and mechanical angles are 120° out-of-phase (three-phase field). On May 1, 1888 Nikola Tesla, with the Westinghouse company, patented an electric motor which exploited the rotating field invented by Ferraris.

In practice, Ferraris succeeded to obtain the two quadrature currents from a unique current source; to do this, he used a small alternator and the two coils of the transformer he had been studying: the primary current was injected in a coil, while the secondary current was injected in the other coil. By regulating resistances and inductances, he eventually obtained two currents 90° out-of-phase.

Rotating field realization (courtesy of Museum of Electrical Technology, University of Pavia)

Having set the two-phase field, he placed a hollow cylinder made of copper in the field: the cylinder, free to move, started to rotate. This way, the transition from conduction machine to induction machine took place.

Ferraris published his results only in the year 1888, when he presented a lecture entitled *"Electrodynamical rotations produced by AC"* at the Science Academy in Turin. The reason of the delay might be attributed to the methodological approach followed by Ferraris: he was not keen of publishing new results based just on experimental evidence; in turn, he aimed at proving their theoretical evidence in the context of the Maxwellian electromagnetism he had been a major interpreter of.

From Santa Radegonda to Lauffen and Frankfurt

The system proposed and implemented by Edison had solved the problem of delivering electric energy at a short distance by means of electromechanical conversion in DC regime; then, the next step was to transfer the electric energy to a remote end user. At this point in time, however, a new problem arose, i.e. the independence of generation voltages with respect to transmission voltages. In fact, at the level of generator machine and user circuit, voltages had to be kept in the limits dictated by sizing criteria and safety, respectively. In contrast, the increased generator-to-user distance made it necessary to increase more and more the voltage values in order to mitigate transmission Joule's losses.

In the year 1891 a crucial experiment took place. An AEG-type three-phase alternator rating a power equal to 230 kVA at the voltage of 95 V with speed of 150 rpm was installed in Lauffen, Germany. The alternator was connected to a hydraulic turbine installed in a cement factory located next to the Lauffen falls on the Neckar river. In the same time, two AEG-type transformers were operating: a step-up transformer increased the voltage to 15 kV for transmission along a three-cable line, connecting Lauffen to Frankfurt am Main over a distance of 175 km. A step-down transformer decreased it to 113 V at the line end, located in the exhibition area in Frankfurt, where a three-phase induction motor rating 75 kW was supplied by the line. The motor made a pump work, eventually activating an artificial water fall. The successful experiment determined the advent of three-phase AC systems for transmission. From the scientific viewpoint, it was the triumph of Ferraris' ideas.

The modern synthesis of Pacinotti's field and Ferraris' field

In the year 1873, Pacinotti published another paper, again in the journal *Il Nuovo Cimento*, in which he showed how to generate a rotating magnetic field by making the brushes of the collector ring rotate (Pacinotti's field). The condition to do this was simply the injection of a suitable AC current in the brush terminals, with a current frequency dependent on the speed of the brushes with respect to the field speed. Thanks to this new interpretation of the dynamo and, more generally of a collector machine, the Pacinotti's field and the Ferraris' field could have been magnetically equivalent at the air gap.

Accordingly, in the year 1925 Riccardo Arnò, former assistant of Ferraris and professor at the Technical University in Milan, put forward the following remark during a conference at the AEI (Italian Society of Electrotechnics): *"Pacinotti succeeded to keep an ideal magnet fixed in space, despite the relevant electric circuit was rotating. Conversely, Ferraris succeeded to make an ideal magnet rotate, despite the relevant electric circuits were fixed. All DC and AC motors nowadays working rely upon these two principles"*.

This view suggested the transformation of a three-phase and, more generally, a poly-phase machine into a collector machine, energetically equivalent as far as air-gap effects and terminal effects are concerned. In particular, it would have been conceptually possible to derive poly-phase fields from multi-sector collectors equipped with a number of rotating brushes suitably located: in a sense, the collector machine became a universal model. The subsequent unified theory of electromechanical conversion, dating back to the contribution of Gabriel Kron, exploited just this principle.

Eventually, the modern development of the power electronics field, based on semiconductor-controlled valves, made it possible to activate static switches according to a sequence of width-modulated pulses. Applied in the control of electrical machines, electronic converters generated a rotating magnetic field at the air-gap of brushless permanent-magnet motors and generators; this way, the synthesis of DC and AC devices took place. There was a long history behind; Nobel laureate (1985) Carlo Rubbia said: *"Volta was the first to introduce electricity as a usable form of energy, but the subsequent developments in the use of electric power originated from two other Italian contributions: the dynamo by Pacinotti and the rotating magnetic field by Galileo Ferraris"*.

References

G. Ferraris: Lezioni di Elettrotecnica. Torino, STEN, 1899. In Italian.

G. Ferraris: Opere. Milano, Hoepli, 1902. In Italian.

M. Guarnieri, *"The Beginning of Electric Energy Transmission: Part One,"* IEEE Industrial Electronics Magazine, pp. 50-52, vol.7, no.1, March 2013.

M. Guarnieri, *"The Beginning of Electric Energy Transmission: Part Two,"* IEEE Industrial Electronics Magazine, pp. 52-54 and 59, vol.7, no.2, June 2013.

J. Kron, *"Generalized Theory of Electrical Machinery,"* AIEE Transactions, 1930.

A. Pacinotti, *"Descrizione di una macchinetta elettromagnetica,"* Il Nuovo Cimento, 1863. In Italian.

A. Pacinotti, *"Sopra una piccola macchina dinamo-elettrica,"* Il Nuovo Cimento, 1870. In Italian.

A. Pacinotti, *"Sulla elettrocalamita trasversale ruotante adoprata come elettro-motore,"* Il Nuovo Cimento, 1873. In Italian.

C. Rubbia, *"From the Voltaic pile to the fuel cell,"* in Alessandro Volta nel Bicentenario dell'invenzione della Pila 1799-1999, Grafica Marelli, Como, 2002, pp. 330-339.

A. Volta, *"On the Electricity excited by the mere Contact of conducting Substances of different kinds,"* Letter to Sir Joseph Banks, Royal Society of Physics, 20th March 1800.

V. Cantoni[1]*, P. Di Barba*[1,2]

1- Dept. of Elect., Computer and Biomedical Engineering, University of Pavia, Italy.
2 - Research Centre for the History of Electrical Technology - CIRSTE, University of Pavia, Italy

History of Electronic Devices

Passive components enable many processing functions used in circuits and systems. However, the addition of active functions, mainly the ones granted by the use of operational amplifiers, transconductors and comparators, significantly expands the processing capability and offers substantial benefits. The historical evolution of devices used in active building blocks is essential for the history of circuit and systems. The first active device were the electronic tubes, then they were replaced by the bipolar transistors followed by an astonishing sequence of technological versions of the MOS transistor.

The Electronic Tube

The appliance that paved the way for developing the first electronic device was invented more than two centuries ago. Its function was not to obtain amplification but for producing light. It was around 1800 when a British scientist, Humphry Davy, discovered that a current flowing in a strip of metal (a filament) can increase its temperature to a very high level and possibly reaches the point of incandescence. However, the filament was quickly consumed by combustion. The invention was then improved by Thomas A. Edison who placed the filament into a glass the bulb with the vacuum inside.

Thomas A. Edison [Wikipedia]).

That allowed the filament to glow for long periods without being consumed. While doing his experiments in 1883, Edison discovered that inserting a second metal strip inside the bulb there was a current flowing from the hot filament to the second electrode. The reason was that hot metals emit electrons that travel toward the second electrode. The flow of current was enhanced when applying a battery between the two terminals but it became almost zero when the polarity of the battery was reversed. Edison saw no practical benefit of that

observation, called Edison effect, and concentrated his efforts on refining his 1880 filed patent for an incandescent lamp.

The Edison effect drew the attention of John Fleming who after his graduation in Cambridge and a short period as Professor of Physics and Mathematics at University College Nottingham, took a position at the Edison Telephone Company. Fleming had the chance to see many of the Edison's inventions and became aware of possible benefits. Some years later after he had joined University College London (1904) Fleming developed a device based on the Edison effect and called it *"valve"*. Its function is the one of a diode or one-way valve for electric current - as it only allows current to flow in one direction. The intended use of the device was as a detector in radio receivers. Because of the possible uses of

Fleming   valve [Wikipedia]).

electronic apparatus, the Fleming valve is considered the first invention in the long history of electronic devices.

John Fleming (left) Lee de Forest (right) [Wikipedia]).

Fleming was an excellent researcher and lecturer. When teaching, he put a lot of care in demonstrations and frequently used examples to remembering things. Among them, there are the left-hand and right-hand rules for relating magnetic field, current and force in electrical machines. Even today, these rules are called Fleming's right and left hand rule.

The Fleming invention and the Edison effect were the bases of the discovery of the American Lee de Forest who some time before 1906 started inspecting the phenomenon for gaining more benefits. He placed a metal screen between the hot filament and the metal strip and found that the current flowing from filament to the other electrode could be regulated by the application of a small voltage between the metal screen, that he called grid, and the filament. De Forest called his invented device *Audion* tube. He found that the control of the current both depends on the voltage between grid and filament and its polarity. A negative voltage opposes the flow of electrons, whereas a positive voltage enhances the flow. Actually, there was a current through the grid, but it was very small when compared to the current through the plate. That means that with small efforts it is possible to control a much larger power.

De Forest was a prolific inventor; he totally filed over 300 patents. However, almost none of them were successful. The Audion tube, instead, being the first device able to perform amplification, was successful and used by the telephone company for transcontinental wired phone calls. The Audion, known now as the triode, was not only a brilliant invention but also the source of trouble and financial turbulences. De Forest was a poor businessman and after a first company was forced to close its operation because of insolvency, he founded the *"De Forest Radio Telephone Company"*. Around 1909, the company was in trouble because of federal charges of misconduct for seeking to promote a *"worthless device"* – the Audion tube. De Forest was later cleared but because of the financial difficulties he reluctantly had to sell the rights of his Audion patents at very low price to a lawyer that was acting for the American Telephone

& Telegraph Company (AT&T). The Audion was then largely used by AT&T in telecommunication systems as an essential amplification component for repeater circuits in long-distance communication. A modified version of the de Forest Audion was then used by Western Electric

Electronic tubes [Wikipedia]).

Co.(WECo) (1915) to build an amplifier for the coast to coast telephone communication.

The anode-grid capacitance of the early valves was very high and this caused unwanted oscillations especially when the frequency of operation was at hundreds of kHz. The solution to the problem was found by H. J. Round who in 1916 had the idea of decoupling the anode from the grid by passing the anode connection out of the top of the glass container instead than going inside the envelope.

In the attempt to increase the amplification factor more grids were added to the triode. An electronic tube with two grids, named tetrode, was patented by William S. Schottky of Telefunken Company, Germany in 1916. The four electrodes are one inside the other with the thermionic cathode in the center, first and second grids and the

H. J. Round (left), W. Schottky (right) [Wikipedia]).

plate. There were distinct types of tetrode tubes with different functions of the two grids. In the screen grid electrode tetrode proposed by Albert Hull, the first grid is the control grid and the second grid is the screen grid. In others, like the space-charge grid tube or the bi-grid valve the role of the control grid and second grid changes according to the expected use. The screen-grid valve was used for medium-frequency, small signal amplification, the beam tetrode was for high power radio-frequency transmissions.

The pentode has an additional grid. It was invented by G. Holst and B. D.H. Tellegen in 1926. The five electrodes inside the evacuated glass container are the heated cathode, the control grid, the screen grid, the suppressor grid, and the anode (or plate). The role of the suppressor grid placed before the anode was to prevent secondary emission electrons emitted by the anode from reaching

A. Hull (left), B. D. H. Tellegen (right) [Wikipedia]).

the screen grid, thus avoiding possible instability and consequent oscillations.

The bulky vacuum tubes have been gradually replaced by solid state devices-However, there are niches of applications that still require tubes and not solid state devices. These are, for example, the cases of high power radio frequency generators for particle accelerators, and broadcast transmitters.

The Bipolar Transistor

The bipolar transistor was the answer to the need of replacing the amplifier made by vacuum tubes because their implementations were bulky, used a lot of power and had a short life. The transistor demonstrated on December 1947 by John Bardeen and Walter Brattain as the point-contact type was anticipated by several studies and discoveries with solid state materials and devices.

Preliminary researches on solid state amplifiers lead to a point-contact zincite (ZnO) crystal diode able to obtain signal amplification up to 5 MHz. It was designed in 1922-23 by Oleg Vladimirovitch Losev, a Russian engineer that worked at the Nizhny Novgorod, Nizhegorod.

Another early discovery was the one of the Austro-

O. Losev (left), J. E. Lilienfeld (right) [Wikipedia]).

Hungarian born physicist, Julius E. Lilienfeld; in 1926, he filed a patent for a *"Method and Apparatus for Controlling Electric Currents"* based on a copper-sulfide three-electrode amplifying device. Lilienfeld was not able to build the amplifier but the patent description had a strong similarity to the later field effect transistor and this prevented future scientists from patenting the field effect structure.

At the beginning of the 20th century radio operators tried to obtain signal amplification by using different materials and modifying the point contact rectifier made by the junction between a fine metallic wire (the so-called 'cat's whiskers') and the surface of certain crystalline materials. A reported example is some amplification achieved using a silicon carbide (carborundum) crystal and two cat's whisker. Another news concerns professional radio operators crystal detectors made by carborundum instead than galena (lead sulfide). They used carborundum in the harbour because being close to a transmitter that material prevented burnout. They also used carborundum at long distances using two cat's whiskers, one of them was excited with a battery for getting higher wave oscillations.

Around 1940 the possibility to have some form of solid state devices was close to reality. At that time Mervin Kelly, Bell Labs Executive Vice President, created a solid state device group. Kelly thought that the reliability and size of the vacuum tube were such that something needed to be done, besides making that device more efficient and smaller. *"Employing the new theoretical methods of solid state quantum physics and the corresponding advances in experimental techniques, a unified approach to all of our solid state problems offers great promise."* he wrote in a memo, *"Hence, all of the research activity in the area of solids is now being consolidated ..."*.

Mervin Kelly. [University of Chicago Photographic Archive]

The persons in charge of the Solid State Physics Group were William Shockley and Stanley Morgan. Shockley with a solid background in solid state physics theory started studying field effect amplifiers but the experiments failed despite the use of many different device configurations and materials. The reason for the failures was found by John Bardeen, a theoretical physicist at the University of Minnesota who joined Bell Labs. His calculations demonstrated that a relatively low concentration of surface states screens the voltage applied to the electrode operating the gate. Another member of the team, Walter Brattain, partnered with Bardeen in charge to build and run experiments that Bardeen, as theoretician suggested and interpreted. They made many attempts until, without saying this to Shockley, they built a device made from strips of gold foil placed on the faces of a plastic triangle. The triangle was pushed down into a slab of germanium to obtain two very close contacts. That was the invention of the point of contact transistor.

Bardeen and Brattain told Shockley of the invention who was at the same time pleased with the results and disappointed because he was not involved in the experiments. He considered the point contact transistor a very good result but not completely satisfactory because the device was fragile and difficult to manufacture.

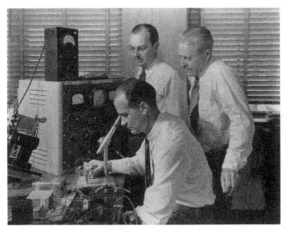

Bardeen, Brattain and Shockey at Bell Labs [Wikipedia]).

The Bardeen and Brattain achievement galvanised Shockey into action. Bardeen supposed that the amplification was due to a change of conductivity of the surface layer. Shockley was thinking of an effect caused by the bulk of the crystal. In just two months, after the point of contact invention, Shockley developed the bipolar junction transistor theory, based on previous studies on minority carriers that Shockley had already made. The observation that the distance between the two contacts of the point contact transistor was very small suggested that having a very thin layer of material in between two silicon pieces with complementary doping could have the similar effect but with a more solid and reliable implementation. Based on that Shockley defined a new type of transistor that he called *"junction"* transistor.

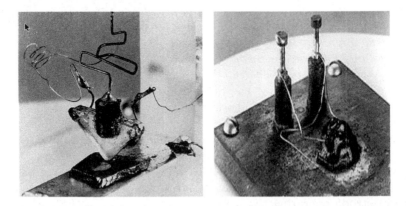

The point contact transistor (left) and the bipolar transistor (right) [Wikipedia]).

The experimental verification of the idea was done shortly after by John Shive who used a very thin piece of germanium with emitter and collector on the two sides of the crystal. The experiment verified the idea that the effect was due to minority carrier injection: the theory of Shockley was correct.

The three scientists had a stormy relationship because of the issue of who deserved the credit of the first invention. Since Shockley gave the initial idea for the transistor, he started the process to patent the transistor under his own name but that obviously disappointed Bardeen and Brattain. The consequent fight also motivated the spark of creativity that led Shockley to develop the bipolar junction transistor theory.

The frequency response of point-contact devices was superior than the one of junction-based transistors and because of this Bell Labs delayed announcing the Shockley achievement until 1951. Silicon was used few years later, in the mid-1950s. It is the semiconductor material still used in today's integrated circuits. Indeed, the use of germanium has advantages, among them the lower melting temperature and the higher mobility of electrons and holes that leads to higher frequency response. However, the key problem is that germanium has a low band gap (0.67 eV while silicon has 1.12 eV) and the leakage current in the junction is much higher than for silicon. Because of that limit the temperature range of operation is limited, so at more than 60-70°C it is almost impossible to have a working device.

Bardeen, Brattain, and Shockley were jointly awarded the 1956 Nobel Prize for inventing a device capable of obtaining solid state amplification. Bardeen and Brattain invented the point contact transistor; Shockley the bipolar transistor. Despite the joint prestigious award and the well known Bell Labs photo that shows Shockley seated at a microscope and the other two behind him, the three scientists did not continue working together. Only Brattain remained at Bell Labs. Bardeen moved to the University of Illinois where he was working on superconductivity. For that activity, he was awarded by a second Nobel Prize in Physics in 1972. Shockley left Bell Labs in 1955 to found Shockley Semiconductor in Mountain View, California.

The Nobel prize Medal.

Shockley hired the best talented scientists. Among them, Gordon Moore and Robert Noyce. The partnership lasted a short time. Eight engineers decided to quit and start a new company, Fairchild Semiconductor. The company focused the fabrication of silicon transistor based on a new process called *"mesa"*. The success of that small reality was then determined by the interest of IBM that ordered mesa silicon transistors for memory drivers.

The MOS Transistor and the Integrated Circuits

The device that is still in use for integrated circuits with annual sales of hundred of billion US dollars (US$ 294.97 billion in 2015) is the MOS transistor. After the precursor works attested by three patents filed by Lilienfeld in 1926 and

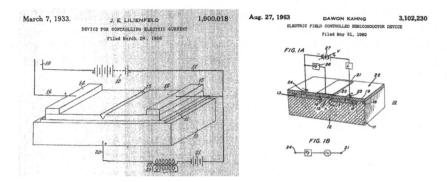

Drawings from the Lilienfeld patent (left) and the Kahng Patent (right).

1928 (granted few year later), many researchers tried experimental verifications of the device. Under Shockley's direction, Walter Brattain and others worked on such a three-electrode device but all the attempts they made were unsuccessful. The problem was the ability to overcome surface states that mask the voltage applied to the gate. Only in 1959 M. M. (John) Atalla and Dawon Kahng at Bell Labs realised the first successful insulated-gate field-effect transistor (FET).

Atalla, the head of a Bell Labs group working on surface problems, focused on thermally grown silicon-dioxide layers and found that the surface states are greatly reduced at the interface between the silicon and its oxide; a sandwich of metal, oxide and silicon was, therefore, the basis of the MOS transistor. Dawon Kahng, a member of Atalla' group, built the device and together with Atalla, presented the experimental results at a conference in 1960. However, the first realisation was 100 times slower than the best junction devices and because of this limit, it was considered almost useless for telephone applications. Kahng, instead, was certain of the benefits of the device because it was easy to fabricate and had a potential use in integrated circuits: the fabrication technology required fewer processing steps than bipolar technology; furthermore, the device admits scaling without compromising performance.

Martin M. (John) Atalla [Wikipedia]).

In 2003 Purdue University awarded Martin M. (John) Atalla the Engineering honorary degree.

Bell Labs did not recognise the value of the silicon MOSFET because of the focus on the bipolar junction-transistor and the development of the epitaxial technology. On the contrary, researchers at Fairchild and RCA perceived the potential advantages. The first time the Atalla passivation technique was used was in 1960 at Fairchild Semiconductor Corp. as an improvement in the fabrication process for bipolar transistors (the planar process). The same year Karl Zaininger and Charles Meuller at RCA and C.T. Sah at Fairchild fabricated

MOS transistors.

Over a few years many other firms fabricated MOS devices for some specific application. However, the low benefit compared to the bipolar device was such that the MOS gained a negligible part of the semiconductor market. It was the use of MOS transistors in integrated digital circuits that boosted use of the device. The technology steadily increased its importance with digital applications and the use of integrated circuits in the computer.

Quantum effects arose in solid state electronics with the invention of the tunnel diode in 1958 by Leo Esaki. This involves electrons moving through a potential barrier when they do not have enough energy to do so. In classical physics terms, they have zero probability of crossing the barrier, but with quantum tunnelling, some can nevertheless appear on the other side. This conduction gives rise to a negative resistance which can be used to make oscillators and amplifiers. Although tunnel diodes still have some special applications, the hopes of the early 1960s that they would become widely used devices did not materialise.

In a 1963 two researchers of the Fairchild R&D Labs, C. T. Sah and Frank Wanlass, discovered that a logic circuit made by a p-channel and n-channel MOS transistors connected as an the inverter requires almost zero power in standby mode. That was the invention of what we call today CMOS technology.

The invention of the integrated circuit, more than one electronic component on the same semiconductor chip, dates back to September 1958 when Jack Kilby at Texas Instruments realised a simple (flip-flop) circuit with two bipolar transistors on a single chip of germanium. For the connections between the transistors he used gold wires. Because of this that integrated technology can be considered 'hybrid'.

Jack Kilby (left) and Robert Noyce (right) [Wikipedia]).

Shortly after (1959) Robert Noyce, nicknamed "the Mayor of Silicon Valley" and Fairchild co-founder, eliminated the wire connections of the Kilby invention by aluminium metal strips. The patterns where produced by photolithography and deposited on top of the oxide used to protect the circuit. The method, called planar process, was more effective than the Kilby approach to realise a complete electronic circuit on a single chip of semiconductor. Noyce filed the patent, *"Semiconductor device-and-lead structure"* in July 1959 and Fairchild fabricated the first monolithic ICs in 1960. Texas Instruments and Fairchild had a long litigation for the right of the integrated circuit and eventually settled with an agreement just before the court decided in favour of Fairchild. Noyce passed away in 1990. Kilby, who died in 2004 received the Nobel prize in 2000 for the invention of the integrated circuit.

Comparison of the μA702 with a 2N1613 planar transistor.

Robert (Bob) Widlar, while at Fairchild, invented the integrated circuit OpAmp, beginning with the 702 and then the 709. These were rather difficult to use because the open-loop gain needed to be shaped carefully in order to avoid instability and oscillation when the feedback loop was closed. However, this was followed by the 741, which had a low-frequency (∼10Hz) dominant real pole which meant that the gain fell by 6 dB per octave over a substantial frequency range, keeping the phase shift at close to 90 degrees. As a result the 741 was almost always stable when the feedback loop was closed. The open loop gain was typically over 100,000, giving a gain-bandwidth product of about 1MHz, so that useful gain on closed loop was still readily available over much more than the audio-frequency range.

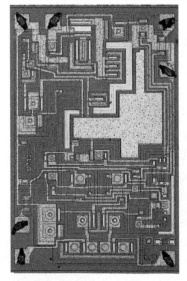

Microphotograph of the 741 OpAmp.

By 1974, it was possible to obtain four 741 OpAmps in a single 14 pin dual-in-line package (two pins for +12V and -12V supply, and three for each of the four amplifiers).

At last the analog filter designer had a nearly-ideal active component to create practically useful active-RC filters.

Moreover, the idealised OpAmp corresponded exactly to the nullor introduced long before, and enabled simple nodal analysis to be used for networks with such idealised OpAmps: each nullor in a non-degenerate location introduces asymmetry into the otherwise symmetric nodal admittance matrix of a passive network, and reduces by one the order of the matrix.

Subsequently much improved OpAmps have been designed, but it was the 702, 704, 741 sequence which pioneered this important development.

Origins of the Microprocessor

The invention of the microprocessor is generally credited to M.E. (Ted) Hoff, employed by Intel. The company was asked to provide a wide range of different integrated circuits for calculators by a Japanese company, Busicom. Hoff proposed making a programmable scheme of three chips, which could be adapted for many different requirements without any re-design, and the four-bit CPU became the Intel 4004, designed by Federico Faggin, Italian by birth and education, naturalized US citizen. It was soon followed by the 8008, an 8-bit version, and then by the 8080 which was followed a year later by the Motorola 6800. Faggin left Intel to form Zilog, creating the very successful Z80. These microprocessors were powerful enough to create a huge range of applications, and to radically change the way electronic systems would be designed. With the aid of the continuing success of Moore's Law, processors of rapidly increasing performance were produced, which has led to the dominance of programmable microelectronics in the modern world. The technology moved from PMOS to NMOS to CMOS, the last enabling very low power operation. Initially there were many very different

F. Faggin [Wikipedia]).

microprocessor designs produced by many different companies, with little standardisation. Intel, AMD and Motorola and Texas Instruments came to dominate the market.

RCA Research Laboratories improved the CMOS technology and achieved very low-power integrated circuits. Gerald Herzog designed CMOS logic and memory circuits in the context of an Air Force program for realising a computer in 1965. RCA fabricated a 288-bit static RAM in 1968 and shortly after produced a family of general-purpose logic devices. As a result, RCA was the first to develop a CMOS microprocessor, the 'COS-MAC 1802, in the mid 1970s, having very low power consumption, for use in control of car engines. Intel introduced CMOS versions of their 8049 and 8051 microcontrollers later, in 1983.

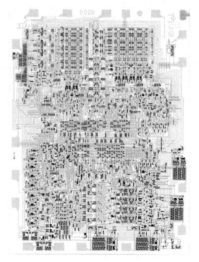

The Intel 4004 was in an 18-pin dual-in-line package, but the requirement for more connections led to most subsequent microprocessors being in 40-pin dual in line packages.

Layout of the Intel 4004 microprocessor. (source: Intel)

Being digital devices, time-multiplexing (e.g. of data and address lines) could be used to reduce pin-count. Texas Instruments put their 9900 microprocessor in a 64-pin dual-in-line package, but that was the upper limit of this configuration, and many other package formats were developed, including pin-grid arrays for up to several hundred pins, and formats to support various flow-soldering, surface mount and high-reliability applications as well as increasing miniaturisation.

Progress in solid state memory has been equally remarkable. Early computers used ferrite-core based components for realising memory. An initial alternative was with bipolar technology but the circuits required many transistors per bit (and a continuous refresh of stored data), and so was possible only for small fast data stores.

Magnetic-core memory storing 1024 bits of data.

Initially, Intel produced PMOS static random access memory chips of less than 1Kbit, but the storage capacity has continued to increase rapidly. Each cell is basically a bistable flip-flop and therefore requires several transistors. For much higher density, the Dynamic RAM was developed, which stores the state of each cell on a capacitor and requires only one transistor per cell. Single DRAM chips with a huge storage capacity are now readily available.

In addition, many other forms of storage have been developed, from programmable ROMs, erasable ROMs (initial versions requiring ultraviolet light erasers, in contrast to electronically re-writeable ROMs). Solid state storage of many Gigabytes is now economically possible in laptop computers.

The application of MOS based integrated circuits to computers and microprocessors gave the MOSFET the importance it enjoys today.

Franco Maloberti Life Fellow IEEE,
University of Pavia, Italy

Anthony C Davies, Life Fellow IEEE,
Emeritus Professor, King's College London, England

The History of

The Resonant Circuit

Bells and whistles have been accessories of humankind since the beginnings of civilization and the key physical property involved in the sounding of these instruments is *resonance*. All musical instruments without exception rely on resonance. The string of a guitar would sound unimpressive if strung between two nails on a stick whereas the mouthpiece of a clarinet would sound no better than a cheap whistle without the body of the clarinet. Yet, resonance is often a nuisance, for example, in cars and aeroplanes, and on occasion it can cause disasters of unbelievable magnitudes, for example, when the occurrence of unanticipated resonance of unbelievable strength destroyed the Tacoma Narrows Bridge soon after it was built in 1940 [1].

The most familiar manifestation of resonance occurs in a pendulum of any size or shape, including a kid's swing. By bending the knees at the appropriate moment during the swing cycle, boys and girls can swing for hours. What makes this possible is the periodic interplay between kinetic and potential energy and, similarly, electrical resonance in a resonant circuit is enabled by the periodic back-and-forth movement of energy from a capacitor and an inductor. As far as this article is concerned, resonance has facilitated the discovery of electromagnetic waves and led several years later to wireless communications, broadcasting, and long-distance telephony. These technologies led to the formation of the American Institute of Electrical Engineers and the Institute of Radio Engineers and after their merger in 1963 to the formation of the Institute of Electrical and Electronics Engineers we know today. Over the years, IEEE and its founding organizations has spawned numerous technical societies, one of the early ones being the IRE Circuit Theory Group.

This article is a modest attempt to trace the evolution of the resonant circuit and its application to wired and wireless communications over the past two and a half centuries. Some great men, for example, Hertz and Tesla, made contributions to its evolution and are mentioned here. Many of these men made other contributions that are not specific to the resonant circuit and a few of those contributions are discussed in detail elsewhere in this volume. On the other hand, the resonant circuit gave rise to entire technologies, for example, passive filters. These technologies are mentioned in passing here but interested readers can refer to the more specialized articles in this volume.

The Leyden Jar

The beginnings of the resonant circuit go back to the invention of the Leyden jar by Pieter van Musschenbroek at Leyden University in the Netherlands in 1746 [2, 81-95]. Its basic construction, illustrated in Fig. 1a, immediately reveals that this is the very first implementation of a capacitor, also commonly referred to as a *condenser* in the past. This was more of a laboratory apparatus than an electrical component, which was used to study the wonders of electricity. The basic function of the Leyden jar, namely, that it stores electrical charge, was known quite early in the history of electricity but a puzzling phenomenon manifested itself some eighty years later. In 1826, Felix Savary observed that

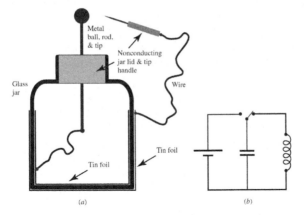

Fig. 1 - (a) Leyden jar, (b) equivalent circuit

a steel needle magnetized by the discharge current of the Leyden jar did not always have the same polarity [3, p. 54]. This was deemed to be a strange behaviour. The Leyden jar was charged by a direct-current source and, naturally, one would expect a direct current discharge. Savary accurately predicted the presence of *oscillations* of some sort. New discharge experiments were carried out in 1842 by Joseph Henry some 15 years later by magnetizing steel needles using the current induced in a conductor placed near the discharge conductor of the Leyden jar. In a presentation to the American Philosophical Society, he concluded that "... *the phenomena require us to admit the existence of a principal discharge in one direction, and then several reflex actions backward and forward, each more feeble than the preceding, until the equilibrium is obtained* ...". In 1847, Hermann von Helmholtz who is well known for numerous achievements in physiology and physics presented a paper on the conservation of energy at the Physical Society of Berlin and the Leyden jar was one of his illustrations of the principle of the conservation of energy. He concluded, among other things, that ".... *the discharge of a battery*[1] *is not a*

---

[1] A battery was an array of interconnected Leyden jars in those days (see [4]).

simple motion of the electricity in one direction but a backward and forward motion between the coatings, in oscillations which become continually smaller until the 'vis viva'[2] is destroyed by the sum of the resistances. ...". (See [5] for details.)

In the early history of the resonant circuit before the early 1850s, the capacitor's partner in crime, the inductor, was hidden in the metal rod with the ball at the end and the connecting wires. Including the sum total of the stray inductances, the equivalent circuit illustrated in Fig. 1b is immediately revealed. In 1853, Sir William Thomson, known as Lord Kelvin to all scientists, published a seminal paper titled *"On transient electric currents"* [6]. In it, he provided a mathematical characterization of the discharge of a capacitor through a conductor. In the process, he identified the missing link, the *'electrodynamic capacity'* or *inductance* as we refer to it today. What he did, was to formulate a second-order differential equation with constant coefficients, in today's language, whose solution was consistent with the observed behaviour of the resonant circuit. He found out that the solution depends critically on the relative values of the capacitance, inductance, and resistance values in today's terminology and that under certain circumstances there is no oscillation at all, the *over-damped* case.

Discovery of Electromagnetic Waves

The most dramatic use of electrical resonance occurred in 1887 when Hertz demonstrated by experiment that electromagnetic waves that behave very much like light exist as proposed by Maxwell in two presentations delivered to the Cambridge Philosophical Society in 1855 and 1856. The experimental set-up of Hertz comprised a *transmitter* made up of an induction coil of that time, known as a *Ruhmkorff* coil, and two metal spheres connected by copper wires to a spark-gap mechanism, as illustrated in Fig. 2a. There is no doubt that Hertz was familiar with resonance in mechanical systems, e.g., the pendulum, as he was a recent physics graduate from the University of Berlin having studied under Gustav Kirchhoff and having been appointed Professor of Physics at Karlsruhe University two years after graduation. He was also very knowledgeable about ongoing research with the Leyden jar and knew about the presence of oscillations in his experimental set up [7, pp. 241-245].

The metal spheres provided a capacitance C and the wiring to the spheres provided inductances L_1 and L_2 as shown in the modern schematic in Fig. 2b. The induction coil served as a step-up transformer, not unlike a modern transformer, and its main purpose was to charge the capacitor formed by the spheres. Ignoring secondary effects, when the spark gap began to conduct, capacitance C along with inductances L_1 and L_2 form a *parallel resonant circuit*[3]. The set-up would produce a damped sinusoidal oscillation when the

---

[2]Latin for 'living force' which can be interpreted as "the energy stored in the Leyden jar" in the present context.

[3]Or a series resonant circuit as the induction coil at the left presented a very high impedance at the resonant frequency and the spark gap presented very low resistance when conducting.

switch at the primary of the induction coil was opened. He also constructed a *receiver* using a loop of copper wire and a spark-gap mechanism similar to that of the transmitter, as shown in Fig. 2c. In to-day's language, Hertz's receiver was a simple antenna which, as is well known, achieves its performance through the interplay between capacitance and inductance.

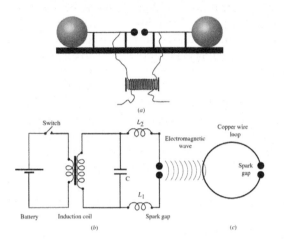

Without the metal spheres and resonant circuit, Hertz's transmitter would have produced a 'wideband' electromagnetic click of very low intensity which would not be easily picked up by a receiver that did not involve electrical resonance. Even with his 'sophisticated' apparatus, it was no simple matter for Hertz to detect that fleeting spark. However, his persistence paid off eventually. He played around with the size of the spheres, the length of the copper wire between the spheres, the size of the wire

Fig. 2 - Experimental setup of Hertz: (a) Actual setup without the induction coil, (b) modern schematic of transmitter, (c) receiver.

loop on the receiver side, and the size of the spark gaps at the transmitter and receiver. It is also known that he performed his experiments in a dark room and he may have used a magnifying glass. It is not known, how many attempts or how long it took but eventually he hit pay dirt when the resonant circuit was 'tuned' sufficiently to produce that magnificent fleeting spark. (See [7], [8] for more details.)

Wireless Telegraphy

Nikola Tesla of Serbian origin emigrated to the US early in 1884 at the age of 28. He dedicated his life to the generation, transmission, and utilization of electrical energy. In 1881, he invented the *Tesla coil*, illustrated in Fig. 3, which he used to generate spectacular sparks for the amazement of everybody and which was to be used soon after as a crucial component in many of the early wireless transmitters for telegraphy. The induction coil was of the same kind as that used by Hertz whereas the output transformer often referred to as a *resonant* transformer was specially designed to radiate energy. The windings of the resonant transformer would possess inductances L_1 and L_2 as well as capacitance C_2 as shown in Fig. 3. Thus the winding inductances and capacitance along with capacitance C_1 would form two *coupled* resonant circuits or *tank* circuits as they used to be called in the past. The end result was that when the spark gap began to conduct, a strong electromagnetic wave began to be radiated at the output of the Tesla coil, whose frequency was determined by the values of the capacitances and inductances.

The quest of Tesla's life was to transmit electrical energy, huge amounts, over wireless systems. In this respect, he filed a patent for a wireless system for the transmission of electrical energy on September 2, 1897, which was eventually granted by the US Patent Office in 1900 (see [9]). The system comprised a transmitter, basically a step-up transformer driven by an alternator, and a receiver, basically

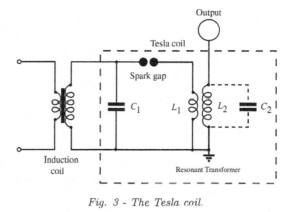

Fig. 3 - The Tesla coil.

a step-down transformer loaded by a series of electrical lamps and motors connected in parallel. A modern schematic of the system that includes the wiring stray capacitances is shown in Fig. 4. As can be seen, the system uses four parallel resonant circuits and for this reason it came to be known as Tesla's system of *four tuned circuits*. In modern terminology, the coupled tuned circuits at the transmitter and receiver were, in effect, *bandpass filters*, the first equipped with a transmitting antenna and the second equipped with a receiving antenna.

Tesla was prolific in other ways just as well. He invented single-phase and multi-phase alternators and induction motors. AC current was chosen for power generation and transmission from the start only because Tesla's AC system won over Edison's DC system. Tesla died of heart failure and in debt in a New York hotel room he used to call home, having sold or given away his many patents in previous years. He failed

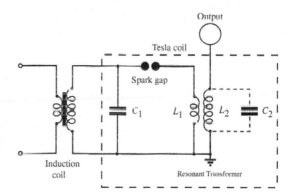

Fig. 4 - Modern schematic of Tesla's wireless system with stray capacitances included.

to fulfil his greatest ambition, namely, to transmit large amounts of power through wireless systems. See article on Tesla for more details.

Inspired by the work of Hertz, Guglielmo Marconi began experimenting with spark transmitters in the attic of the family home in Pontecchio near Venice while still a teenager. He explored ingenious ways that would increase the distance over which effective transmission could be achieved. Soon he was able to transmit signals over an impressive distance of about 1.5 km. At the age of 21, Marconi traveled to London with his wireless system determined to make his fortune. While in London, he gained the attention of a certain

William Preece, Chief Electrical Engineer of the British Post Office (part of
which is now British Telecom). In a landmark presentation on December 2,
1896, Preece demonstrated Marconi's invention. When a lever was operated at
the transmitting box, a bell was caused to ring in the receiving box across the
room, the first *remote control*. Through a series of demonstrations, Marconi
transmitted signals of Morse code over a distance of 6 km and after that 16 km.
In due course, he was able to send Morse signals across the Atlantic. Marconi
was a smart system designer and a clever entrepreneur who would readily adopt
and modify ideas reported by his peers, including Tesla's coil. (See [10] for
more information.)

A spark-gap wireless transmitter illustrating the principle of operation of
transmitters used during the late 1890s and early 1900s by Marconi and others
is illustrated in Fig. 5a. Basically, the transmitter consisted of an induction coil
in series with a relay, a parallel resonant circuit, and a spark gap constructed
from two metal balls similar to those used by Hertz. When the Morse key was
depressed, a voltage was induced in the primary as well as the secondary of the
induction coil and a spark was initiated at the spark gap. The electromagnetic
field of the primary opened the relay switch which interrupted the current but
when the field collapsed, the relay was reset and if the Morse key was kept
depressed a second cycle would begin. Thus as long as the Morse key was
kept depressed, a series of damped oscillations of the type shown in Fig. 5b
was generated in the loop of the secondary thereby sustaining a continuous
oscillation at the resonant frequency.

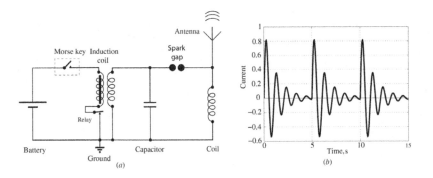

Fig. 5 - Early wireless transmitter: (a) Schematic, (b) typical transmitted signal.

From the start, the pioneers of the time realized that the higher the voltage
of the transmitter source and the taller the antenna tower, the further the
electromagnetic wave would travel, and to achieve more accurate transmission
over larger distances, they began to use larger and larger supply voltages,
eventually in the range of kilovolts. This imposed unusual requirements on the
design of the components used. Just to put things into perspective, the vital
statistics of some of the components used by Marconi in 1902 in his first North
American telegraph station at Glace Bay, Nova Scotia, and the corresponding

station across the Atlantic at Clifden, Ireland, should be mentioned. The transmitter voltage source comprised three 5000-volt DC generators in series connected in parallel with a 12,000- volt battery made up of 6000 2-volt batteries in series.[4] On the other hand, the capacitor comprised 1800 sheets of metal each measuring 9 x 3.6 m (yes, meters) hanging 30 cm apart from the ceiling in a huge room specially constructed to house the capacitor (see [10, pp. 395-397]). The spark itself, on the other hand, grew to the gigantic size of 4 inches [11].

Thanks to Marconi's many innovations, wireless telegraphy became a major industry and other ancillary applications soon emerged. The British Royal Navy, for example, developed *wavemeters* whose purpose was to set up and maintain transmitters and receivers to their correct wavelengths. The wide bandwidth of spark transmitters could be controlled by adjusting the rate of decay of the damped oscillations generated (see Fig. 5b) and instruments called *decremeters* were developed [11]. All these accessories made good use of the resonant circuit.

The 20th Century

By the end of the 19th century, engineers and scientists of all sorts began to examine the possibility of transmitting voice over the air and that technology took quite a while to perfect. Resonant circuits along with high-voltage dc generators and huge antennas served wonderfully well when a series of clicks were to be transmitted over the air but transmitting voice was another matter. The real breakthrough came about in 1906 when an American inventor by the name of Lee de Forest added another electrode to Fleming's vacuum-tube diode to invent the so-called *audion*, eventually called a *triode*, as an amplifying device. However, the resonant circuit did not disappear from the scene. A new partnership evolved where the large dc generators and tall antenna towers were made redundant by electronic amplification but the resonant circuit maintained its hold on frequency selectivity.

Eventually wireless voice transmission became possible. Along with it, other processes evolved such as modulation, demodulation, and the superheterodyne principle but the resonant circuit persisted. Before too long, the whole world was populated by broadcasting and TV stations and large sections of the world population became initially *listeners* and in due course *viewers*. When electromagnetic waves of different frequencies are beamed across the land by different broadcasting stations, it is prerequisite that the frequencies be well defined and free of harmonics while the radio and TV receivers need to be capable of differentiating among many different frequencies. The required selectivity has been achieved over the years through the use of resonant circuits in conjunction with power amplification devices.

The telephone system grew quite rapidly after the innovations of Alexander Graham Bell during the late 1870s. Things worked out well within cities but

---

[4]Simple application of Kirchhoff's voltage law would suggest that the 12,000-volt battery was charged continuously presumably to be ready to continue transmission in the event of malfunction in one of the DC generators.

eventually connections had to be established between cities and long-distance telephony through cables was born. Laying a telephone cable over a distance of hundreds of miles would cost a fortune and having just two persons using such a cable at any one time would not have been an economical proposition.

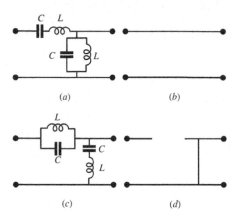

Fig. 6 - Resonant circuits: (a) bandpass filter, (b) equivalent circuit at resonance, (c) bandstop filter, (d) equivalent circuit at resonant frequency.

A parallel resonant circuit has a high impedance, ideally infinite, and its dual, the series resonant circuit, has a low impedance, ideally zero, at the resonant frequency $f_0 = 1/(2\sqrt{LC})$. Thus the circuits in Figs. 6a and 6c can be represented by the idealized circuits in Figs. 6b and 6d, respectively. In the first case, there is a straight through connection and in the second there is no connection at all between input and output. Consequently, the circuit in Fig. 6a will pass signal components whose frequencies are at or close to the resonant frequency whereas the circuit in Fig. 6c will stop such signal components. In effect, the first circuit behaves as a *bandpass* filter and the second one as a *bandstop* filter.

Through the clever use of resonant circuits, engineers and scientists developed a great variety of *passive*[5] filters, largely over the first half of the 20th century, and through the use of large numbers of passive filters along with new processes like frequency translation in so-called filter banks it became possible for numerous subscribers at one end of a long-distance cable to talk to as many subscribers at the other end at the same time without interfering with each other.

With the emergence first of the transistor and later the integrated circuit a surge towards circuit miniaturization began primarily to achieve cost reduction and improved reliability although the savings in space achieved did not hurt at all. The twin constituent elements of the resonant circuit, the capacitor and inductor, are in theory regarded as *dual* elements. However, in practice, they are anything but dual. The capacitor tends to improve by providing more capacitance when the distance between the electrodes is reduced. Unfortunately, the inductor does not scale down so well. As the inductor size is reduced, it becomes necessary to use thinner and thinner wire and as a consequence, the inherent resistance of the inductor is increased thus resulting in an inductor of poor quality. So a drive began during the 1950s to banish the inductor from circuits. Electrical filtering had always relied on electrical resonance and new innovations were called for. Through the use of amplifiers, resistors, and capacitors, engineers began to develop so-called *RC-active filters*. One

---

[5]So-called because they do not require energy sources, e.g., batteries.

way to deal with the absence of resonance was to create negative impedances through the use of *negative-impedance converters*. Resonance-like performance was achieved by subtracting negative from positive impedances. However, that approach immediately caused a serious side effect. Early types of *RC*-active filters were very sensitive to element variations that could be caused by environmental changes or by element tolerances.

The one branch of *RC*-active filters that attracted the author's attention soon after receiving his doctorate was the class of *inductorless* filters. The impedance of a capacitor is $1/sC$ and if a circuit can be constructed that can invert the impedance inversion of a capacitor, an impedance sC can be obtained which represents a simulated inductance equal to C. A circuit that can accomplish impedance inversion, called a *gyrator*, was proposed by Tellegen in 1948 [12]. He also proposed an implementation for such a device based on the piezoelectric effect but it did not catch on. Successful gyrator circuits were eventually developed using transistors or amplifiers. It is for this reason that inductorless filters are usually included in the class of active filters although they are actually passive filters in principle like their prototypes. In a short paper published in 1966 [13], Orchard provided a rationale that due to their passivity property, passive filters unlike their active counterparts are inherently relatively insensitive to element variations, which encouraged many, including the author, to pursue more research on inductorless filters. The author's modest contribution to the field was a family of gyrator circuits using operational amplifiers. These circuits along with the state-of-the-art in 1969 are documented in [14] and the references therein. The gyrator brought about renewed interest in the amazing resonant circuit.

In more recent times, new branches of filtering emerged that are more amenable to miniaturization as detailed in the articles that follow.

Acknowledgements

I would like to thank Anthony C. Davies of King's College London, England, and Antonio C. M. de Queiroz and Paulo S. R. Diniz of the Federal University of Rio de Janeiro, Brazil, for bringing certain crucial publications to my attention and for reading the article and providing feedback.

References

[1] "*Tacoma Narrows Bridge (1940).*" [Online]. Available: https://en.wikipedia. org/wiki/Tacoma Narrows Bridge (1940)

[2] J. Priestley, "*The History and Present State of Electricity with Original Experiments,*" London, 1767, printed for J. Dodsley, J. Johnson, B. Davenport, and T. Cadell. [Online]. Available: https:// archive. org/details/ historyandprese00priegoog

[3] F. Savary, "*Mémoire sur l'aimantation,*" vol. 34, Oct. 1827, pp. 557. [Online]. Available: http://gallica.bnf.fr/ark:/12148/bpt6k65691629/f11.image

[4] "*Electrical battery of Leyden jars,*" 1760-1769. [Online]. Available: http:// www.benfranklin300.org/frankliniana/result.php id=72&sec=0

[5] J. Blanchard, "*The history of electrical resonance,*" Bell System Technical Journal, vol. 20:4, pp. 415433, Oct. 1941. [Online]. Available: https://archive.org/details /bstj20-4-415

[6] W. Thomson, "*Transient electric currents,*" vol. XXXIV, June 1853, pp. 393404. [Online]. Available: https://books.google.com.br/books-id=3Ov22-gFMnEC

[7] O. Darrigol, "*Electrodynamics from Ampère to Einstein,*" New York: Oxford University Press, 2000. [Online]. Available: http://www2.pv.infn.it/ boffelli/PAS/ Darrigol.pdf

[8] "*Heinrich Rudolf Hertz,*" [Online]. Available: http://en.wikipedia.org/ wiki /Heinrich Hertz

[9] N. Tesla, "*System of transmission of electrical energy,*" Mar. 10 1900, U.S. Patent 645,576. [Online]. Av.: http://keelynet.com/tesla/00645576.pdf

[10] T. K. Sarkar, R. J. Mailloux, A. A. Oliner, M. Salazar-Palma, and D. L. Sengupta, "*History of Wireless,*" New York: Wiley, 2006.

[11] A. C. Davies, "*The right tunes? Wavemeters for British Army and Air Force uses in World War One time,*" 2014, Paper written in support of a presentation at a conference on Making Telecommunications in the First World War on 24th January 2014 at Oxford University.

[12] B. D. H. Tellegen, "*The gyrator, a new electronic network element,*" Philips Res. Rep., vol. 3, pp. 81101, 1948.

[13] H. J. Orchard, "*Inductorless filters,*" Electron. Lett., vol. 2, pp. 224225, 1966.

[14] A. Antoniou, "*Realisation of gyrators using operational amplifiers, and their use in RC-active-network synthesis,*" Proc. Inst. Elec. Eng., vol. 116, no. 11, pp. 18381850, Nov. 1969.

Andreas Antoniou, Life Fellow IEEE

Professor Emeritus, University of Victoria,

British Columbia, Canada

Filters

Around the year 1890 several people worked with the idea to improve the properties of long-distance transmission lines by inserting coils at regular intervals in these lines. Among those people were Vaschy and Heaviside. The results were discouraging at that time, and no real progress was made, until in 1899 M. I. Pupin investigated these cables. He found that a line which contains coils at regular intervals can be represented by an equivalent uniform cable if the coils are spaced closely enough. The equivalence decreases if the distance between two adjacent coils is increased, and disappears altogether if this distance is larger than half the wave length of the signal that is propagated in the cable. By his thorough mathematical and experimental research, Pupin found that the damping in cables for telegraphy and telephony can be substantially reduced by judiciously inserting these coils, which has resulted in a widespread use of these so-called 'Pupin lines' throughout the world.

Michael I. Pupin
(http://www.electrical4u.com)

The technical roots of the passive filters were on the one hand, the loaded lines and on the other hand, the resonant circuits. Already before World War I the first applications of wave filters in telephone systems were described but it was not until 1915 that the use of wave filters in telephone systems were described in patents and publications of K. W. Wagner from Imperial Physical Technical Institute (PTR) in Germany and G. A. Campbell from Bell Laboratories in the United States, written independently. However, due to the war, these patents and papers could not appear in print. The works of Wagner and Campbell were based on their own research on the damping behavior of transmission lines, which were provided with Heaviside-Pupin coils. After World War I the results of both engineers were published and became extremely essential for the new carrier method of multiplexing telephone. Therefore, intensive research on wave filters and other aspects of carrier multiplexing systems were taken in particular at Bell Laboratories, where O. Zobel developed a first systematic and very powerful concept of filter design in 1923. In order to construct an electrical wave filter Zobel combined filter sections, Wagner-Campbell sections and Zobel sections, where resonant circuits are included, under the condition that the image impedances were matched between sections. In practice, the pass and stop bands of different electrical wave filters can be calculated, so that a *"catalog"* of filter characteristics was available, which is suitable for many applications.

Zobel's concept of filter design is basically an analytical procedure. In 1924 the reactance theorem of R. M. Foster opened the door to a synthetic concept of the filter design, where a desired pass and stop band characteristic is the starting point and a filter circuit can be obtained by means of a systematic procedure.

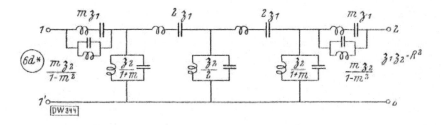

Zobel filter circuit (W. Cauer: Siebschaltungen 1931).

Cauer immediately recognized the potentialities of Foster's result and in his doctorate thesis (1926) *"The realization of impedances of specified frequency dependence"* he presented the first steps towards a scientific program which converted empirical approaches of the electrical filter design into a mathematical concept. This program addressed three distinct classes of problems where the following questions for a network or a transfer function arise: 1) Realization conditions for admitted network functions. 2) Interpolation and approximation problems as well as circuit realization. 3) Equivalence of circuits. This synthesis program focused the view to the realization techniques where Foster and Cauer already had contributed. Cauer also developed a synthetic design theory of wave filters and reconstructed Zobel's filter circuits. In 1931 Cauer published a monograph that included the theory and a filter catalog for wave filters. Unfortunately, Cauer's method is restricted in its applications since he used bridge circuits which are very sensitive with respect to parameter variations.

Theory and Design of Uniform and Composite Electric Wave-filters

By OTTO J. ZOBEL

THE electric wave-filter, as regards its general transmission characteristics and its extremely important rôle in communication systems, is well known. Its physical theory was discussed in detail in the preceding number of this Journal by its inventor, G. A. Campbell. In the present paper it is proposed to present systematic general methods of wave-filter design, together with representative designs, which have been developed in connection with the practical utilization of this device in the plant of the Bell System.

First is considered a general theory of design combining physical and analytical considerations which gives explicitly the structure of a uniform type of wave-filter having any preassigned transmitting and attenuating bands as well as desirable impedance and quite arbitrary attenuation characteristics. Next, this theory is applied to the design of a low-and-band pass wave-filter from which are derived design formulæ for all the practical uniform wave-filter structures in present use, belonging to the classes low pass, high pass, low-and-high pass, and band pass. Then the subject of composite wave-filters is taken up, these being non-uniform wave-filter networks

First page of the Otto J. Zobel paper on theory of electric filters: Bell System Technical Journal, January 1923

The Zobel method to use 'composite wave filters' bore the heritage from the transmission-line theory and was expressed in terms of characteristic impedances that should be matched if stages were cascaded, and wave-propagation constants that were used to describe the attenuation characteristics of the filter. As a first-order approximation, it was usually assumed that the filter was terminated by its characteristic impedance at all frequencies, which is not practically possible, because the characteristic impedance varies with frequency and a matched termination cannot be achieved with any finite number of lumped components. An approximate constant frequency termination was achievable by using an m-derived section with m=0.6, but this could not be described as an exact process in the sense that the Insertion Loss method of Darlington and Cauer achieved it. In theory, this assumption always gave rise to flat passbands in which – apart from parasitic losses – no damping at all occurred.

The theory was advantageous for designing complicated filters with a large number of sections. However, the exact attenuation characteristic of a filter that consisted of a large number of stages, and was terminated in a fixed resistance, was very difficult to determine, and still much more difficult to design. Therefore one had little control over the transmission characteristics in general and the transfer irregularities near the band limits in particular.

S. Butterworth split the filter in sections of order maximally equal to four, and separating these sections by amplifiers (constructed with electronic tubes), so that there was no interaction between parts, and the transfer function of the filter could be well designed. In this way he was able to construct filters with the famous and well defined maximally-flat transfer functions that were named after him. His work was published in 1930.

In 1931 Cauer's Ph. D. student O. Brune from MIT, Mass., constructed a realization technique for RLC one-ports with mutual inductances. Later on, a synthesis approach for RLC one-ports without mutual inductances were presented by R. Bott and R. Duffin in 1949. Until the end of World War II several authors contributed to the filter synthesis problem of 2-ports mostly in the United States and Germany, where image-parameter theory of constant impedance filter pairs. g. H. W. Bode, W. Cauer, R. Feldtkeller, A. Fialkow, I. Gerst, C. M. Gewertz, F. Glowatski, E. A. Guillemin, and H. Piloty but also from other countries A. C. Bartlett and R. Julia.

Although the Cauer's synthesis program contains some restrictions he illustrated for the first time that a circuit synthesis is very useful. Actually, the importance of Cauer's concept of filter synthesis resulted in the development of the modern theory of filter design which based on insertion-loss functions. The foundation a general insertion loss theory was published by E. L. Norton from Bell Laboratories in 1937 where he introduced a constant resistance filter pairs. Based on Norton's concept G. Cocci (1938-1940), S. Darlington (1939), W. Cauer (1939-1941), and H. Piloty (1931-1941) developed the insertion-loss theory independently, where a desired function which can be interpreted as power-ratio is realizable as a filter circuit. Especially, the paper of Darlington *Synthesis of Reactance 4-Poles Which Produce Prescribed Insertion Loss*

October, 1930   536   EXPERIMENTAL WIRELESS &

On the Theory of Filter Amplifiers.*

By S. Butterworth, M.Sc.
(Admiralty Research Laboratory).

Title of the Butterworth paper publishes on "Experimental wireless & the wireless engineering," 1930.

Characteristics published in 1939 influenced circuit theorists strongly.

As already mentioned, the three steps of Cauer's program of filter synthesis were implemented within the procedure of the new design theory of LC filters. A first detailed discussion about this theory was presented in Cauer's monograph *Theory of Linear AC Circuits* from 1941 (written in German: *Theorie der linearen Wechselstromschaltungen*) and since 1958 also available in English . Additional work was done by W. Bader about polynomial filter and LC filters with small resistors by and N. Ming (a former Ph.D. of W. Cauer).

If a power-ratio function is prescribed and the insertion-loss theory is used, to calculate the parameters of circuit elements of a filter, tedious calculations arise. In addition, we obtain ill-conditioned numerical problems when we use the polynomial coefficients of the prescribed filtering function and the calculations must be performed with a large number of decimals. This is possible only with modern digital computers. Therefore, most publications and books about the realization of filters by means of the insertion-loss theory have appeared after the World War II. The majority of publications came from the United States, Germany and Japan as well as a few researchers from other countries; without being complete, the following authors should be mentioned: N. Balabanian, M. Bayard, H. J. Carlin, E. A. Guillemin, H. J. Orchard, M. E. van Valkenburg, L. Weinberg, D. C. Youla; G. Fritzsche, E. Lüder, R. Saal, E. Ulbrich, R. Unbehauen; T. Fujisawa, H. Ozsaki, H. Takahasi, D.F. Tuttle as well as B. D. H. Tellegen, T. Laurent and others. As a final step, V. Belevitch (1951) as well as Y. Oono and K. Yasuura (1954) developed a complete synthesis of reciprocal and nonreciprocal lumped passive n-ports in terms of the scattering matrix.

Although beginning with the early 1950s a new area of electrical filters began, where transistor and later in operational amplifiers are included and the area of active filter circuit arose, the classical filter synthesis theory of passive filters is until now one of the most elegant mathematical theories in electrical engineering that influenced many other areas. For example, the concepts of positive real functions and passivity are now alive in control theory and numerics.

Finally, it should be emphasized that much more information about the history of design of passive filters can be found in the excellent papers of V. Belevitch, *Summary of the History of Circuit Theory* Proc. IRE, vol. 50, no. 5, pp. 848-855, May 1962, W. Klein, *The Historical Development of Filter Theories* (in German), Frequenz, Vol. 7, No. 11, 326-331, 1953, and E. Cauer, W. Mathis, R. Pauli, *Life and Work of Wilhelm Cauer (1900 1945)*, Proc. MTNS'2000, Perpignan, France, June 19 - 23, 2000 (available in internet) as well as an article of S. Darlington, *A History of Network Synthesis and Filter Theory for Circuits Composed of Resistors, Inductors, and Capacitors* , IEEE Trans. Circ. Syst., Vol. CAS-31, N0. 1, pp. 3-13, January 1984, with many personal reminiscences.

Wolfgang Mathis
Leibniz University Hannover, Germany

Adaptive Filters

The concept of adaptation is inherent to the learning process of any living animal. In general the animal probes some actions and according to the responses to these actions, next times new forms of actions are tried and the reactions are observed again and again. A very simple example is when a human being is adjusting the volume of a radio, in this process the human increases and decreases the volume up to a point the brain judges the volume as pleasant. This is a natural adaptive action where some desired volume is stored in the human brain, serving as a reference, and the human is adapting the volume up to the point where maximum pleasure is reached. This adaptive system consists of a single adaptive parameter that is changed through a twist in the volume button determined by the feedback of a pleasure measurement error, consisting of the difference between the human desired volume and the volume he/she is actually hearing.

In a discrete-time representation the adaptive filtering problem can be described in a simple form as the interaction among three distinct sequences, namely: $x(k)$ defining the input signal; $d(k)$ defining the desired (or reference) signal; and $y(k)$ representing the adaptive filter output signal. The variable k might represent the time index or the relative position of an entry of the sequence. The action to be taken to perform the adaptation is determined by how satisfied one is with the

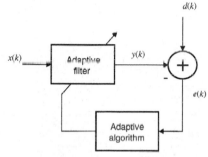

Fig. 1 - *General Adaptive Filtering Configuration.*

comparison between the desired signal and the filter output signal. The degree of satisfaction is measured through a function of the error signal usually defined as $e(k) = d(k) - y(k)$. The error is used in the learning process inherent to adaptive filters. This process normally entails the minimization of a cost function represented as $F(e(k)) - F(e(x(k); d(k)))$, where in most cases a collection of input signals assembled in a vector $\mathbf{x}$ generates the adaptive filter output through a weight vector w as $\mathbf{w}^T\mathbf{x}$. The typical configuration of an adaptive filter is depicted in Fig. 1, where one can observe the mapping from the input signal to the output signal, the formation of the error signal and its role in the adaptation process.

The ingenuity in choosing the signals to form the input signal vector and the reference signal determines the effective action performed by the adaptive filter. Typical applications are: prediction of the future or past behavior of a time series; identification of an unknown system model; noisy signal enhancement; and communication channel equalization. All these applications are pervasive in our daily lives.

This brief introduction helps us clarify the historical motivations for the development of this area as we see today. The description below highlights key developments that led to most of the adaptive filtering algorithms currently employed in uncountable products. Omissions are unavoidable, particularly in such a brief account of history.

Least Squares Estimation

The origins of the theory of adaptive filtering can be traced back to the eighteen century when there was a growing interest to bring about a solution to the linear estimation problem. In today's matrix formulation this would entail finding a good solution $\hat{\mathbf{w}}$ to the problem $\mathbf{X}^T \hat{\mathbf{w}} + \mathbf{n} = \mathbf{d}$, where the entries of vector $\mathbf{d}$ represents the set of observations, each row of $\mathbf{X}^T$ represents another set of observations related to the inputs in our adaptive filter configuration, and each entry of $\mathbf{n}$ represents measurement noise (or disturbance). The key problem is how to select a set of parameters $\mathbf{w}$ such that $\mathbf{y} = \mathbf{X}^T \mathbf{w}$ and such that a norm of the error vector $\mathbf{e} = \mathbf{d} - \mathbf{y}$ is minimized. Since that time the attempt is to minimize the l_1, l_2, or the l_∞ norms of the error, representing the least absolute deviations, the least squares, and the minimax absolute errors, respectively.

The least squares (LS) method consists of choosing the parameters leading to minimum sum of the squared errors. In some cases the entries in the summation can be weighted according to a particular interest, giving less or more importance to a predefined error measurement.

The LS method was originally proposed by the German mathematician Carl Friedrich Gauss (1777-1855) [6] around 1795; see Fig. 2. In the early 1800's there was some debate regarding the authorship of the LS method owing to the fact that Gauss did not publish his work until 1809. In between the French mathematician Adrien-Marie Legendre (1752-1833) published the LS method in 1805, in an independent way and in a didactic form; see the book by Gauss and particularly Prof. G. W. Stewart afterword

Fig. 2 - Carl Friedrich Gauss (1777-1855) [Wikipedia])

[1]. It is a fact that Gauss was not very keen on keeping the details of derivations of important results, and in some cases he seemed to delay the publication of his new results. Today it is widely accepted that both mathematicians should receive credit for the proposal of the LS method, and in both cases they were motivated by their astronomy studies. Therefore, we can infer that the LS method originated from the practical problem of estimating the movement of

---

[6] The man who proposed the least-squares solution in 1794 or 1795. Gauss worked as Professor of Astronomy and Director of the astronomical observatory in Göttingen, Germany, most of his life. He has made so many contributions to science so that anyone working in engineering, mathematics and statistics comes often across his name through the Gaussian distribution and the LS method, for example. In signal processing the popular fast Fourier transform (FFT) was actually first used by Gauss.

celestial bodies. It is worth giving credit to Gauss and Pierre Simon Laplace (1749-1827) for their many contributions to the LS theory, whereas Legendre was a key figure in its popularization. The competing objective functions to combine observations based on the l_1 and l_∞ norms were also addressed by the Croatian mathematician Rogerius Josephus Boscovich (1711-1787) followed by Laplace, and Laplace, respectively. The main stumbling block to use these cost functions were the lack of mathematical and computational tools to solve more involved problems. In this matter the LS was more amenable to analytical solution.

The early versions of the LS solution were meant for batch computations involving the solution of sets of linear equations known as normal equations. A possible tool to solve the LS problem was also proposed by Gauss: the so-called Gauss elimination algorithm (it should be noted that Isaac Newton (1643-1727) has employed a simplified form in 1707). As it is widely known the LS solution becomes ill-conditioned when the number of parameters is high. In reality, the adaptive filters are meant to work in non-batch setups where previously accessed data do not need to be stored, although their combined contribution is still taken into account whenever new measurements become available.

For typical online applications of adaptive filters, the solution employed is the recursive least squares (RLS) strategy, where new incoming measurements are smartly incorporated to the available solution obtained with all previous measurements. The RLS capitalizes on the description of the normal equations in recursive form. An important tool that played a key role in the popularization of the RLS is the so-called matrix inversion lemma, attributed in one of its versions to American mathematician Max A. Woodbury [3], whose feature is to enable the calculation of the inverse of a matrix after it has been perturbed by a matrix with small rank. The actual origin of the matrix inversion formulas is debatable and a good discussion about this issue can be found in [4]. Another key contribution included in the traditional RLS algorithm is the recursive description of the least squares problem attributed to British Statistician Robin L. Plackett [5], which, as quoted by him, brings the benefit that: *the estimates of parameters, their covariance matrix, and the sum of squares of residuals all require to be adjusted, with minimum of fresh calculation, due to the appearance of new observations.* Plackett admits his work uses a different approach and is a generalization of an early solution by Gauss. These two ingredients, namely the matrix inversion lemma and recursive description of the the LS problem, gave rise to the online RLS algorithm that we find in modern textbooks on adaptive filters such as [16], [17]. The RLS algorithm can be considered the first adaptive filtering algorithm that is still in use today [5], but it should be mentioned that its recognition as an online adaptive filtering algorithm comes much later.

The development of variants of the RLS algorithm gained some momentum in the 1980's trying to address some drawbacks such as stability problems and computational complexity. Pioneer works tried to achieve reduction in computation counts by exploiting special structures of the input data vector such as [6], [7], whereas some others were concerned with numerical stability

Fig. 3 - Bernard Widrow (1929), left, Marcian E. Hoff (1937), right.

[8]. The RLS is also related to the celebrated Kalman filter, a tracking method for nonstationary environment widely used in practical problems [9].

Least Mean Square Estimation

The proposal of the least-mean-square (LMS) algorithm by Professor Bernard Widrow (1929) and his student Marcian E. Hoff (1937) in 1960 [10] is a landmark in the history of adaptive filtering (Fig. 3). The LMS is so relevant that its history and the consequences following its discovery importance are so great to justify a companion article. A few facts are worth mentioning. The LMS algorithm minimizes the instantaneous squared error $e^2(k)$ that can be interpreted as an unbiased estimate of the expected value of the squared error. The minimization of the expected value of the squared error with respect to the filter coefficients leads to the widely known Wiener filter [12]. The Wiener filter cannot be applied in nonstationary environments, a drawback not shared by the LMS algorithm.

The LMS algorithm originated many other very important algorithms such as the normalized LMS and the family of transform-domain LMS algorithm. Even the conventional RLS algorithm can be interpreted as an LMS-Newton algorithm through an adequate change of variables. The unbeatable low computational complexity of the LMS algorithm is a huge asset that attracted its use in most embedded applications of adaptive filtering.

The LMS algorithm is by far the most widely used algorithm in adaptive filtering for several reasons, such as: low computational complexity; proof of convergence in stationary environment; unbiased convergence in the mean to the Wiener solution in ergodic environments; and stable behavior when implemented with finite-precision arithmetic; just to mention a few [11].

Affine Projection Estimation

Another important family of algorithms is the so-called affine projection (AP), proposed by Kuzuhiko Ozeki and Tetsuo Umeda in 1984 [14]. It is relevant to mention that some indication of the data reuse concept utilized in the AP formulation can found in [13]. The AP algorithms find the solution of the adaptive filter coefficients at the intersection of the hyperplanes leading to set of a posteriori errors equal to zero, where the number of hyperplanes is defined by the user and represents how many of the recent input and reference data are reused in the most recent updating. Their drawbacks are the increased computational complexity as compared to the LMS algorithm, and higher algorithm misadjustment, which can be, as usual, controlled through the introduction of a convergence factor sacrificing the convergence speed.

The general technical characteristic that has attracted attention to the AP algorithms is the range of flexible choices for its setup, that can in many cases allow a tradeoff in the features of the classical LMS and RLS algorithms. By no means the popularity of the AP algorithms matches those of the LMS and RLS algorithms.

Some Recent Happenings

In many applications the reference signal is not available so that one has to employ alternative objective functions based on some a priori knowledge related to the nature of the signals involved. These types of algorithms are known as *blind adaptive-filtering* algorithms. They are also called *training-less* or *unsupervised algorithms*, since their learning does not include any reference or training signal. The blind adaptive filter has been receiving quite a lot of attention since the 1980's.

It has been also a noticeable interest in data-selective adaptive filters, which avoid utilizing data carrying limited innovation. This family of algorithms usually allows some computational savings and a reduction in the misadjustment, without compromising the convergence speed [15], [16].

Another current trend is the re-development of the entire adaptive filter theory to deal with distributed systems where data originates from different sensors, in which the way to combine the data plays a crucial role. These techniques have been mostly addressed in the last ten years. Similarly, there is a growing interest in adaptive filtering algorithms that take into consideration sparsity properties of the estimated parameters.

Today there are over fifty books addressing adaptive filtering and many of its applications. To a certain extent, the early books by Prof. S. Haykin [17], [18] and by Profs. B. Widrow and S. D. Stearns [11] stimulated many formal adaptive filtering courses in many universities, and sparked research in this field around the world.

Acknowledgment

We would like to thank Marcello L. R. de Campos of the Federal University of Rio de Janeiro (UFRJ) for bringing key publications to our attention and allowing access to his books. Profs. Luiz W. P. Biscainho and Wallace A. Martins of UFRJ kindly read the manuscript and provided crucial comments. Prof. A. Antoniou of University of Victoria inspired this contribution.

References

[1] C. F. Gauss, *"Theoria Combinationis Observationum Erroribus Minimis Obnoxiae:*

Pars Prior, Pars Posterior, Supplementum,"* Classics in Applied Mathematics, translated as *Theory of the Combination of Observations Least Subject to Errors: Part one, Part two, Supplement,"* by G. W. Stewart, SIAM, Philadelphia, PE, 1995.

[2] R. W. Farebrother, *"Fitting Linear Relationships: A History of the Calculus of Observations 1750-1900,"* Springer-Verlag, New York, NY, 1999.

[3] M. A. Woodbury, *"Inverting Modified Matrices,"* Memorandum Rept. 42, Statistical Research Group, Princeton University, Princeton, NJ, 1950.

[4] W. W. Hager, "Updating the inverse of a matrix," *SIAM Review*, Vol. 31, pp. 221-239, June 1989. June 1989.

[5] R. L. Plackett, "Some theorems in least squares," *Biometrika*, Vol. 37, pp. 149-157, June 1950.

[6] D. D. Falconer and L. Ljung, "Application of fast Kalman estimation to adaptive equalization," *IEEE Trans. on Communications*, vol. COM-26, pp. 1439-1446, Oct. 1978.

[7] D. L. Lee, M. Morf, and B. Friedlander, "Recursive least squares ladder estimation algorithms," *IEEE Trans. on Acoust., Speech, and Signal Processing*, vol. ASSP-29, pp. 627-641, June 1981.

[8] J. G. McWhirter, "Recursive least-squares minimization using a systolic array," *Proc. of SPIE, Real Time Signal Processing* VI, vol. 431, pp. 105-112, 1983.

[9] R. E. Kalman, "A new approach to linear filtering and prediction problem," *Trans. ASME Journal of Basic Engineering*, vol. 82, pp. 34-45, March 1960.

[10] B. Widrow and M. E. Hoff, "Adaptive switching circuits," *IRE WESCON Conv. Rec.*, pt. 4, pp. 96-104, 1960.

[11] B. Widrow and S. D. Stearns, *"Adaptive Signal Processing,"*, Prentice-Hall, Englewood Clifs, NJ, 1985.

[12] N. Wiener, *"Extrapolation, Interpolation, and Smoothing of Stationary Time Series with Engineering Applications,"* Wiley, New York, NY, 1949.

[13] T. Hinamoto and S. Maekawa, "Extended theory of learning identification," *Electrical Engineering in Japan*, Vol. 95, pp. 101-107, 1975.

[14] K. Ozeki, T. Umeda, "An adaptive filtering algorithm using an orthogonal projection to an affine subspace and its properties," *Electronics and Communications in Japan*, Vol. 67-A, pp. 19-27, 1984.

[15] S. Gollamudi, S. Nagaraj, S. Kapoor, and Y.-F. Huang, "Set-membership filtering and a set-membership normalized LMS algorithm with an adaptive step size," *IEEE Signal Processing Letters*, vol. 5, pp. 111-114, May 1998.

[16] P. S. R. Diniz, *"Adaptive Filtering: Algorithm and Practical Implementation,"*, 4th edition, Springer, New York, NY, 2013.

[17] S. Haykin, *"Adaptive Filter Theory,"*, 5th edition, Pearson, Harlow, England, 2014.

[18] S. Haykin, *"Introduction to Adaptive Filters,"*, Macmillan, New York, NY, 1984.

Paulo S. R. Diniz
Universidade Federal do Rio de Janeiro, Brazil

Bernard Widrow, IEEE Life Fellow
Emeritus Professor, Stanford University, USA

Adaptive Signal Processing: Widrow's Group

There a 60-year history of development of adaptive signal processing concepts and applications by B. Widrow and his students, some of which is highlighted in this chapter. The work started in 1956 when Widrow was an Assistant Professor of Electrical Engineering at MIT. The work was continued in 1959 at Stanford. About 15 Master's students at MIT and 88 Ph.D. students at Stanford completed theses in a field that was to become adaptive signal processing.

Introduction

In the summer of 1956, Widrow and his colleague Ken Shoulders spent a week attending a seminar at Dartmouth college on the subject of "artificial intelligence" [1]. The pioneers in the field were there, John McCarthy, Marvin Minsky, Claude Shannon. The discussion was fascinating.

Returning to MIT, Widrow became interested in thinking and in the possibility of building a thinking machine. Given the computer hardware that existed then, a more modest goal was set however.

Digital filters and Z-transforms were well understood. Starting with a transversal filter, a tapped delay line, the coefficients or weights could be adjusted to create an optimal filter. But optimal in what sense? Wiener filter theory and Wiener filtering in discrete form were known and understood. The impulse response of a digital filter could be controlled by adjusting the weights. Typical Wiener problems such as prediction and noise filtering were being considered, adjusting the weights to minimize mean square error (MSE). The MSE was discovered to be is a quadratic function of the weights, with statistically stationary inputs. The weights were adjusted by following the gradient of the MSE function, a quadratic bowl. The objective was to drive the weights to the bottom of the paraboloidal (hyperparaboloid in many dimensions) bowl in order to minimize MSE. This became a Wiener filter that learned. The idea was tested on MIT's IBM 701 computer. It worked.

The algorithm was based on the method of steepest descent, the ball rolling down the hill to reach the bottom. The gradient was measured one component at a time, rocking each weight forward and backward, measuring MSE with the forward and backward settings. The difference in the MSE measurements divided by the weight change gave an estimate of the respective gradient component, in accord with elementary calculus. The MSE measurements were taken with a finite number of error samples and were noisy. When adapting, the noise of the gradient vector caused a loss in performance. When the adaptive algorithm "converged," the weight vector would take a random walk, undergo Brownian motion, about the bottom of the bowl. The algorithm never converged on the bottom of the bowl. The MSE was always greater than the minimum MSE that one would get with the true Wiener solution. The "misadjustment" was defined to be the average excess MSE, due to the Brownian motion, divided by the minimum MSE of the Wiener solution [2]. This ratio is a dimensionless measure of how far the adaptive solution is from the ideal Wiener solution which is based on perfect knowledge, i.e., from an infinite amount of data. Formulas for misadjustment and stability were derived.

The Invention of the LMS Algorithm

A few weeks after arriving at Stanford, in the Autumn of 1959, Widrow received a phone call from John Linvill. John was the identical twin of his MIT thesis adviser, Bill Linvill. (John Linvill started solid state research at Stanford and was surely one of the fathers of what was to become Silicon Valley.) John suggested talking to a brilliant new student, Ted (Marcian) Hoff, to see if he would be interested in research on learning systems. Widrow and Hoff met for the first time on a Friday afternoon in the Autumn of 1959.

Widrow was at the blackboard explaining to Ted what he knew about adaptive filters and, incidentally, about how learning could take place in an artificial neural element. The learning process was the same. He explained how the gradient was measured, one component at a time. Somehow, in the back and forth discussion, an idea popped out about a simpler way to measure the gradient, getting all of its components at the same time from the same piece of data. The existing method required separate data for each gradient component. We had a new gradient that did not require squaring, averaging, or differentiation. It could be obtained directly from the data and used very much less data.

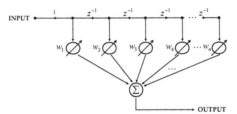

Fig. 1: *An adaptive FIR digital filter.*

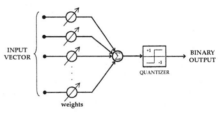

Fig. 2: *An adaptive neuron and synapses.*

The FIR digital filter shown in Fig. 1, and the neuron shown in Fig. 2 were being discussed. The linear combiner of Fig. 3 is common to both and turned out to be the essential adaptive element. This element is in all learning systems to this day.

Referring to Fig. 3, the input vector at the kth instant is X_k, the output is $y_k = X_k^T W_k = W_k^T X_k$. The desired response input is d_k. The error is the difference between the desired response and the actual response, i.e.,

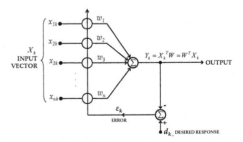

Fig. 3: *An adaptive linear combiner.*

$\epsilon_k = d_k - X_k^T W_k$. The method of steepest descent for adapting the weights can be written as:

$$W_{k+1} = W_k + \mu(-\hat{\nabla}_k). \tag{5.1}$$

The next weight vector equals the present weight vector plus μ (a constant) multiplied by the negative of the gradient $\hat{\nabla}_k$. This is actually a gradient estimate obtained from physical measurement. The LMS gradient estimate was obtained by simply approximating the MSE= $E[\epsilon^2]$ with ϵ_k^2, the square of a single error sample. This is a very crude approximation! Accordingly, the LMS gradient is

$$\hat{\nabla}_k = \frac{\partial \epsilon_k^2}{\partial W_k} = 2\epsilon_k \frac{\partial \epsilon_k}{\partial W_k} = -2\epsilon_k X_k. \tag{5.2}$$

Substituting into equation (5.1), the LMS algorithm [3] is

$$W_{k+1} = W_k + 2\mu\epsilon_k X_k. \tag{5.3}$$

$$\epsilon_k = d_k - X_k^T W_k \tag{5.4}$$

The parameter μ is chosen to control the rate of convergence and stability. The gradient estimate is very noisy, but keeping μ small, the small adaptation steps average and follow the negative gradient. It can be shown that the average of the gradient estimates equals the true gradient and thus the LMS gradient is an unbiased estimate of the true gradient. A sampling of research related to the LMS algorithm is given [2][3][4][5].

Having written the LMS algorithm on the blackboard, within a half hour Ted had it working on an analog computer located across the hall from Widrow's office. We were training a linear combiner as a single neuron performing pattern classification.

An input pattern was presented as a vector X_k, and the desired response d_k was the class of that pattern, either +1 or 1. We trained in several patterns and were very pleased with the results. We were very exited. We called the adaptive neuron Adaline, for adaptive linear neuron. Hoff designed the circuits for a portable Adaline box. A photo of the portable Adaline is shown in Fig. 4.

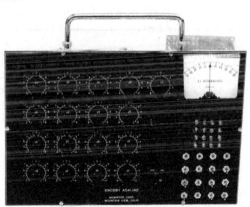

Fig. 4: *Knobby Adaline, a single neuron. A learning machine from 1959.*

It was not clear if we were engineers having invented a learning algorithm or if we were physicists having discovered a natural phenomenon. Secretly, Widrow thought that we had discovered a key element of learning in the brain. About a year later, he became skeptical and held that skepticism for a long time. A few years ago, he re-examined the issue and is now thinking that the LMS algorithm may actually be used by nature to perform learning at the synaptic and neural level. This is reflected in his latest work [6].

Adaptive Antennas

In 1967, Widrow, students, and colleagues published a paper in the Proceedings of the IEEE entitled Adaptive Antenna Systems [7]. The ideas resulted from working with sonar signals. The problem had to do with a submarine attempting to detect a distant quiet submarine while noisy surface ships nearby were making this task next to impossible. The submarine has an array of hydrophones for listening in the sea. A beam is formed, but interference can be received by the beam's sidelobes. By connecting the inputs from the sensors to adaptive weights, it became possible to form a beam in a desired direction and to simultaneously create nulls in the sidelobe structure in the directions of the interference without knowing their directions a priori.

When submitted to IEEE, the paper was examined by three reviewers. One said absolutely do not publish. Another said there were interesting things in the paper and it should be published. The third reviewer was neutral and could not decide. Fortunately, the editor decided to publish with some minor revisions. A few years later, Widrow and his group were informed that the paper had become a citation classic. At that time, Stanford had more Nobel Prizes than citation classic papers. This paper started a whole field of adaptive arrays. Today, adaptive antennas are called "smart antennas." They are being researched for use with mobile phones. Adaptive arrays on the phones and on the cell tower will allow different callers to use the same frequency in the same cell at the same time by nulling out the unwanted signals. The economic advantage from this technology will be huge. The best way to implement this is with the adaptive beamformer of Griffiths and Jim [8]. Griffiths was one of Widrow's Ph.D. students. Jim was one of Griffiths' Ph.D. students.

Adaptive Noise Canceling

At Stanford, the medical school and hospital are located near the school of engineering, within a few minutes walk. The proximity of these two schools greatly facilitated joint research.

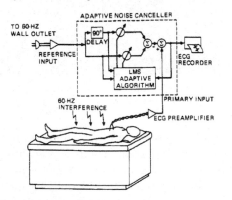

Fig. 5: *Canceling 60 Hz interference in electrocardiography. From [9].*

In the late 1960's, one of Widrow's Ph.D. students, Donald F. Specht, who was doing research on neural networks for classification of electrocardiograms as related to disease states, introduced him to Dr. Jobst Von Der Groeben, a cardiologist from whom Don was obtaining EKG data. This meeting started a long-term interest in biomedical engineering problems.

Dr. Von Der Groeben had

a filtering problem. When recording an EKG, the patient's body acted like an antenna, picking up 60 Hz interference that was making EKG's difficult to interpret. To eliminate the 60 Hz interference, he used a Buttterworth low pass filter, cutting off at 50 Hz, with a 60 db loss at 60 Hz. This filter eliminated the 60 Hz interference, but cut out all EKG components above 50 Hz. Dr. Von Der Groeben wanted to see the EKG out to 200 Hz. This was a problem.

In the early 1970's, Widrow thought of a possible solution: fight fire with fire. Take a 60 Hz signal directly from the electric wall outlet, filter it by controlling magnitude and phase, and subtract it from the EKG signal. With the right magnitude and phase, it would be possible to completely subtract out the 60 Hz interference. A group of his students recorded the data at the hospital and brought this back to our lab. The idea worked. A block diagram of the resulting system is shown in Fig. 5.

In 1963, a daughter, Deborah, was born to Mrs. Widrow at the Stanford Hospital. A young resident on duty in the maternity ward invited Professor Widrow to visit the nurses station where he had a Tektronix scope connected through a rotary selector switch to the dozen or so labor rooms. He was monitoring fetal electro-cardiograms. It was possible to see Debbie's electro-cardiogram before she was born.

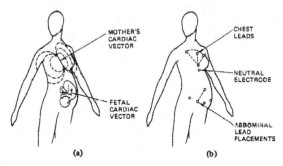

Fig. 6 - *Canceling maternal heartbeat in fetal electrocardiography. (a) Cardiac electric field vector of mother and fetus. (b) Placement of leads. From [9].*

Abdominal electrodes pick up signals from the baby's heart, but strong interference from the mother's heart, present on her abdomen, make the fetal EKG difficult to interpret. The baby's heart beats independently of the mother's heart. A high-pass filter could not sep-arate these signals since the second harmonic of the mother's signal is often close to the fundamental of the baby's signal. The young doctor asked, "what can be done?". It looked like an impossible problem. Many years went by. Long after solving the 60 Hz interfer-ence problem, the fetal EKG problem came up with a pos-sible solution. Place chest

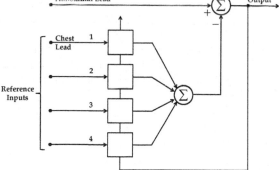

Fig. 7 - *Multi-reference noise canceller used in fetal EKG experiment. From [9].*

electrodes on the mother to pick up her electrocardiographic signal, loud and clear. Process this signal with an adaptive filter and subtract from the abdominal signal to eliminate the maternal interference [9][10]. This involves wideband adaptive noise canceling. Canceling 60 Hz interference was a narrow band problem. The maternal EKG was canceled and the doctor was happy. The lead placements are shown in Fig. 6, and a block diagram of the resulting system is shown in Fig. 7. This experiment demonstrated that wideband noise canceling was possible [10][11]. LMS is the most widely used learning algorithm in the world today.

References

[1]   J. Moor, *"The Dartmouth College Artificial Intelligence Conference: The Next Fifty Years,"* AAAI AI Magazine, vol. 27, no. 4, pp. 87-91, 2006.

[2]   B. Widrow and S. D. Stearns, *Adaptive Signal Processing*, Prentice-Hall, 1985.

[3]   B. Widrow and M. E. Hoff Jr., *"Adaptive Switching Circuits,"* IRE WESCON Convention Record, pp. 96-104, 1960.

[4]   B. Widrow, *"Rate of Adaptation in Control Systems,"* American Rocket Society (ARS) Journal, pp. 1378-1385, September 1961.

[5]   B. Widrow and E. Walach, *"On the Statistical Efficiency of the LMS Algorithm with Nonstationary Inputs,"* IEEE Transactions on Information Theory, vol. IT-30, no. 2, pp. 211-221, 1984.

[6]   B. Widrow, Y. Kim, and D. Park, *"The Hebbian-LMS Learning Algorithm,"* IEEE Computational Intelligence Magazine, vol. 10, pp. 37-53, 2015.

[7]   B. Widrow, P. E. Mantey, L. J. Griffiths, and B. B. Goode, *"Adaptive Antenna Systems,"* Proceedings of IEEE Transactions on Antennas and Propagation, vol. 55, pp. 2143-2159, 1967.

[8]   L. Griffiths and C. Jim, *"An Alternative Approach to Linearly Constrained Adaptive Beamforming,"* Proceedings of IEEE Transactions on Antennas and Propagation, vol. 30, pp. 27-34, 1985.

[9]   B. Widrow, J. M. McCool, J. Kaunitz, C. S. Williams, R. H. Hearn, J. R. Zeidler, J. E. Dong, and R. C. Goodlin, *"Adaptive Noise Cancelling: Principles and Applications,"* Proceedings of the IEEE, vol. 63, no. 12, pp. 1692-1716, 1975.

[10]   E. R. Ferrara and B. Widrow, *"Fetal Electrocardiogram Enhancement by Time-Sequenced Adaptive Filtering,"* IEEE Transactions on Biomedical Engineering, vol. BME-29, no. 6, pp. 458640, June 1982.

[11]   M. Yelderman, B. Widrow, J. M. Cioffi, E. Hesler, and J. A. Leddy, *"ECG Enhancement by Adaptive Cancellation of Electrosurgical Interference,"* IEEE Transactions on Biomedical Engineering, vol. BME-30, no. 7, pp. 392-398, June 1983.

Bernard Widrow
Stanford University, CA

Dookun Park
Stanford University, CA

Active Filters

An *electrical filter* processes an electrical input signal such that the output signal has desirable properties according to the specifications, which are often expressed in terms of frequency-selective performance.

For about 50 years from their conception, which dates back to 1915 when independently Wagner in Germany and Campbell in the U.S. introduced passive electronic wave filters, the LC passive filter technology dominated most applications. But in the mild 60's the rapid growth of the telecommunication industry along with the need of filters with more accurate performance and together with a higher volume production at low cost asked for newer filter technologies, thus the *Active Filter* came out. Soon filters with higher precision in tuning and stability over time were available [1]-[3].

A considerable support in the advance of Active Filters design came also from the development of silicon integrated circuit technology. An extensive search for inductorless filters suitable for very-low frequency applications (where inductors tend to be bulky and expensive) started in the 60's as well and, when the monolithic operational amplifier (*op-amp*) was introduced in the mid 60's, active filters using op-amp and only discrete resistors and capacitors, called active RC filters, were feasible and became an attractive alternative to passive filters.

Active RC filters provided the following additional advantages with respect the passive-LC and RLC filters [1]:

- Increased circuit reliability because all processing steps can be automated;
- Huge reduction of cost in large quantities production;
- Better performance due to the feasibility of high-quality components;
- Reduction in parasitics due to smaller size.

Additional to the above advantages which arise from the physical implementation, other advantages derive from the circuit-theoretic nature:

- Both design and tuning process are simpler than those for passive filters;
- A wider class of filtering functions can be implemented;
- Gain can be provided, in contrast to the passive filters which exhibit a significant loss.

Of course together with their advantages RC active filters exhibit some drawbacks compared to the passive filters. In particular, they:

- are typically limited to the audio-frequency applications due to the finite bandwidth of the active components;
- are generally more affected by component drift in manufacture and environment sensitive, i.e. their sensitivity is in general higher that the passive filters one;
- require a power supply.

From the historical point of view, the realization of complex poles of transfer function by means of active RC network was known since 1938. Indeed, selective circuits were implemented through a passive network in feedback with a voltage

Fig. 1 - John G. Linvill (News. Stanford.edu).

amplifier realized with vacuum-tubes [4]-[5]. However, the community had to wait for about fifteen years to formally open the active RC filters era. Indeed, even if there was an U.S. patent by Dietzold (dated 1951) from which several active filters were experimentally tested and built at Bell Laboratories, only with the paper of Linvill [6] we can consider the RC active filter topic to be formally born in 1954, as clearly suggested by the Linvill paper title itself.

The aim of [6] was to overcome the RC passive filters drawback of high in-band loss and poor economy elements, but allowing to implement inductorless filters. Moreover, unlike the patent of Dietzold, where an amplifier is used as the active element, Linvill proposes the use of the active component as a negative-impedance converter in [6]. One year later, in 1955, another relevant paper authored by Sallen and Key was published on this topic [7]. The paper presents an alternative method to realize sharp cut-off filters at very low frequencies by using an active element which can be a simple cathode-follower circuit. In particular, even if the realization of any order filter is considered, the synthesis procedure is based on the cascade of passive first-order sections and active second-order sections. Hence, it was possible to realize second-order sections with just one active element per section, which was important because active stages were expensive. A well-known block scheme of a typical Sallen and Key filtering section (with a generic amplifier of K gain) is shown in Fig. 2.

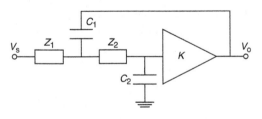

Fig. 2 - The classical Sallen and Key filter.

Up to 1960 only other three main contributions were published in this topic. The first, one year after the Sallen and Key paper, was a brief from Margolis which presented a Butterworth active filter using an RC network and vacuum tube amplifiers [8]. The other in 1957 by Yanagisawa introduced a novel active-filter realization through the current-inversion type converter combined with the RC network [9], and the last by Horowitz developed an optimized method to design RC active filters with negative impedance blocks [10].

Again Horowitz opened the next decade where a more intense research on active RC Filters was carried out. He authored two papers submitted in 1960, but published one in the same 1960 [11] and the other in the next year [12]. In both papers the non ideality of active elements (particularly critical using transistors instead of vacuum tubes in active filters) were taken into account for the first time and optimization to minimize the number of components was also pursued.

Since complex poles cannot be realized by means of a passive RC structure, to design active RC filters the desired polynomials are achieved by following two possible synthesis procedures. One takes advantage of polynomial subtraction and makes use of the various negative impedance converters available [5], [8] and [11], the other is based on polynomial addition and exploits the insertion of a selective network in the feedback path of an amplifier [1], [4] and [6]. In general the method based on polynomial subtraction is more sensitive to active parameter variation (the classical method exploiting polynomial addition has lower sensitivity to active parameters, but passive parameter sensitivity for large-Q pole pairs is just as large as in the negative impedance converter).

As already mentioned, an intensification of active RC filters research started around 1964 even thanks to the availability of inexpensive small operational amplifier modules which allowed straightforward implementation of filters by means of integrated amplifiers combined with discrete resistors and capacitors [13]; thus, since

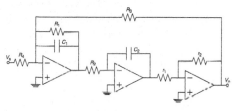

Fig. 3 - The "Biquad" active RC filter.

1965 several papers and letters were published. Few years later, taking inspiration from the interesting paper of Kerwin, Huelsman and Newcomb [14], Tow proposed in paper [15] the popular active RC filters state space realization (see Fig. 3), which he then termed "Biquad", [16] name confirmed by papers [17]-[18] and since then generally accepted.

To complete the description of the most popular and useful structures to realize active RC filters, Girling and Good [19] introduced in 1970 the leap-frog topology (named also ladder) amenable for high order filters, which adds to the Sallen and Key and the Biquad famous topologies. A typical block scheme of a ladder filter is shown in Fig. 4.

At the end of the sixties educational papers and books on active RC filters started to appear [20]-[21], the topic quickly moved to become mature, and a collection of the main papers was published by the IEEE [22]. Convenience in the realization of active RC filters was also favored in the seventies by the thin-film hybrid integrated circuits techniques. Then, when a considerable

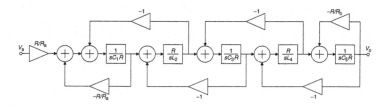

Fig. 4 - Block diagram of third order active filter.

progress was made, a second collection of papers with established and mature results on active RC filters was printed by the IEEE ten years after the former collection [23].

References

[1] M.S. Ghausi, K. R. Laker, Modern Filter Design (Active RC and Switched Capacitor), Prentice Hall, 1981.

[2] M. E. Van Valkenburg, Analog Filter Design, Holt, Rinehart and Winston, 1982.

[3] S. K. Mitra, C. F. Kurth, (Eds.), Miniaturized and Integrated Filters, J. Wiley, 1989.

[4] G. H. Fritzinger, "Frequency Discrimination by Inverse Feedback", Proc. of the IRE, Vol. 26, pp. 207-225, January 1938.

[5] H. H. Scott, "A New Type of Selective Circuit and Some Applications", Proc. of the IRE, Vol. 26, pp. 226-235, February 1938.

[6] J. G. Linvill, "RC Active Filters", Proc. of the IRE, Vol. 42, pp. 555-564, March 1954.

[7] R. P. Sallen, E. L. Key, "Practical Method of Design RC Active Filters", IRE Transaction on Circuit Theory, CT-1, pp. 74-85, March 1955.

[8] S. G. Margolis, "On the Design of Active Filters with Butterworth Characteristics", IRE Transactions on Circuit Theory, Vol. 3, p. 202, September 1956.

[9] T. Yanagisawa, "RC Active Networks Using Current Inversion Type Negative Impedance Converters", IRE Transactions on Circuit Theory, Vol. 4, pp. 140-144, September 1957.

[10] I. M. Horowitz, "Optimization of Negative-Impedance Conversion Methods of Active RC Synthesis", IRE Transactions on Circuit Theory, Vol. 6, pp. 296-303, September 1959.

[11] I. M. Horowitz, "Exact Design of Transistor RC Band-Pass Filters with Prescribed Active Parameter Insensitivity", IRE Transactions on Circuit Theory, Vol. 7, pp. 313-320, September 1960.

[12] I. M. Horowitz, "Optimum Design of Single-Stage Gyrator-RC Filters with Prescribed Sensitivity", IRE Transactions on Circuit Theory, Vol. 8, pp. 88-94, June 1961.

[13] E. J. Foster, "Active Low-Pass Filter Design", IEEE Transactions on Audio, Vol. AU-13, pp. 104-111, Sept./Oct. 1965.

[14] W. J. Kerwin, L. P. Huelsman, R. W. Newcomb, "State-Variable Synthesi for Insensitive Integrated Circuit Transfer Function", IEEE Journal of Solid-State Circuits, Vol. SC-2, pp. 87-92, Septerber 1967.

[15] J. Tow, "Active RC Filters A State Space Realization", Proc. IEEE, Vol. 56, pp. 1137-1139, June 1968.

[16] J. Tow, "Design Formulas for Active RC Filters Using Operational-Amplifiers Biquad", Electonic Letters, Vol. 5, pp. 339-341, July 1969.

[17] L. C. Thomas, "The Biquad: Part I Some Practical Design Considerations", IEEE Transactions on Circuits and Systems, Vol. CAS-18, pp. 350-357, May 1971.

[18] L. C. Thomas, "The Biquad: Part II A Multipurpose Active Filtering System", IEEE Transactions on Circuits and Systems, Vol. CAS-18, pp. 358-361, May 1971.

[19] F. E. Girling, R. F. Good, "Active Filters 12. The leap-frog or active-ladder synthesis", Wireless World, Vol. 76, pp. 341-345, July 1970.

[20] J. Tow, "A step-by-step active-filter design", IEEE Spectrum, Vol. 6, pp. 64-68, December 1969.

[21] L. P. Huelsman, Theory and Design of Active RC Circuits, Mc Graw Hill, 1968.

[22] S. K. Mitra (Ed.), Active Inductorless Filters, IEEE Press, 1971.

[23] R. Schaumann, M. A. Soderstrand, K. R. Laker (Eds.), Modern Active Filters Design, IEEE Press, 1981.

Gaetano Palumbo

University of Catania, Italy

Multirate Filters

Multirate Filtering

A linear time-invariant system usually operates with a single sampling rate both at the input and the output. Systems using multiple sampling rates at different stages are called multirate systems. Although they provide different sampling rates the original signal components are never destroyed. A filter normally alters some characteristics or components of a signal and a filter bank is an array of filters that separates the input signal into multiple components, each one carrying a single frequency sub-band of the original signal. In filter banks the process of decomposition is called analysis and the reconstruction process is called synthesis. A multirate filter bank can be analyzed at different rates corresponding to the bandwidth of the frequency bands and multirate filter design uses decimation (downsampling) by a factor of M that keeps every Mth sample of an original sequence, and interpolation (upsampling) by a factor of L that generates the original L samples necessary to restore the sampling rate. In multirate digital filters, decimators and interpolators allow the reduction of computations by using lower sampling rates. On the other hand, multirate analog filters reduce the overall capacitance, thus saving chip area, allowing lower power consumption, as well. Moreover, in multirate filtering the signal should be filtered before decimation and after interpolation, otherwise aliasing and imaging will occur, respectively.

Multirate Digital Filters (1973-current)

Multirate Filters have evolved originally from the research area of signal processing, in particular, digital signal processing, and they were designated as multirate digital filters which can be organized in multirate filter banks where the signal is separated in different frequency domains. These multirate digital filters emerged in the beginning of the 1970's, where they found numerous applications, whenever it was necessary to change the sampling rate of a digital signal, in digital modulation, digital waveform coding, speech processing, image processing, voice

R. W. Shafer (left), L. R. Rubiner (right) [http://ethw.org]

or video codecs, digital audio and antennas' systems. One of the pioneering works in this area dealt with interpolation and was entitled "*A digital signal processing approach to interpolation*," published in the Proceedings of IEEE, in June 1973, by Ronald W. Schafer and Lawrence

R. Rabiner, both with a Ph.D. from MIT, in 1968 and 1967, respectively, and that have worked together initially in the AT&T Bell Laboratories.

Other works followed, like for example "*New results in the design of digital interpolators*," in IEEE Transactions on Acoustics, Speech, and Signal Processing, in June 1975, by G.

Front page of the book "Multirate Digital Signal Processing" (left), Ronald E. Crochiere (right)

Oetken, T. W. Parks and H.W. Schussler, but it was the seminal work of Ronald E. Crochiere and Lawrence R. Rabiner, both working at AT&T Bell Labs, first published, in March 1981, in the Proceedings of IEEE and entitled "*Interpolation and decimation of digital signals: A tutorial review*," followed later by the book "*Multirate Digital Signal Processing*," published by Prentice Hall in 1983, that constituted the reference for future developments in this field.

The continuous and fast development in this area led to numerous applications in the most diverse communication systems, some already mentioned above, but, in particular in wideband digital communications, namely, digital television and audio broadcasting, DSL Internet access, wireless networks, power-line networks and 4G mobile communications. Then, it would be important to highlight here, as well, a recent book by Y-P Lin, S-M Phoong and P. P. Vaidyanathan, entitled "*Filter Bank Transceivers for OFDM and DMT Systems*," published by the Cambridge University Press in 2010, where the

The front pages of two cited seminal books and one of the authors of both, P. P. Vaidyanathan.

fundamentals of multirate signal processing and the multirate formulation of communication systems are presented and embedded in the composition of advanced communication systems.

Another example of the proliferation and up-to-date importance of multirate filters can be found in many new circuit implementations, as it can be outlined here with a recent work from MIT, by Michael Price, James Glass, Anantha P. Chandrakasan, published in the IEEE ISSCC 2014 Digest and entitled *"A 6mW 5K-Word Real-Time Speech Recognizer Using WFST Models"*.

Multirate Analog Filters (1990-2005)

Based in the same theory of Multirate Digital Signal Processing and by utilizing Switched-Capacitor (SC) Filters, several multirate analog filtering architectures were developed throughout the 1990's and until mid-2000's. Since the integration of high-frequency analog filtering into the system analog front-end was highly required for the fast growing high-speed communications and signal processing solutions and although these front-ends represented a small portion of the total mixed signal system, they are usually its speed and performance bottleneck. In particular, the contin-

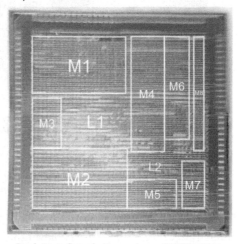

Real-time Speech Recognizer, ISSCC 2014.

uous reduction of the power supply and the increase of the operation speed, as well as the digital noise driven by the constantly growing digital signal processing core created additional challenges. Developing trends in high-speed communications and signal processing imposed the integration of high frequency analog filtering, traditionally implemented by external analog components, as much as possible on a system-chip to obtain better performance and reliability at low cost.

Even with the signal processing systems appearing to be mostly entirely digital, they still always required to contain one or more integrated analog filtering functions internally or as interface with the natural analog world. Moreover, filtering requirements at very high frequencies, where ultrafast sampling and digital circuitry with its associated data conversion may not be realistic and economical, usually imposes the utilization of analog techniques. Besides, although SC filters still need Continuous-Time (CT) front Anti-Aliasing and post smoothing or Anti-Imaging filters, if multirate techniques are embedded in their structures, they will allow a significant relaxation of the CT front-end filtering. Then, several multirate SC decimating and interpolating filters were designed on chip for high-speed applications, namely, for a high-order and high-selectivity multirate SC lowpass decimating filter of a fast modem, in 1.8μm CMOS, with

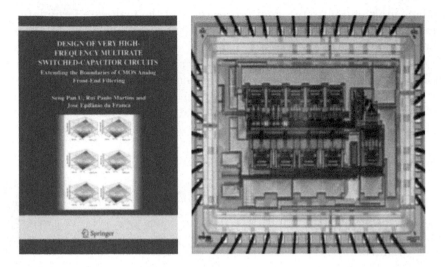

Cover page of "Design of Very High-Frequency Multirate Switched-Capacitor Circuits" (left) and, microphotograph of a bandpass Interpolating Filter - ISSCC 2002.

significant savings in area and power consumption, as reported by R. P. Martins, J. E. Franca and F. Maloberti, from Universities of Lisbon and Pavia, in IEEE Journal of Solid-State Circuits, in September 1993, under the title *"An Optimum CMOS Switched-Capacitor Antialiasing Decimating Filter,"* and a multirate SC bandpass interpolating filter which attained, by then, the fastest ever frequency of operation in a filter, in this case, in its output at 400MS/s, fulfilling the specifications of a direct digital frequency system (DDFS), in 0.35μm CMOS, published also in the IEEE Journal of Solid-State Circuits, in January 2004, under the title *"A 2.5-V 57-MHz 15-Tap SC Bandpass Interpolating Filter With 320-MS/s Output for DDFS System in 0.35μm CMOS"*. This work was later reported in a book authored by Seng-Pan U, R.P.Martins (University of Macau) and J.E.Franca (University of Lisbon), by Springer, in September 2005, entitled *"Design of Very High-Frequency Multirate Switched-Capacitor Circuits Extending the Boundaries of CMOS Analog Front-End Filtering."*

Rui Martins

University of Macau, Macao, China
and Instituto Superior Técnico,
Universidade de Lisboa, Portugal

Switched Capacitor Filters

James C. Maxwell, *A Treatise on Electricity and Magnetism*, Oxford: Clarendon Press, 1873, vol. 2, pp. 374-375,

If the magnet of a galvanometer included in the circuit is loaded, so as to swing so slowly that a great many discharges of the condenser occur in the time of one free vibration of the magnet, the succession of discharges will act on the magnet like a steady current whose strength is $2EC/T$.

If the condenser is now removed and a resistance coil substituted for it, and adjusted till the steady current through the galvanometer produces the same deflexion as the succession of discharges, and if R is the resistance of the whole circuit when this is the case,

$$\frac{E}{R} = \frac{2EC}{T}$$

$$or \quad R = \frac{T}{2C}$$

We may thus compare the condenser with its commutator in motion to a wire of a certain electrical resistance, ..." [1][2]

It is almost always the case that our most important innovations were conceived long ago by the great geniuses who preceded us. Such is the case with the switched-capacitor (SC) *"resistor"* concept, a fundamental idea that underpins modern CMOS switched-capacitor circuits, which was described by James Clerk Maxwell in 1873. Maxwell's brilliant idea lay dormant for almost a century before the emergence of a *"killer application"*; i.e., the conversion to digital transmission and switching networks from the analog telephone systems that had persisted for nearly a century following Alexander Graham Bell's seminal patent [3]. Maxwell's epiphany began to reappear in the late 1960's when A. Fettweis disclosed the wave digital filter concept [4] and further in the early 1970's when D.L. Fried described circuit ideas for low-pass, high-pass and n-path SC filters [5]. Key contributions which made MOS a mainstream digital technology were the invention of the microprocessor [6], which led to the release of the 4004 4-bit CPU by Intel in March 1971, and the first 1,024 bit DRAM, which replaced magnetic core memories, in October 1970 [7].

Prof. David A. Hodges joined the EECS faculty at UC Berkeley in 1970 after working for four years at Bell Telephone Laboratories. His broad knowledge of the analog telephone system was a key factor in defining the overall vision and setting the research directions along with Profs. Paul R. Gray and Robert W. Brodersen who joined him in 1971 and 1976, respectively.

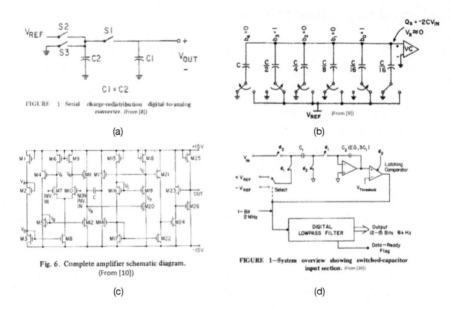

Fig. 1 - (a) *Charge-redistribution DAC by Suarez (1974). (b) Successive-approximation ADC (McCreary and Gray, 1975). (c) The first ΣΔ ADC with switched-capacitors (Hauser, 1985). (d) The first MOS operational amplifier by Tsividis and Gray (1976).*

The big picture was to embrace Gordon Moore's 1965 prediction of exponential growth for digital MOS circuits, and challenge the conventional wisdom which at the time held that bipolar technology was the only good analog technology, to design and demonstrate precision analog circuitry on the same MOS IC (i.e., mixed-signal integrated circuits). Their landmark research enabled the rapid development and worldwide adoption of PCM digital telephony, and after more than four decades, SC circuits continue to have tremendous social and economic impacts. Emerging applications areas include ultra-low-power biomedical circuits and systems, ultra high-efficiency and energy-scavenged RF communications devices and networks, etc.

Prior to the early 1970's, most of the world's best circuit designers thought analog circuits simply could not be implemented in MOS, especially in the metal-gate PMOS and NMOS technologies of the day. This attitude was easy to understand given the incredible performance available from conventional discrete bipolar operational amplifiers (e.g., open-loop output voltage to input voltage ratio $> 10^6$ V/V, etc.). When used with discrete resistors and capacitors, those amplifiers enabled active-RC filters that were the backbone of many analog signal processing applications including the existing analog telephone systems. Adding inertia to that approach was a massive amount of available literature on the analysis, design, and optimization of active-RC circuits. For other key functions such as data conversion, resistor-based (e.g. R-2R ladders) and current-steering approaches were ubiquitous, although high precision was often achieved using somewhat expensive trimming techniques. In addition to their limited matching accuracy, the integrated resistors used in those designs were also shown to be inferior to MOS capacitors in terms of temperature and voltage coefficients and deleterious piezoelectric effects.

In the face of considerable skepticism from the established bipolar analog integrated circuits design community, MOS analog and switched-capacitor design techniques were launched in the mid 1970's; several seminal design techniques are depicted in the four figures above. Arguably, perhaps, the most important idea was the first and the simplest – the all-MOS two-capacitor charge-redistribution digital-to-analog converter designed and demonstrated by Suarez, et al. [8] (Fig. 1-a). This work was of paramount importance for several reasons. It demonstrated that high-precision data conversion could be achieved by exploiting the native devices of MOS technologies; namely, precision MOS capacitors, and clock-controlled zero-offset MOSFET switches. Resistors, which have always been an anathema to MOS technologies, were not needed. Moreover, the resulting charge-based signal processing technique using MOSFETs with no gate currents were attractive compared to current-steering techniques in bipolar. An 8-bit A/D converter was also demonstrated in [8], but mitigation of the top-plate parasitic capacitance effects limited the minimum capacitance to about 25 pF, which, in turn, limited the maximum speed of operation.

McCreary and Gray described an all-MOS successive-approximation weighted-capacitor A/D converter in 1975 [9] (Fig. 1-b). It used a 10-bit binary-weighted capacitor array wherein the smallest and largest capacitors were 0.24 pF and 120 pF, respectively. Thus, the minimum capacitance was reduced by $\sim$100X compared to the two-capacitor design because the new topology was insensitive to parasitic capacitances. Consequently, a 10-bit conversion was completed in only 20 μs. This approach remains ubiquitous in integrated systems that require high-speed data conversion.

In 1976, Tsividis and Gray developed a two-stage charge-redistribution A/D converter, which showed that the required μ-law companding for PCM telephony was realizable in MOS [11].

The first delta-sigma oversampled data converter that used switched-capacitor techniques was introduced by Hauser (Fig. 1-c), et al. in 1985 [20]. The approach is widely used today because it exploits the relentless scaling of CMOS digital signal processing circuitry to enable low-power and very high-accuracy signal conversion at moderate conversion rates.

Another seminal contribution by Tsividis and Gray was the first MOS operational amplifier in 1976 [10] (Fig. 1-d). It exploited the high matching accuracy achievable with MOSFET devices by using replica biasing techniques to achieve low offset voltages, good common-mode and power-supply rejection ratios, etc. The open-loop voltage gain was only 51 dB, but it was adequate for the emerging mixed-signal data conversion and sampled-data filtering applications. It showed, once and for all, that high-gain bipolar amplifiers were not essential for high precision.

Another important development was the time-interleaved A/D approach of Black and Hodges [12]. It is widely used today in applications that require very

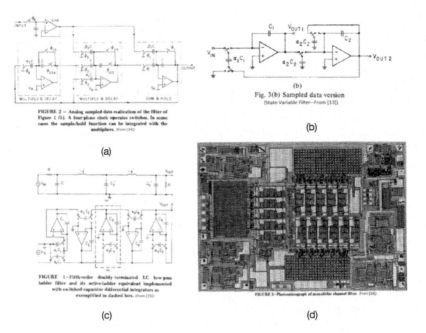

(a)

(b)

(c)

(d)

Fig. 2 - (a) A second-order SC filter by Young, et al. (1977). (b) Hosticka, et al. second-order switched-capacitor (1977). (c) The first MOS switched-capacitor (fifth-order) ladder filters by Allstot (1978). (d) The first silicon-gate CMOS PCM filter (Black, et al., 1980

high-speed A/D conversion in CMOS.

The development of an all-MOS approach to replace active-RC filters was the next challenge. In 1977, Young, et al. described a second-order SC filter derived from a direct form second-order digital filter prototype [14] (Fig. 2-a). About the same time, Hosticka, et al., presented a second-order switched-capacitor filter derived from an active-RC state-variable prototype [13] (Fig. 2-b). It used less chip area with lower sensitivity to capacitor ratio variations than the direct form version. Caves, et al., described a similar approach in 1977 [19].

Implementation of a PCM codec chip for digital telephony required a precision anti-aliasing filter preceding the A/D converter and a precision reconstruction filter following D/A converter. Fifth-order lowpass filters were required with stringent passband accuracies (< 0.1 dB variation error across process, voltage and temperature corners). Orchard showed in 1966 that the low sensitivity properties of passive doubly-terminated RLC ladder filters were retained in active implementations [22]. Thus, the double-terminated ladder structure was adopted for the first MOS switched-capacitor (fifth-order) ladder filters by Allstot, et al. [15] (Fig. 2-c). Black, et al., designed the first silicon-gate CMOS PCM filter chip in 1980, which became the industry standard [16] (Fig. 2-d). SC ladder filters are still widely used in high-performance CMOS mixed-signal processing systems.

As described above, switched-capacitor circuits played a key role in the

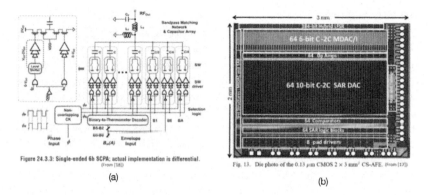

Figure 24.3.3: Single-ended 6b SCPA; actual implementation is differential.
(From [18])
3 mm
2 mm
64 6-bit C-2C MDAC/I
64 10-bit C-2C SAR DAC
8 pad drivers
Fig. 13. Die photo of the 0.13 μm CMOS 2 × 3 mm² CS-AFE. (From [17])

(a) (b)

Fig. 3 - (a) The first switched-capacitor RF power amplifier (Yoo,2001). (b) 64-channel compressed sensing analog front-end (Gangopadhyay, et al., 2014

late 1970's and early 1980's in the worldwide conversion of the old analog telephone systems into modern digital switching networks. It is probably not surprising then that SC techniques played critical roles in the development of radio frequency integrated circuits that have enabled ubiquitous wireless connectivity. A.A. Abidi pioneered radio research in MOS technologies and made seminal contributions to advanced radio architectures that employed CMOS SC techniques including the first design and implementation of direct-conversion transceivers for digital communications in 1995 [21].

Switched-capacitor circuit techniques are continuing to expand into many important emerging markets. In 2011, Yoo, et al., demonstrated the first switched-capacitor RF power amplifier which featured record power efficiency [18] (Fig. 3-a). This approach is attractive for massive MIMO transceivers because of its high efficiency at typical back-off power levels. CMOS chips with the capability to drive more than 100 antennas are under development. At the other frequency extreme, switched-capacitor circuits are finding many applications in ultra-low-power biomedical sensor systems. One example of this trend is the 64-channel compressed sensing analog front-end described by Gangopadhyay, et al, in 2014 [17] (Fig. 3-b). The future is bright for SC circuits.

References

[1] James C. Maxwell, "*A Treatise on Electricity and Magnetism*," Oxford: Clarendon Press, vol. 2, 1873.

[2] Yannis P. Tsividis, Letter to the Editor, "*James Maxwell and switched-capacitor circuits*," IEEE Circuits and Systems Magazine, p. 16, June 1984.

[3] Alexander Graham Bell, "*Improvement in Telegraphy*," U.S. Patent #174465, 1876.

[4] A. Fettweis, "*Digital filters related to classical structures*," AEÜ: Archive für Elektronik und Übertragungstechnik, vol. 25, pp. 79-89, Feb. 1971.

[5] D.L. Fried, "*Analog sample-data filters*," IEEE J. Solid-State Circuits, vol. 7, pp.

302-304, Aug.1972.

[6] F. Faggin, M. Hoff, and S. Mazor, *"Memory System for a Multi-chip Digital Computer,"* U.S. Patent #3,821,715, June 28, 1974.

[7] Intel 1103 1,024 bit dynamic random access memory.

[8] R.E. Suarez, P.R. Gray, and D.A. Hodges, *"An all-MOS charge-redistribution A/D conversion technique,"* IEEE Intl. Solid-State Circuits Conf., 1974, pp. 194-195, 248.

[9] J. McCreary and P.R Gray, *"A high-speed, all-MOS, successive-approximation weighted capacitor A/D conversion technique,"* IEEE Intl. Solid-State Circuits Conf., 1975, pp. 38-39, 211.

[10] Y. Tsividis and P.R. Gray, *"An integrated NMOS operational amplifier with internal compensation,"* IEEE J. Solid-State Circuits, vol. 11, pp. 748-753, 1976.

[11] Y. Tsividis, P.R. Gray, D.A. Hodges, and J. Chacko *"An all-MOS companded PCM voice encoder,"* IEEE ISSCC, 1976, pp. 24-25, 211.

[12] W.C. Black, Jr. and D.A. Hodges, *"Time interleaved converter arrays,"* IEEE Intl. Solid-State Circuits Conf., 1980, pp. 14-15, 254.

[13] B.J. Hosticka, R.W. Brodersen, and P.R. Gray, *"MOS sampled data recursive filters using switched capacitor integrators,"* IEEE J. Solid-State Circuits, vol. 12, pp. 600-608, Dec. 1977.

[14] I.A. Young, D.A. Hodges, and P.R. Gray, *"Analog NMOS sampled-data recursive filter,"* IEEE Intl. Solid-State Circuits Conf., 1977, pp. 156-157.

[15] D.J. Allstot, R.W. Brodersen, and P.R. Gray, *"Fully-integrated high-order NMOS sampled-data ladder filters,"* IEEE ISSCC, 1978, pp. 82-83, 268.

[16] W.C. Black, Jr., D.J. Allstot, S. Patel, and J. Wieser, *"CMOS PCM channel filter,"* IEEE ISSCC, 1980, pp. 84-85, 262.

[17] D. Gangopadhyay, E.G. Allstot, A.M.R. Dixon, K. Natarajan, S. Gupta, and D.J. Allstot, *"Compressed sensing analog front-end for bio-sensor applications,"* IEEE J. Solid-State Circuits, vol. 49, pp. 426-438, Feb. 2014.

[18] S.M. Yoo, J.S. Walling, E.C. Woo, and D.J. Allstot, *"A switched-capacitor power amplifier for EER/polar transmitters,"* IEEE ISSCC, 2011, pp. 428-430.

[19] J.T. Caves, S.D. Rosenbaum, M.A. Copeland, and C.F. Rahim, *"Sampled analog filtering using switched capacitors as resistor equivalents,"* IEEE J. Solid-State Circuits, vol. 12, pp. 592-599, Dec. 1977.

[20] M. Hauser, P. Hurst and R.W. Brodersen, *"MOS ADC-filter combination that does not require precision analog components,"* IEEE Intl. Solid-State Circuits Conf., 1985, pp. 80-81, 313.

[21] A.A. Abidi, *"Direct-conversion radio transceivers for digital communications,"* IEEE J. Solid-State Circuits, vol. 30, pp. 1399-1410, Dec. 1995.

[22] H.J. Orchard, *"Inductorless filters,"* Electron Letters, vol. 2, pp. 224-225, 1966.

David J. Allstot
Dept. of EECS
University of California, Berkeley

Nonlinear Circuits

Linear circuits represent a small, albeit very important, subset of physical systems. Even the simplest circuit element, a piece of wire, exhibits nonlinear behavior. When the voltage V applied across its terminals is small, Ohm's law ($I = V/R$), which dates from the early nineteenth century, predicts that the current I through the wire will be linearly related to this voltage, the constant of proportionality being the resistance R. The resistance of a real wire depends on the current flowing in it. A large current will cause the wire to heat up and its resistance to change. Thus, more generally, the current is a function ($\hat{I}(\cdot)$) of the applied voltage, i.e. $I = \hat{I}(V)$.

A two-terminal circuit element whose current is a function of the voltage across its terminals is called a voltage-controlled nonlinear resistor; the earliest electronic device, the thermionic diode, is of this type. If the voltage is a function of the current, the element is said to be current-controlled.

Multiterminal electronic circuit elements, such as thermionic valves and transistors, also exhibit nonlinear behavior. For example, the output driving-point (DP) characteristics of a triode in the common cathode configuration can be described at DC by the nonlinear function

$$I_A = \hat{I}_A(V_A, V_G),$$

where I_A is the anode current and V_A and V_G denote the anode and grid voltages, respectively, with respect to the cathode.

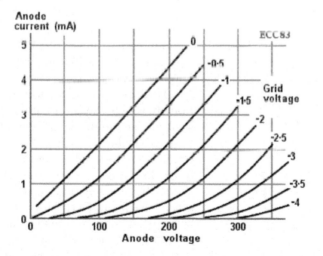

Output driving-point characteristics of a triode in common cathode configuration

Dynamic elements, such as inductors and capacitors, may also exhibit nonlinear behavior. For example, a voltage-controlled capacitor (varactor) is usually described by

$$I = \hat{C}(V)\frac{d}{dt}V$$

or, more appropriately,

$$Q = \hat{Q}(V),$$

where the stored charge Q depends on the terminal voltage V.

The term *"nonlinear circuits"* refers to electrical circuits that contain one or more nonlinear elements. Chua, Desoer and Kuh present an excellent overview of the subject in their 1987 texbook *"Linear and Nonlinear Circuits."*

Rectification

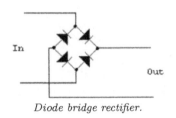

Diode bridge rectifier.

The first electronic circuits, which emerged in the early twentieth century, were powered by electro-chemical batteries. Even today, most electronic circuits assume that a DC voltage source is available to power them. However, mains electricity is distributed globally as an alternating signal. The conversion of AC to DC—a process called rectification—exploits nonlinearity to convert a signal which has zero DC component, such as a sine wave, to another which has a non-zero DC component. The required input-output transfer function of the rectifier, namely $V_{OUT} = |V_{IN}|$, where $|\cdot|$ denotes the absolute value, is usually implemented as a bridge of electronic diodes. These can be modeled as voltage-controlled nonlinear resistors.

Oscillation

The growth of channelized radio communications in the 1920s gave rise to the need for a range of electronic circuits, including oscillators and amplifiers, to generate carrier and modulation signals and to send them over great distances.

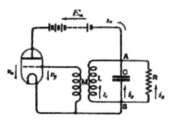

Van der Pol oscillator.

Oscillation requires a feedback configuration in which the loop gain is sufficient to sustain the amplitude of a signal circulating at a particular frequency. Barkhausen's criterion for sinusoidal oscillation, which dates from the 1930s, requires that the loop gain should be precisely unity at the frequency of oscillation f_0. In fact, Barkhausen's criterion, which is based on a simplified linear model, is necessary but not sufficient to produce oscillation. A more precise bifurcation analysis, based on nonlinear considerations, shows that (i) the small-signal (linearized) loop gain should be greater than unity to produce local instability and (ii) that a nonlinear amplitude-limiting mechanism is required to maintain a stable oscillation. The Dutchman Van der Pol is credited with introducing, in the late 1920s, a simplified cubic model of the nonlinearity in a triode-based LC oscillator which could predict the condition for oscillation, as well as estimating the shape, amplitude and frequency of the resulting periodic signal. The so-called Van der Pol equation is:

$$\frac{d^2}{dt^2}X - \epsilon(1 - X^2)\frac{d}{dt}X + X = 0.$$

For small values of the parameter ϵ, the oscillation is almost sinusoidal; for large ϵ, it exhibits jumps, and is called a relaxation oscillation.

Around the same time, in the 1930s Soviet Union, Andronov and his colleagues wrote the classic monograph "*Theory of Oscillators*," which summarized the state of the art in oscillator design. Although the circuits described in the book are based on the only three-terminal electronic devices which were available at the time, namely thermionic valves, the authors explain both qualitatively and quantitatively the basis of sustained oscillation from a mathematical point of view, independently of the details of the constituent electronic devices. Their approach, which is

Alexandr Andronov [Wikipedia]).

based on theoretical investigations of nonlinear dynamics originating with Poincaré in the late nineteenth century, has been hugely influential in the understanding of oscillation phenomena, especially since it was republished in English in the 1960s.

Feedback Linearization

As audio communications and sound reproduction grew in popularity between the First and Second World Wars, with the widespread update of voice telephony, "*talkies*," and records, the quality of amplifiers drew increasing attention. Nonlinearity causes interference between nearby communication channels and distortion in sound quality. A range of circuit techniques were introduced, the goals of which were to maximize the linear range of amplifier elements. Most of these exploit feedback to linearize amplifier blocks.

Transient and Stability Analysis

Demands from the military during the Second World War spurred the development of control strategies for anti-aircraft guns, bomb sights, and aircraft themselves. The subsequent Cold War and Space Race pushed the field of automatic control even further, resulting in the emergence of effective control strategies for nonlinear systems. These ideas were applied in circuits to ensure stability and to minimize settling times in response to switched inputs.

Analog Computation

Control systems based on analog computers required subsystems to perform both linear and nonlinear mathematical operations. While addition and subtraction can be performed by passive linear circuits, integration, differentiation, and amplification require active circuits. The operational amplifier was developed as the core active building block for these operations, its high gain being exploited, via feedback linearization, to minimize the adverse effects of nonlinearity.

Analog computers also required nonlinear operations such as multiplication and division. The inherent exponential characteristic of the bipolar junction transistor became indispensable for calculations using logarithms. For example, multiplication and division could be evaluated in terms of logarithm, addition, subtraction, and exponentiation:

$$X \times Y \ = \ \exp\big(\ln(X) + \ln(Y)\big)$$
$$X/Y \ = \ \exp\big(\ln(X) - \ln(Y)\big).$$

Log domain computation also inspired the translinear principle which underpins the Gilbert cell mixer that has become ubiquitous in radiofrequency electronics, as well as low-voltage current-mode circuits using bipolar junction and subthreshold MOS transistors.

Digital Circuits

The advent of binary computation using digital circuits in the 1940s and 1950s motivated the development of techniques for the analysis and design of fast binary logic circuits and storage elements. A key nonlinear phenomenon exploited in these applications is regeneration resulting from local positive feedback. As with many nonlinear phenomena, circuit implementations preceded theoretical analysis. It took several more decades before Willson *et al.* proved why cross-coupling transistors results in latching.

Transistors and Integrated Circuits

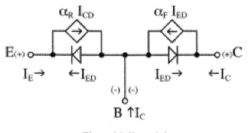

Ebers-Moll model.

Transistors are inherently nonlinear elements. As more transistors were combined in the 1950s to produce complex electronic systems, it became necessary to develop simple but accurate models that exploited advances in semiconductor physics. The Ebers-Moll model represents the bipolar junction transistor (BJT) at DC as a combination of nonlinear resistors (diodes) and current-controlled current sources. The dynamic behavior of the BJT is modeled by parasitic capacitors and diodes.

The introduction of integrated circuits in the 1960s made it necessary to be able to simulate huge networks containing many transistors. Hand analysis of such large nonlinear circuits was no longer feasible so attention turned to developing efficient numerical modeling and simulation methods.

The Golden Years

During the 1960s and 1970s, many of the key theoretical results underpinning nonlinear circuits were elaborated. The core knowledge of the subject was collected in textbooks such as Hayashi's "Nonlinear Oscillations in Physical Systems" (1964), Stern's "Theory of Nonlinear Networks and Systems: An Introduction" (1965) and Chua's "Introduction to Nonlinear Network Theory" (1969). In 1974, Willson edited a collection of state-of-the-art research publications under the auspices of the IEEE Circuits and Systems Society. This book included influential papers from academics and industrial research laboratories, most notably Bell Labs. These adressed issues such as the existence and uniqueness of solutions of nonlinear systems of equations, as well as efficient numerical methods for computing the DC solutions and dynamic behavior of large transistor networks.

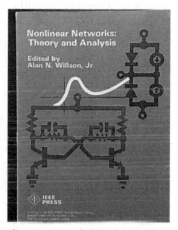

Cover page of "Nonlinear Networks: Theory and Analysis.

Computer-Aided Design

State equation descriptions of circuits, a legacy from nonlinear system theory, were proving cumbersome and inflexible for simulating large circuits. Two alternative methods were developed to describe networks of electronic circuit elements: Sparse Tableau Analysis (STA) and Modified Nodal Analysis (MNA).

STA, which was developed by Brayton *et al.* at IBM, determines all node voltages and branch currents in a network, while MNA uses a smaller set of equations, and therefore less computer memory. Even though MNA calculates only a limited set of circuit variables, these include node voltages, which makes it very useful for comparing simulations with physical implementations.

Circuit simulation was subdivided into four key tasks in Rohrer's MNA-based CANCER software at UC Berkeley in the early 1970s, and this continued into SPICE: DC analysis, small-signal analysis, transient analysis, and noise analysis.

The goal of DC analysis is to determine the DC solution of a circuit, the so-called operating point. Small-signal analysis is concerned with the behavior of the circuit in the vicinity of the operating point. Transient analysis returns voltages and currents as a function of time. Noise analysis uses the linearized small-signal model to determine the noise spectral density as a function of frequency.

The general representation of a circuit using MNA is as follows:

$$H(X) + \frac{d}{dt}Q(X) = U,$$

where X is a vector of unknown node voltages and branch currents, $H(\cdot)$ and $Q(\cdot)$ are (possibly) nonlinear functions capturing the branch equations and connectivity, and U is a vector of independent voltage and current sources.

Nonlinear DC Analysis The first step is to solve for the DC solution, namely when X and U are constant. Setting $dX/dt = 0$ and $U = U_Q$ gives

$$H(X) = U_Q.$$

This equation is solved using a nonlinear root-finding algorithm such as Newton-Raphson. In practice, the nonlinear elements are iteratively replaced by affine equivalents (a combination of linear elements and independent sources). The resulting matrix inversion problem is solved by Gaussian elimination.

Small-Signal (AC) Analysis Once the DC solution has been obtained, the circuit is linearized at the operating point. In particular, each nonlinear element is replaced by a linearized (affine) equivalent at the operating point. Energy storage elements are replaced by complex conductances. The resulting circuit is solved for one frequency at a time.

Transient Analysis An implicit integration scheme, originally based on the trapezoidal rule, but later Gear's method for stiff systems, reduces transient analysis to a succession of DC analyses at each time point.

Frequency Domain Noise Analysis The thermal and shot noise characteristics of a circuit are calculated in conjunction with the AC small-signal analysis. The adjoint matrix of the linearized circuit is determined and then solved for one frequency at a time.

Nonlinear Dynamics

An oscillator is an example of a nonlinear dynamical system. The exact form of the nonlinearity is often unknown, or complicated, making analytical solution of the circuit equations difficult, if not impossible, in general. The advent of low-cost computational power in the 1980s sparked interest in studying the more complex behaviors which can be exhibited by nonlinear dynamical systems, including complicated periodic and non-periodic oscillations.

As early as 1927, Van der Pol had noted a phenomenon which he called *"frequency demultiplication"* as well as strange noise-like behavior in a forced oscillator. With the evolution of understanding in the field of nonlinear dynamics, these phenomena have been identified as what are now commonly known as injection-locking and chaos, respectively. In forced oscillatory systems, these phenomena are closely related to synchronization.

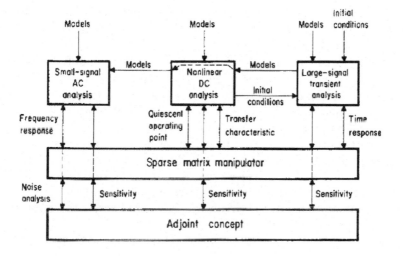

Computational approach to circuit simulation used in CANCER.

Unforced oscillators, such as that introduced by Lorenz in the 1960s as a greatly simplified model of climate dynamics, can also exhibit chaotic behavior. Chua's oscillator, invented in 1984, is a relatively simple electronic circuit described by three ordinary differential equations which has been studied extensively. It contains three dynamic energy storage elements, a linear resistor, and a nonlinear resistor. By changing the parameters of a two-terminal nonlinear resistor and/or the values of the linear circuit elements, Chua's oscilllator can exhibit a plethora of behaviors including periodic oscillation, subharmonics, and chaos.

Local and Global Behavior Nonlinear dynamics takes a geometrical approach to circuits and systems. The state of a system is represented as a point in a multi-dimensional state space. As time passes, the state follows a trajectory through the state space. A DC solution corresponds to a fixed point in state

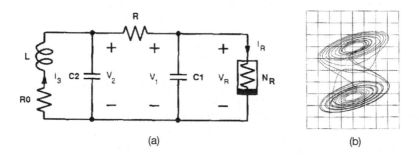

(a)   (b)

(a) Chua's oscillator and (b) a chaotic attractor.

space, i.e. the state does not change with time. A periodic solution corresponds to a closed trajectory: a trajectory returns to the same point in space once every fundamental period T.

The behavior of a nonlinear circuit might be different in different parts of its state space. Therefore, one must distinguish between local and global behavior.

If trajectories starting near a fixed point converge to that point, it is stable; otherwise, it is unstable. For example, a bistable flip-flop has three equilibrium points, two of which are locally stable; the other is locally unstable. No matter where it starts in the state space, the flip-flop will eventually converge to one or other of the stable fixed points. Thus globally, the circuit goes to one stable fixed point or the other. Because trajectories move *towards* these fixed points, they are called attractors. Attractors are separated in state space by separatrices. With an appropriate input, the circuit can be pushed across the separatrix and attracted to the other fixed point.

An LC oscillator is typically characterized by an unstable equilibrium point surrounded by a stable periodic trajectory (a so-called limit cycle). Multiple solutions could coexist, for example a simple periodic solution and a complex aperiodic solution. If properly designed, the only attractor in the state space is a stable limit cycle.

The local stability of a fixed point (equilibrium point) is determined by the eigenvalues (equivalently poles) of the linearization of the circuit (the small-signal model) at this point.

The idea of stability can also be applied to periodic solutions. If trajectories close to a periodic trajectory (a cycle) converge towards that cycle, it is orbitally stable. Orbital stability can be determined by sampling the system approximately once per period when a trajectory crosses a so-called Poincaré section. The fixed point in the resulting discrete-time system (the Poincaré map) inherits the stability of the periodic trajectory in the underlying continuous-time system.

The local stability of a periodic trajectory is determined by the eigenvalues of the fixed point of the linearization of the Poincaré map (the small-signal model of the equivalent discrete-time system).

The stability of a periodic trajectory can also be determined by perturbation analysis along the trajectory, using methods pioneered by Floquet in the late nineteenth century. The so-called monodromy matrix maps trajectories forward by one period close to the cycle. The eigenvalues of the monodromy matrix are called characteristic multipliers.

Bifurcations and Chaos

The concept of structural stability refers to the dependence of the steady-state behavior of a nonlinear dynamical system on the parameters of the system. For example, changing the negative resistance in an LC oscillator can make the fixed point unstable so that the only attracting solution is a limit cycle, corresponding to an oscillating state. This qualitative change in behavior is called a bifurcation. Common bifurcations in electronic circuits include the Hopf bifurcation associated with most LC oscillators (where an equilibrium point changes stability), and fold bifurcations in Schmitt triggers and latches (where a stable and an unstable fixed point merge and disappear).

Chaos is associated with special types of trajectories in nonlinear dynamical systems: a heteroclinic trajectory connects one fixed point to another, a homoclinic trajectory connects a fixed point to itself. Chaotic trajectories are confined to a bounded region of the state space but they are not periodic. In the frequency domain, they have a continuous spectrum, and appear noiselike. The real parts of the Floquet exponents are called Lyapunov exponents. Chaotic behavior is characterized by a positive Lyapunov exponent.

The subject of bifurcations and chaos is covered in detail elsewhere in this book.

Synchronization A number of applications in electronics require signals which have precise phase relationships. These include modulation schemes for digital communication systems which require In-phase (I) and Quadrature (Q) signals, and clock recovery in data communications. Injection-locking is used for synchronization and in frequency division. Synchronization is also central to the operation of electrical power systems.

These applications exploit the inherently nonlinear phenomenon of synchronization, first reported in coupled mechanical clocks by Huygens in the seventeenth century.

Neural Networks

Hodgkin and Huxley were awarded the Nobel Prize in Physiology or Medicine in 1963 for their electrical model of the squid giant axon. The observation that neurons can be modelled by electrical components, supported by efficient numerical methods, and an improved understanding of nonlinear dynamics, stimulated research into biologically-inspired computation. Neural networks contain a number of similar subcircuits (analogous to neurons) connected together to form complex systems. Widrow and Hoff developed an early electrical neural network in 1960 which was based on the McCulloch-Pitts neuron and used memistors as programmable elements. In the late 1980s, Chua and Yang in-

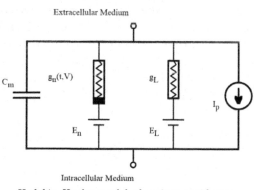

Hodgkin-Huxley model of a giant squid axon.

troduced locally-connected cellular neural networks. Today, neural networks based on this architecture can outperform conventional digital machines in applications such as near real-time image processing.

Delta-Sigma Modulation

The 1970s saw digital signal processing techniques spreading into domains which had heretofore been the preserve of analog circuits. In particular, a combination of oversampling and delta-sigma modulation (DSM) simplified the linearity requirements for data conversion circuits. Inose *et al.* described DSM-based coding for telecommunications. Candy introduced DSM-based ADCs. DSM-based DACs became ubiquitous in audio playback of digital recordings, beginning with CDs in the 1980s. DSM has underpinned frequency synthesizers in wireless communications since its introduction in fractional-N PLLs is the 1990s.

DSM theory assumes that the *"noise"* introduced by the quantization process has a white power spectral density and that it is uncorrelated with the signal being processed. It soon became clear that this is not the case. Complex nonlinear phenomena in DSMs include idle tones in DACs and ADCs and fractional spurs in synthesizers. Techniques such as mismatch-shaping and dithering have been developed to alleviate these problems.

Noise Analysis

The noise analysis method which was implemented in CANCER and later SPICE, and its variants, assumes a linear model of the circuit at the operating point and white noise sources. In practice, circuit designers need to predict the noise in large classes of switching and oscillating circuits, including switched-capacitor circuits, oscillators, and phase-locked loops. These circuits are characterized by complex nonlinear interactions with noise sources, some of whose power spectra (such as $1/f$ noise in MOS transistors) are not white. Two main linear time-variant methods have been developed to solve this problem in the case of oscillator phase noise.

The Impulse Sensitivity Function (ISF) was introduced by Hajimiri and Lee in 1998. This characterizes the sensitivity of the circuit to impulse perturbations introduced at different time instants. In addition to being time-consuming, this method is prone to error if the user chooses impulses that are too large. A more accurate technique, introduced by Demir *et al.* in 2000, is based on perturbation analysis along a limit cycle. The Perturbation Projection Vector (PPV) method is based on calculating the eigenvalues of the monodromy matrix associated with a periodic solution.

Michael Peter Kennedy

University College Cork, Ireland

Chaos

Deterministic dynamical systems are those whose state is uniquely determined by the knowledge of an initial condition and of the mathematical model describing them. Depending on whether they evolve in continuous- or discrete-time, such model assumes the form of a set of differential or difference equations.

As an example, lumped circuits composed by *resistive elements* –such as resistors, independent and controlled voltage and current sources[7], and *reactive element* –capacitor and inductors– are continuous-time deterministic dynamical system with respect to state variables expressed as voltages, currents, electric charges and magnetic fluxes. Similarly, switched capacitors/current circuits, digital filters, sigma-delta modulators, or circuits containing sample-and-hold in their structure, are examples of discrete-time deterministic dynamical systems.[8]

For any electrical engineer, *linear* circuits play a fundamental role. They are, in fact, the starting point for tackling the study of circuits, and, as a consequence, *nonlinear* circuits are typically considered a "distorted" version of them. From this point of view, phenomena like signal distortion, generation of harmonics, etc., are a straightforward consequence on the nonlinear characteristics of (some of) the circuits elements. As such, natural methods for their analysis are, for instance, the *harmonic balance* or the *describing function technique*, and more in general, all those methods which are based on a series expansion aiming to capture the deviation of the circuit from a linear behavior.

Such an approach is fully justified for "weakly" nonlinear circuits, but it is not able to describe the influence of "strong" nonlinear circuit elements. Furthermore, it is not possible to draw a line between weak and strong nonlinear circuits by simply looking at the characteristics of the individual circuits elements. Their interactions plays, in fact, a fundamental role to determine the final behavior of the circuits. On the other hand, one needs to consider that all electronic components, both active and passive, are *intrinsically* nonlinear, as are (some of) the circuits interconnecting them. For many practical applications, a circuit designer exploits suitable techniques and circuit topologies (differential configurations, feedback, inverse-function-based cancellation techniques, etc.) aiming to reduce the influence of the nonlinear elements. Despite effective, these *linearization techniques* cannot fully cancel nonlinearities. Consequently, any circuit designer is constantly faced with the problem of evaluating the limits of validity of the particular linearization method used and of how the residual nonlinear effects degrades circuit performance. For instance, effects like harmonic distortion and intermodulation must always been considered in the design of "quasi-linear" electronic circuits.

In other applications, the goal is not to cancel the nonlinearity, but to exploit it to achieve the desired signal processing application, as it is the case for analog multipliers, logarithmic amplifiers and oscillators and multivibrators. Also in

---

[7]The model of more complex elements such as bipolar junction or MOS transistors can be suitably expressed by using the former elements.

[8]From now on we will drop the term determinist and use simply the expression "dynamical system".

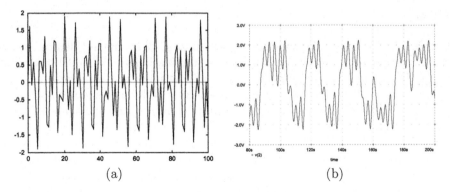

Figure 5.1: Time evolution of a quasi-periodic solution (a) and of a chaotic solution (b).

this case, a designer must be well aware of the intrinsic limitations of the design techniques used to optimize performance of the nonlinear circuit at hand.

One of the most spectacular example of a strong nonlinear behavior is a steady-state random-like behavior which is exhibited by relatively simple nonlinear circuits. This phenomena is called *chaos* and possess an interest which goes well beyond simple scientific curiosity, since it has fund useful applications in the area of communication, electromagnetic interference (EMI) reduction and random number generation.

Features of Chaos

There is no single adopted definition of *chaos* and one may simply rely on a phenomenological one by calling *chaotic behavior* the aperiodic, stead-state, random-like, bounded solution of a continuous-time dynamical system $\dot{x}(t) = F(x(t), u(t))$, with $x(t)$, $u(t) \in \mathbb{R}^N$ e $t \in \mathbb{R}^+$, or discrete-time one $x(n+1) = F(x(n), u(n))$ where $x(n)$, $u(n) \in \mathbb{R}^N$ e $n \in \mathbb{N}^9$. To fully appreciate the peculiar features of this behavior, we need to compare it, both in terms of time- and frequency domain behavior, with the other possible steady-state solutions of a dynamical system: equilibrium points or DC solutions, periodic solutions, and quasi-periodic solutions. An equilibrium point is a solution which is constant in time and whose power density spectrum (PDS) has a single component at DC. A solution is periodic if it repeats in time after a minimum period $T_0 > 0$ and whose PDS Is composed of harmonics at frequencies multiple of $f_0 = 1/T_0$. A quasi-periodic solution is characterized by the fact that it can be expressed as the sum of a finite or infinite number of periodic signals with *incommensurable* frequencies [1]. For example $\phi(t) = \sin t + \sin \sqrt{5}\, t$ is a quasi-periodic signal. Waveforms corresponding to this kind of solutions can have a quite complex time evolution as shown, for example, in Fig. 5.1 (a). The PDS of a quasi-periodic signal is composed by a discrete set of lines at incommensurate frequencies.

To understand chaos, let us refer to a practical example of a systems capable to exhibit a chaotic solution, namely Chua's circuit [2], shown in Fig. 5.2 (a).

---

[9]When $u(t)) = 0$ ($u(n) = 0$) the systems is called autonomous, otherwise is said non-autonomous.

This is an third-order autonomous circuit with 2 capacitors of capacitance C_1 and C_2 and one inductor of inductance L as reactive elements, oNe resistor of conductance G, and a single nonlinear resistor –called Chua's diode– whose linear piecewise characteristic is $g_b(V_b) = G_b V_b + 1/2(G_a - G_b)(|V_b + B_p| - |V_b - B_p|)$. The equation describing the circuits are $C_1 dV_{C_1}/dt = G(V_{C_2} - V_{C_1}) - g_b(V_{C_1})$, $C_2 dV_{C_2}/dt = G(V_{C_1} - V_{C_2}) + I_L$ and $L dI_L/dt = -V_{C_2}$.

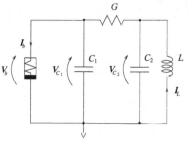

Figure 5.2: *Chua's circuit*

Assuming the following values for the parameters of the circuit[10] $C_1 = (1/9)$ F, $C_2 = 1$ F, $L = (1/7)$ H, $G = 0.7$ Ω^{-1}, $G_a = -0.8$ Ω^{-1}, $G_b = -0.5$ Ω^{-1}, $B_p = 1$ V, one can obtains the waveforms for $V_{C_2}(t)$ and the projection of the circuit state trajectory on the plane $V_{C_1} - V_{C_2}$ reported in Fig. 5.1 (b) and Fig. 5.3 (a), respectively.

It is evident from these figures that the system trajectory is bounded and aperiodic, but also that it is clearly difficult to determine which are the difference between the waveform of Fig. 5.1 (b) and the one of characterizing a quasi-periodic behavior of Fig. 5.1 (a). To distinguish among the two solutions one needs to consider the signals PDS. As it is shown in Fig. 5.3 (b), the PDS of a chaotic signal is clearly different with respect to the one of a periodic or quasi-periodic signal, since it is *continuous*, *broadband* and *noise-like*. This is one of the most important distinct features of a signal generated by a chaotic system. The trajectory of chaotic system in the state space concentrates, in its aperiodic motion, in a bounded region called *strange attractor* (see Fig. 5.3 (a)) [3], whose description and properties are however too complicated to be described here.

An important question to ask is about the minimum complexity that a dynamical system must have to be able to exhibit chaotic behavior. The answer to this question relies on the Poincaré-Bendixon theorem [4], which, roughly speaking, states that any trajectory of a second-order continuous-time autonomous dynamical system can only converge either to an equilibrium point or to a limit cycle. This obviously excludes the possibility of an irregular, random-like behavior at steady-state. Consequently, to exhibit a chaotic solution, a continuous-time dynamical system must be at least *non-autonomous of the second-order* or *autonomous of the third-order* (as it is the case for Chua's circuit).

On the contrary, if one considers the case of discrete-time autonomous dynamical systems, the minimum complexity reduces to one-dimensional assuming that function $F(\cdot)$ in non-invertible: in this case, one typically speaks of *one-dimensional maps*[11]. typical example is is the *Tent Map* $x(n+1) = M(x(n))$, where $M : X = [0, 1] \mapsto [0, 1]$ and $M(x) = 1 - 2|x - 1/2|$.

---

[10] This is of course not the only possible choice.

[11] This indicates both the nonlinear function and the dynamical system based on this

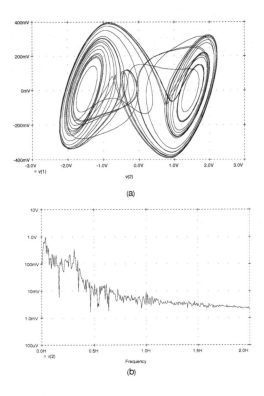

Figure 5.3: Projection of the trajectory on the plane $V_{C_1} - V_{C_2}$ (a) and broadband PDS of signal V_{C_2} (b), for a Chua's circuit operating in chaotic regime.

H. Poincaré [Wikipedia]).

The most peculiar and distinctive property of chaotic systems is the so-called *sensitive dependence on initial conditions*. This phenomenon was originally discovered by Henri Poincaré in a particular case of the three-body problem, who also suggested that it could be common also in other areas such as meteorology. Interesting enough, meteorology is an area that played indeed an important role in chaos history. In the 1960s the meteorologist Edward Lorenz experimented with primitive computer simulations of weather prediction using a program based on 12 recursive equations. When doing so, he tried to recreate an interesting solution, one he had simulated previously, by re-entering the values the computer had previously calculated. To speed up the process, he used a truncated version of these values as intial condition, and, to his suprise the obtained solution was very different from the one he had previously seen and was changing dramatically when the initial condition was even minimally changed.

To more concretly highlight this phenomenon, let us consider two different trajectories generated by a Tent map starting from two slightly different initial conditions, namely $\pi/10$ e $\pi/10 + 10^{-4}$. The plot of the two time series is shown in Fig. 5.4 (a): as it can be seen, despite the two initial conditions are very close, a few iterations are sufficient to make the two time series almost uncorrelated from each other. This phenomena is very important. In fact, in practice, due to measurement errors and noise (physical or computational), it is not possible to know the state of a dynamical system with absolute precision. Due to the sensitive dependence property, such errors in the knowledge of the system state will be somehow "amplified" by the system dynamics and *cause the*

Edward Lorenz [Wikipedia]).

inability to predict the system behavior in long terms, despite the fact that the system is totally deterministic [5], [6]. In other words, we may conclude that the study of single trajectories brings little or no information on the general properties of a chaotic system and that attempting to characterize typical behaviors from them is doomed to fail. In fact, any error in the knowledge of the state gets amplified by the system evolution till a point when it becomes completely impossible to predict the orbit position in the state space. This is the reason why deterministic architectures that exhibit chaotic behaviors, are sometimes considered to lay on the borderline between *deterministic systems* and *stochastic processes*.

To cope with the problem of characterizing the typical behavior of a chaotic map, we may take advantage of tools originally developed for random process theory [7], [8]. To understand the main idea grounding this approach, let us consider one of the two time series mentioned in the last example, that is $\{x(n)\}$ $0 \leq n \leq N - 1$ generated by a Tent map starting from $x_0 = \pi/10$. Divide the interval X into $m = 20$ subintervals $X_i = [(i-1)/m, i/m)$, $i = 1, \ldots, m$ and construct the histogram with the values f_i indicating the fraction of the N elements which belongs to X_i defined as

G.D. Birkhoff [Wikipedia]).

$$f_i = \frac{m}{N}\{\# \text{ di } M^k(x_0) \in [(i-1)/m, i/m), \ k = 1, \ldots, N\}. \qquad (5.5)$$

The result is shown Fig. 5.4 (b). What is remarkable is that the result remains the same also by considering the time series starting from $\pi/10 + 10^{-4}$ and others initial conditions. The conclusion is that the state distribution in a long trajectory seems not to be affected by the sensitive dependence on initial conditions! In other words, a better characterization of the properties of a

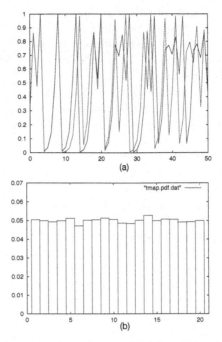

Figure 5.4: Example of sensitive dependence on initial condition for the Tent map (a) and Histogram constructed according to (5.5) with $m = 20$, $N = 5000$ and $x(0) = \pi/10$..

chaotic system seems possible in terms of a statistical approach (by computing f.i. the probability distribution of the state).

Historically the two most important contributions which grounds this theory (called *Ergodic theory*) were made in the 1930s and are due to George David Birkhoff and John Von Neumann who proved that under certain conditions, the time average of a function along each trajectories exists almost everywhere and is related to the statistical average computed with respect to the probability distribution of the state. The area has advanced significantly since the late 1970s and this set of tools has also entered those used by engineers to study chaos. In particular, if the chaotic maps are piecewise linear Markov, one can develop a methodology to fully characterize the quantized sequences generated by these maps [8]. In other words, these maps are generator of stochastic process with statistical features that can be tuned by changing the map parameters (such as the slopes) and can also easily implement as CMOS integrated circuits, and this is the key to many of the engineering applications of chaos.

Application of Chaos

In this section we briefly highlight some of the most interesting results of the application of chaotic circuits and systems to problems of engineering interest.

Chaos-based Communication

Over the past two decades, there has been significant interest in exploiting chaotic dynamics in communications. Topics which have been explored include chaotic encryption for security, chaotic spreading codes for multi-user access in spread-spectrum systems, and chaotic modulation for the transmission of analog and digital information (see [9] and references therein).

At signal level, continuous-time chaotic systems can be used to generate wideband carriers for digital modulation schemes. The best digital chaotic modulation scheme in the literature is FM-DCSK, which has been shown to offer some advantages over conventional digital communication systems in a multipath environment [10].

At code level, discrete-time chaotic systems can be used to generate spreading codes for Direct Sequence Code Division Multiple Access (DS-CDMA) communication systems. Actually, it can be proven that by substituting classical pseudo-random (i.e. m-, Gold, Kasami, ...) spreading sequences with those generated by a suitable chaotic map, one is able to achieve the absolute minimum multiple-access interference [11]. Hence, when multiple-access interference is the main cause of errors in the communication, chaos-based spreading allows to obtain the optimum system performance, which can be computed to have an average 15.47% (and optimized 60% peak value) increase in users capacity with respect to systems adopting random or pseudo-random spreading. This is an optimum result in the sense that no other choice of sequence generation policy can result in a lower interference. Theoretical predictions have been also confirmed by a prototype system comprising of 8 transmitters and a receiver matched with one of the transmitters [12], which was the first example of a chaos-based communication system prototype that outperformed the equivalent one employing standard methodology.

Electromagnetic Interference Reduction

Nowadays there are very many sources of electromagnetic emission (ranging from radio transmitters and radars to relays and dc electric motors, as well as digital electronic devices), so that the problem of designing electromagnetic compatible (EMC) systems has become of great and widespread practical concern. High performance computing platforms are a perfect example where the presence of Electromagnetic Interference (EMI) must be handled with particular care in order to design more and more compact systems and architectures that can operate without having any of the subsystems interfering with the neighboring one. Switching power converters also offer an extremely significant example where EMI is a severe problem. For such systems, interference is mainly due to timing signals, such as clock signals, widely employed in digital circuits, or the control pulse-trains used in switching power converters. Due to their periodic nature and to the presence of sharp edges, the power of such signals is in fact concentrated at those frequencies corresponding to the multiples of the timing signal period. This can obviously produce serious problems in terms of EMI.

To solve this issue one can adopt a-priori solutions at the very beginning of the design stage, in order to assure that the implemented electronic equipment generates EMI characterized by a maximally flat power spectral density. With

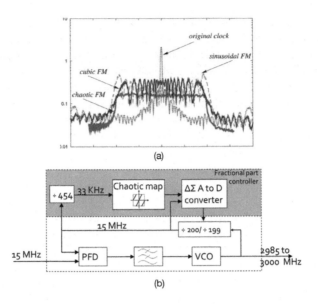

Figure 5.5: (a) Measured EMI reduction when different FM modulating signal are employed, and (b) architecture of a chaos-based spread-spectrum clock generator for EMI reduction in SATA-II applications

this, its integral within any frequency range (and therefore within the sensitivity bandwidth of any potential victim) can be made as low as possible. An appreciable EMI reduction can be achieved by means of a simple quasi-stationary frequency modulation (FM) directed at intentionally perturbing a normally narrowband timing signal, thus spreading the energy associated with each harmonic over a large bandwidth in order to reduce the peak value. From this point of view, classical solutions range from simple sinusoidal FM to the more sophisticated case where the frequency deviation profile is expressed as a family of cubic polynomials patented by Lexmark. Notably, this last solution reduces the peak value of the spectrum by more than 7dB with respect to the unperturbed signal and has been employed in several commercial products from Intel, IBM, and Cypress.

The use of the modulating signal generated by a suitably chosen chaotic map allows to reduce the EMI spectrum by an additional 9dB with respect to the previously mentioned solutions, as it is shown in Fig. 5.5 (a). This methodology has also been applied to reduce EMI for 3-GHz Serial Advanced Technology Attachment II (SATA-II) applications, by using chaos in a spread spectrum clock generator whose architecture is shown in Fig.5.5 (b). The results reported in [13] for a prototype implemented in 0.13 um CMOS technology show a measured peak reduction greater than 14 dB.

Interested readers can refer to [14] for a survey on chaos-based EMI reduction methods.

Random Number Generators

Random Number Generators (RNGs) represent a critical point in the implementation of many security schemes and can be regarded as fundamental cryptographic primitives. It is generally recognized that ideal (or so called true) random number generators (TRNGs) can only be approximated. An ideal source must in fact be capable of producing infinitely long sequences made of perfectly independent bits, with the property that, when restarted, it never reproduces a previously delivered sequence (non-repeatability). Practical implementations of RNGs can be classified into two major categories, namely pseudo-RNGs and physical-RNGs. Pseudo-RNGs are deterministic, numeric algorithms that expand short seeds into long bit sequences. Conversely, physical-RNGs rely on microscopic processes resulting in macroscopic observables which can be regarded as random noise (quantum noise, thermal noise, etc.). Pseudo-RNGs generally depart more from the ideal specifications. Being necessarily based on finite memory algorithms, they exhibit periodic behaviors and generate correlated samples. The same reason also makes them completely repeatable and obviously unsuitable for data security and cryptography. Their substantial advantage is the algorithmic nature which makes them easily embeddable in any digital circuit or system. Physical-RNGs, on the other hand, are the best approximation of TRNGs, but they may require very specialized hardware and/or environmental conditions, which make them expensive to embed. In spite of this liability, security related applications have recently been strongly pushing their development and deployment, so that bigger players in IT (e.g. Intel and Via) are now introducing physical-RNGs in their security platforms. Chaotic maps have always claimed to be capable of implementing TRNGs. The basic idea is that, rather than exploiting natural phenomena like noise that are hardly controllable and mathematically unmanageable, one can rely on artificial and simpler ones, as one-dimensional chaotic maps, that derive unpredictability from complexity. My main contributions in this field have been to tackle this problem both at the theoretical and system design level. By exploiting the statistical approach to chaos, it as been possible [15], [16] to analytically prove that chaotic maps exist that behave as TRNGs in an ideal setting and that robustly maintain such properties with respect to implementation errors, and, furthermore, that such maps can be practically realized out of pipeline analog to digital converters (ADC) parts which are now ubiquitous in any mixed mode systems. With this, one can easily reuse design expertise and even analog Intellectual-Property (IP) blocks to quickly embed true random sources in SOCs and specialized apparatuses.

References

[1] T. S. Parker and L. O. Chua, *Practical Numerical Algorithms for Chaotic Systems*, Springer-Verlag, New-York, 1989.

[2] R. Madan, *Chua's Circuit: a Paridigm for Chaos*, World Scientific, Singapore, 1993.

[3] S. Wiggins, *Introduction to Applied Nonlinear Dynamical Sytems and Chaos*, Springer–Verlag, 1996.

[4] J. Hale, *Ordinary differential equations*, Wiley, New York, 1969.

[5] E. Ott, *Chaos in dynamical systems*, Cambridge University Press, 1993.

[6] R. L. Devaney, *An Introduction to Chaotic Dynamical Systems*, Benjammin/Cumming Publishing Co, 1986.

[7] A. Lasota, M. C. Mackey, *Chaos, Fractals, and Noise*, Springer-Verlag, 1994.

[8] G. Setti, G. Mazzini, R. Rovatti and S. Callegari, "Statistical modeling of discrete-time chaotic processes-basic finite-dimensional tools and applications," *Proceedings of the IEEE*, vol. 90, no. 5, pp. 662-690, May 2002.

[9] M.P. Kennedy, R. Rovatti, G. Setti, eds, *Chaotic Electronics in Telecommunications*, CRC Press, 2000.

[10] M. P. Kennedy, G. Kolumban, G. Kis, Z. Jako, "Performance evaluation of FM-DCSK modulation in multipath environments," *IEEE Transactions on Circuits and Systems I: Fundamental Theory and Applications*, vol. 47, no. 12, pp. 1702-1711, Dec 2000.

[11] G. Mazzini, R. Rovatti, G. Setti, "Interference Minimization by Autocorrelation Shaping in Asynchronous DS-CDMA Systems: Chaos-Based Spreading is Nearly Optimal," Electronics Letters, vol. 35, pp. 1054-1055, 1999.

[12] G. Mazzini, R. Rovatti, G. Setti, "Chaos-Based Asynchronous DS-CDMA Systems and Enhanced Rake Receivers: Measuring the Improvements, *IEEE Transactions on Circuits and Systems- Part I*, vol. 48, n. 12, pp. 1445-1453, 2001.

[13] F. Pareschi, G. Setti, R. Rovatti, "A 3 GHz Serial ATA Spread Spectrum Clock Generator Employing a Chaotic PAM Modulation," *IEEE Transactions on Circuits and Systems Part I*, vol. 57, n. 10, pp. 2577 - 2587 2010.

[14] F. Pareschi, R. Rovatti, G. Setti, "EMI Reduction via Spread Spectrum in DC/DC Converters: State of the Art, Optimization, and Tradeoffs, *IEEE Access*, vol.3, pp.2857-2874, 2015.

[15] J35 S. Callegari, R. Rovatti, G. Setti, "Embeddable ADC-Based True Random Number Generator for Cryptographic Applications Exploiting Nonlinear Signal Processing and Chaos," IEEE Transactions on Signal Processing, pp. 793-805, vol. 53, n. 2, 2005.

[16] J21 F. Pareschi, G. Setti, R. Rovatti, "Implementation and Testing of High-speed CMOS True Random Number Generators based on Chaotic Systems, *IEEE Transactions on Circuits and Systems Part I*, vol. 57, n. 12, pp. 3124 3137, 2010

Gianluca Setti

University of Ferrara, Italy

Neural Networks

A brief scientific historical review of Neural Networks embodies an interdisciplinary summary spanning almost all of the conventional disciplines of psychology, neuroscience, mathematics, computer science, and even physics and philosophy. Stephen Grossberg in the Editorial "Towards building a neural networks community" for his last day as the founding editor–in–chief of the journal *Neural Networks* wrote: "*I gradually began to realize that I was contributing to a major scientific revolution whose groundwork was laid by great nineteenth physicists such as Hermann von Helmholtz, James Clerk Maxwell, and Ernst Mach. These interdisciplinary scientists were physicists as well as psychologists and physiologists. Their discoveries made clear that understanding the Three N's of the brain would require new intuitions and mathematics*" [1]

The Three N's of brain dynamics represent the fundamental principles underlying large networks of biological neurons:

- Nonlinear phenomena
- Non–stationary nature
- Nonlocal interactions

According to S. Grossberg, those principles were the basis for the schism between Physics and Psychology [2] and led the scientific community to introduce biological neural models as a new paradigm to fill the gap between behavioral experience and brain dynamics, i.e. the Neural Networks concept represented a radical break with previous scientific methods. The cornerstones that led to the definition of the Three N's of brain dynamics can be briefly summarized as follow (a full historical treatment can be found in [1,2] and the reference therein contained):

- Helmholtz [3] studies on visual perception made clear that, in contrast to the classical Newtonian approach to color theory, human perception of white light requires the interactions among neurons with long–range interactions. In addition to the *nonlocal* neural process of white light, experimental and theoretical investigations led to unfold *nonlinear* phenomena in a network of retinal neurons stimulated by sources of light with different luminance, brightness and color [4,5,6]

- the doctrine of *unconscious inference* [7] following the original Helmholtz's concept of the process of visual perception, that is as reported in [2] "*[...] Helmholtz realized that we perceive, in part, what we expect to perceive based upon past learning*". Biological learning actions emerge as *non-stationary* processes due to competitive–cooperative interactions between nonlinear and nonlocal networks and spatio–temporal external stimuli.

In addition to Helmholtz's discoveries, Maxwell's trichromatic color theory and Mach's studies on optical illusion (known as Mach bands) revealed that the linear, local and stationary mathematics available during the nineteenth century was not appropriate to cope with mind and brain natural phenomena. To quote

S. Grossberg's manuscript [2] *"The schism between physics and psychology encouraged theorists trained in the physics tradition to believe that no theories of behavior and brain exist"*

This preliminary historical insights presents the scientific context in which Neural Networks have been developed and the pillars that influence contemporary research in terms of architectures and algorithms. Following the approach presented in the [8,9], the temporal advance of the scientific research on Neural Networks during the last eighty years can be organized in four chief periods (see in particular [10] where the four stages are presented in details): Beginning of neural networks, First golden age, Quiet Years and Renewed enthusiasm.

The 1940s: Beginning of neural networks

In 1943 McCulloch and Pitts published a seminal paper [11] in which they showed that any arithmetic o logical function could, in principle, be computed by simple neural networks. In 1949 a further fundamental contribution was made by D.O. Hebb that proposed a specific learning rule for synapses (i.e. the specific junction that allows interactions between biological neurons). In the book entitled *The Organization of Behaviour* Hebb exploited the learning law to describe the adaptation of synapse due to the firing activity of neurons [12]. The theory, that is the precursor of what is now referred to as Spike–Time–Dependent–Plasticity, made an effort to explain association phenomena during the learning process observed in some experimental results from psychology. In this period, real–world applications were pioneered by Bernard Widrow (his *ADELINE* was the first adaptive linear neural networks) and John von Neumann formalized the theoretical framework of brain–inspired computing embedding the Three N's concept [13].

The 1950s and 1960s: First golden age

Fig. 1 - Portion of the Harvard-IBM Mark 1. [Wikipedia])

The great scientific achievements of the 1940s culminated in the *Mark I Perceptron* developed during 1957 and 1958 by the psychologist Frank Rosenblatt. The Mark I Perceptron was a neuronal network based on the Hebbian rule initially implemented on a room–size IBM computer (Fig. 1) but subsequently realized in custom–built hardware. The possibility to learn and classify simple images of triangles and squares gave rise to a huge interest in autonomous self–learning machines mimicking human intelligence. In 1958 the New York Times reported that the Perceptron was *"the embryo of an electronic computer that [the American Navy] expects will be able to walk, talk, see, write, reproduce itself and be conscious of its existence"* and a few years later Rosenblatt wrote a comprehensive book on *Principles of Neurodynamics* [14].

Yet alongside the hype, the researchers' interest faded away due to the overoptimistic expectations that artificial brains were just a few years away and the work published by Marvin Minsky and Seymour Papert in 1969. In [15], Minsky and Papert pointed out that the impossibility to compute the XOR predicate was a serious shortcoming of a single-layer Perceptron. Minky and Papert's analysis effectively slowed down interest and funding of neural network research for more than a decade.

The 1970s: Quiet Years

Many researchers overlooked that Minky and Papert's criticism was specific for two–layer neural networks with linear activation functions and developed more complex models combining multiple layers of Perceptrons with nonlinear activation functions to enhance learning capabilities. Among the researchers that sustained neural network research in the 1970s, Stephen Grossberg played a prominent role introducing self–organizing networks and the adaptive resonance theory [16,17]. The pioneering works of Amari on random selection of neurons [18] and Fukushima on machine vision [19] were also fundamental to pave the way for the renaissance stage.

The 1980s and 1990s: Renewed enthusiasm

During the 1980s, the renewed interest in neural network was fueled by many active researchers that contributed to the development of several applications. Teuvo Kohonen proposed the novel idea of self–organizing maps (i.e. neural networks with unsupervised competitive learning) and John J. Hopfield (Fig. 2) presented the revolutionary crossbar associative networks, referred to as *Hopfield Network*. Hopfield also developed a rigorous mathematical methodology to describe network states by means of a global energy function whose minima (local or global) correspond to optimal solutions and was very active in promoting neural network research through numerous talks and highly readable publications [20,21]. Rumelhart and McClelland's books *Parallel Distributed Processing* [22] sprung backpropagation training algorithms and set the stage for

Fig. 2 - John J. Hopfield, Princeton University, NJ, USA. (Source: www.ieee.org/about/awards/bios /rosenblatt_recipients.html))

the current applications of neural networks including *Cellular Neural/Nonlinear Networks, Radial Basis Function (RBF) Networks, Convolutional Neural Networks and Deep Learning.*

In conclusion, this brief historical review has been written toward a twofold aim: to summarize the chief milestones in Neural Networks and their applications, and to stimulate interested readers to dig for more information into the cited comprehensive references.

References

[1] S. Grossberg, "Towards building a neural networks community," Neural Networks, vol. 23, no. 10, pp. 1135–1138, 2010

[2] S. Grossberg, "Nonlinear neural networks: Principles, mechanisms, and architectures," Neural networks vol. 1, no. 1, pp. 17–61, 1988

[3] James Southall, Cocke Powell, ed. Helmholtz's "Treatise on Physiological Optics," Translated from the Third German Edition. Dover, 1962

[4] L. E. Arend, J. N. Buehler, and G. R. Lockhead. "Difference information in brightness perception," Perception and Psychophysics vol. 9, no. 3, pp. 367–370, 1971

[5] Edwin H. Land, "The retinex theory of color vision," Scientific America., 1977

[6] John D. Mollon and Lindsay Theodore Sharpe, eds. "Colour vision: physiology and psychophysics," Academic Press, 1983

[7] Edwin Garrigues Boring, "History of experimental psychology," Genesis Publishing Pvt Ltd, 1929

[8] Jacek M. Zurada, " Introduction to artificial neural systems," Vol. 8. St. Paul: West publishing company, 1992

[9] Simon Haykin, "Neural Network: A comprehensive foundation," Prentice Hall, 2004

[10] Yadav Neha, Anupam Yadav, and Manoj Kumar. An Introduction to Neural Network Methods for Differential Equations. Springer, 2015

[11] W.S. McCulloch, W. Pitts, "A logical Calculus of the ideas immanent in nervous activity", Bull. Math. Biol. 5, 115133,1943

[12] D.O. Hebb, "The Organization of Behaviour: A Neuropsychological Theory", Wiley, New York, 1949

[13] J.V. Neumann, "'The General and Logical Theory of Automata," Wiley, New York, 1951

[14] F. Rosenblatt, "Principles of Neurodynamics", Spartan Books, Washington, 1961

[15] M. Minsky, S. Papert, "Perceptrons", MIT Press, Cambridge, 1969

[16] S. Grossberg, "Pavlovian pattern learning by nonlinear neural networks,"Proceedings of the National Academy of Sciences, vol. 68, pp. 828–831, 1971

[17] S. Grossberg, "Studies of mind and brain: Neural principles of learning, perception, development, cognition and motor control," Boston: D. Reidel Press, 1982

[18] S. Amari, "A theory of adaptive pattern classifiers," IEEE Trans. Electron. Comput. vol. 16, no. 3, pp- 299–307, 1967.

[19] K. Fukushima, "Visual feature extraction by multilayered networks of analog threshold elements," IEEE Trans. Syst. Sci. Cyber, vol. 5, no. 4, pp. 322–333, 1969

[20] J.J. Hopfield, "Neural Networks and physical systems with emergent collective computational abilities," Proc. Natl Acad. Sci. vol. 79, pp. 2254–2258, 1982

[21] J.J. Hopfield, "Neurons with graded response have collective computational properties like those of two state neurons," Proc. Natl. Acad. Sci. vol. 81, pp. 3088–3092, 1984

[22] D.E. Rumelhart, J.L. McClelland, "Parallel Distributed Processing: Explorations in the Microstructure of Cognition I and II," MIT Press, Cambridge, 1986

Fernando Corinto
Politecnico di Torino, Italy

Circuits and Systems for Fuzzy Logic

Multiple-valued logic, i.e., a logic in which truth values are not only the two commonly identified with the number 0 and 1, has been studied since 1920s, most notably by Jan Łukasiewicz and Alfred Tarski. Yet, it was not until Lotfi Zadeh in 1965 proposed the theory of fuzzy sets that the engineering aspects of this discipline emerged. Since then, fuzzy logic has found applications in many fields like, but by far not limited to, control theory and artificial intelligence. Most of those applications were supported by the development of circuits and systems for practical implementation of the corresponding computational steps.

Lotfi Zadeh.

Conceptually, to arrive at fuzzy systems, one must follow a three-step process starting from relaxing the concept of "belonging to a set" or "satisfying a property" that are no longer seen as propositions that may be either totally true or totally false but as statements that can be true to different degrees, commonly taken as real number between 0 and 1.

This is perfect to model implicitly vague concepts like "small number", "brilliant color", etc., that naturally appear when one tries to model the strategies that human beings adopt to solve problems by reacting to real world inputs. In practice each property or set is mapped into a membership function whose co-domain is [0, 1] and says how much a certain instance belongs to that set of satisfies that property.

The second step is the definition of a set of operations that one may want to do on truth degrees to model conjunction, disjunction, negation, and implication. There is a huge literature that addresses the definition of such operations either from an axiomatic point of view (e.g., what are the mathematical functions that satisfy the properties characterizing proper con-

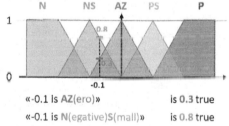

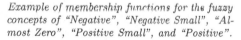

«-0.1 is AZ(ero)»   is 0.3 true

«-0.1 is N(egative)S(mall)»   is 0.8 true

Example of membership functions for the fuzzy concepts of "Negative", "Negative Small", "Almost Zero", "Positive Small", and "Positive".

junction of predicates), or with the aim of optimizing some feature or performance of the overall system. The latter possibility actually depends on the third step that must be taken to unleash the applicative virtues of fuzzy logic. Thanks to the modeling of implications, fuzzy logic is able to model the relationship between premises and and thus from known inputs to outputs that can be inferred from them.

Overall, fuzzy logic gives a quantitative semantic to describe non-linear relationships between inputs and outputs in an almost-natural language, while ensuring that the actual numbers crunched by the machine will behave coherently with the abstract meaning of the rules used in the description. The overall model turns out to be extremely powerful. For example, the model suggested in 1975 by Ebrahim Mamdani implies a global relationship linking a certain number of inputs $x_0, x_1, \ldots, x_{n-1}$ to the output derived by the application of certain number R of fuzzy rules of this form

$$y = \frac{\sum_{r=0}^{R-1} Y_r t_r}{\sum_{r=0}^{n-1} t_r} \qquad (5.6)$$

where Y_r is the output that would be prescribed by the r-th rule if it were the only one to be applied and t_r is the degree-of-truth of the r-th rule, that in a fuzzy context is a proper real number in the range $[0, 1]$. The degrees of truth of the rules are computed depending on the input as in

$$t_r = T\left(x_0, x_1, \ldots, x_{n-1}; \bar{X}_r\right) \qquad (5.7)$$

where $\bar{X}_r$ is the vector of parameter of the r-th rule and the function T is substantially a bell-shaped function in its n arguments. In 1992 Bart Kosko clarified that, under suitable assumptions, such a formulation has the capability or approximating arbitrarily well any continuous function in the inputs.

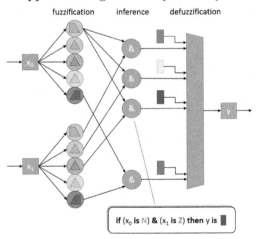

The neural-network view of a fuzzy system.

This originated the idea that what was being developed was a general-purpose method whose reach could go beyond the applications that were already known. Within this framework, the practitioners of the Circuits and Systems Society have been deeply involved in developing design flows for fuzzy systems. The IEEE Transactions on Circuits and Systems (I and II), the IEEE Journal of Solid-State Circuits, and the newly founded IEEE Transactions on Fuzzy Systems began to host a significant number of papers dedicated to the implementation of fuzzy systems both in the analog and in the digital domain, often combined with the neural-network paradigm.

The reason for such a cross-fertilization is that all the models of fuzzy systems can be thought as a sequence of stages of suitably interconnected nodes each of which performs an elementary operation depending on a small number of parameters. If this is done for a Mamdani-type fuzzy system (by far the most common model), the first two stages are in charge of computing the functions in (5.7) depending on the parameters in $\bar{X}_r$ controlling the nodes of the first stage, while the third stage implements the center-of-gravity operation and depends on the parameters Y_r.

This allows learning algorithms to adapt the system parameter in response to examples of correct input-output behavior. Such a learning can be exploited in two ways: to fine tune the parameters of a system that has already been sketched following a set of natural language rules mapped in the network structure, or to exploit available example of desired input-output behavior to infer back the rules describing it. Implementations and design flows for fuzzy and neuro-fuzzy systems have been a hot topic from 1992 to, at least, 2000.

The path taken by analog implementations hinged on two main points: fuzzy systems are naturally nonlinear and they are intrinsically tolerant to inaccuracies, in the

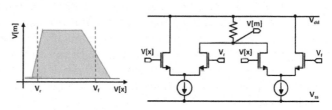

An analog circuit for membership calculation.

sense that their core behavior is encoded in rules and not in the exact values of the parameters. These features clearly stimulated analog designers to find particularly small circuits whose nonlinearity could be mapped into the nonlinearity of, say, membership functions.

The figure above shows one of the classical proposals appeared in 1996 in which the input quantities and the degree of membership are encoded as voltages (V[x] and V[m] respectively), approximately trapezoidal shapes are obtained by proper unbalancing of differential pairs controlled by reference voltages (V_r and V_f). Slopes depend on the sizing of transistors. Regrettably, voltage-domain implementation have problems when it comes to design the normalization loops needed to at least implicitly perform the division in the center of-gravity stage computing the output. This is why, several authors proposed implementations in which the membership degrees and then the truth values t_r of the rules are represented by currents. With this encoding, an array of OTAs can be deployed, each of them producing a current proportional to the truth value t_r of a rule and to the difference between the output y and the output

Though extremely effective in increasing the so-called FLIPS merit figure (Fuzzy Logic Inference Per Second) for systems with a limited number of inputs, all those techniques suffered from a curse of dimensionality preventing them from being applicable to complex models with more than, say, 4 inputs. Intuitively speaking this is due to the fact that is each input may fall in M fuzzy categories and there are n inputs, in principle one should specify

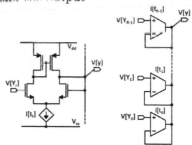

An analog circuit for the computation of (5.1).

$R = M^n$ possible rules each with a corresponding consequence, and all these potentially affect the outpt in (5.6). To counter this, an extension of the algorithms used for the minimization of boolean logic function was proposed in 1995 and applied successfully to the set of fuzzy rules defining a fuzzy system. The rule minimization figure shows an example of 6 rules that describe the same input-output relationship of 25 rules defined for 2 inputs that may be characterized by the 5 fuzzy sets in the first figure on the membership functions.

An even more effective way of approaching digital implementation derived from the equivalence between a specific kind of Mamdani fuzzy systems and piecewise-linear or piecewise-quad-ratic interpolators, established in 1998. On one hand, this new point of view allowed to prove that suitably designed fuzzy systems are able not only to approximate all continuous functions pointwise but are also able to reproduce differential behaviors like first- and even second-order derivatives. On the other hand, piecewise-linear interpolation can be given a computationally convenient formulation that does not suffer from the curse of dimensionality since its computational complexity grows like $Mn + n \log n$ instead of M^n.

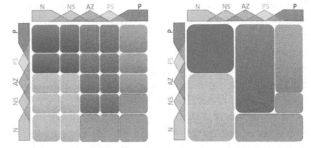

In parallel to all these developments in the implementation of fuzzy systems, the fuzzy paradigm itself grew to embrace an even larger plethora of applications. It is nowadays quite common to consider a fuzzy controller as an obvious option when de- *From 25 to 6 rules describing the same input-output relationship (rule minimization).*

vising the control policy for a new plant or to list a fuzzy classifier among the viable techniques to tackle a new pattern-recognition task. Part of this *ease of use* that is commonly perceived about fuzzy systems is surely due to the efforts devoted to the study of their effective implementation.

References

This list is far from exhaustive and should be taken only as the starting point for a thorough bibliographic and historical search along the issues sketched above.

L.A. Zadeh, "Fuzzy sets", Information and Control, vol. 8, n. 3, pp. 338–353, 1965.

E.H. Mamdani, S. Assilian, "An experiment in linguistic synthesis with a fuzzy logic controller," International Journal of Man-Machine Studies, pp. 1-13, 1975

B. Kosko, "Fuzzy systems as universal approximators," IEEE International Conference on Fuzzy Systems, pp. 1153–1162, 1992.

A. Pagni, R. Poluzzi, G.G. Rizzotto. "WARP: Weight Associative Rule Processor. A dedicated VLSI fuzzy logic megacell," 1992.

T. Miki, H. Matsumoto, K. Ohto, and T. Yamakawa, "Silicon implementation for a novel high-speed fuzzy inference engine: Mega-flips analog fuzzy processor", J. Intell. Fuzzy Syst., pp. 27-42, 1993

R. Rovatti, R. Guerrieri, G. Baccarani, "An enhanced two-level Boolean synthesis methodology for fuzzy rules minimization," IEEE Transactions on Fuzzy Systems, pp. 288-299, 1995.

A. Kandel, G. Langholz, *Fuzzy Hardware: architectures and applications*, Springer Sience+ Business Media, 1998.

P. Echevarria, M.V. Martínez, J. Echanobe, I. del Campo, J.M. Tarela, "Digital Hardware Implementation of High Dimensional Fuzzy Systems," Applications of Fuzzy Sets Theory, Lecture Notes in Computer Science, vol. 4578, pp. 245–252, 2007.

Riccardo Rovatti

University of Bologna, Italy

Signal Processing

As is officially stated on the IEEE CAS website, the area of interest for the IEEE Circuits and Systems Society is defined as:

"The theory, analysis, design (computer aided design), and practical implementation of circuits, and the application of circuit theoretic techniques to systems and to signal processing. The coverage of this field includes the spectrum of activities from, and including, basic scientific theory to industrial applications."

Clearly, the field of *"signal processing"* stands right alongside *"systems"* insofar as being an integral part of the CAS Society. Moreover, research in the signal processing field, and its practical applications, have been a part of the IEEE Circuits and Systems Society even as far back as the time when we were known as the IRE Professional Group on Circuit Theory, starting on March 20, 1951 with the first meeting of the IRE Professional Group on Circuit Theory.

We then might ask: When did the first *"signal processing"* paper appear in the IEEE Transactions on Circuits and Systems? To some extent, the answer depends on what is meant by *"signal processing."* For example, even the Sept. 1954 issue of the IRE Trans. on Circuit Theory has two papers on the use of z-transforms: one by John Truxal of Brooklyn Polytechnic Inst., entitled *"Numerical analysis for net-*

John Truxal [Wikipedia]).

work design," and a short correspondence piece by Bill Huggins of Johns Hopkins University: *"A low-pass transformation for z-transforms."* Truxal points out that *"The increased availability of digital and analog computers has permitted the network synthesist"* [Wow, how often do you see the word *"synthesist"*] *to embrace an entirely new approach to design.* Basically he refers to modulating a signal x(t) by a sequence of unit-impulses at regular time intervals t = T, 2T, 3T, ... Huggins refers to the Truxal article and describes a technique to deal with difficulties that arise with the use of empirically determined inputs and outputs. While it seems that in 1954 CAS members were just beginning to grasp for the best way to employ newly-available fast digital-computation capabilities, it is clear that (like the rest of the world) circuits (and signals) were *"going digital."*

In the December 1956 issue of the IRE Transactions on Circuit Theory we find such papers as: *"Signal Theory,"* by W. H. Huggins, pp. 210-216; and *"Signal theory in speech transmission,"* by E. E. David, pp. 232-244; which are both oriented along the lines of continuous-time signals. But also there is this paper: *"Study of rough amplitude quantization by means of Nyquist sampling theory,"* by B. Widrow, pp. 266-276; in which we can find discrete-time signals and figures of discrete-time systems that use blocks looking much like those

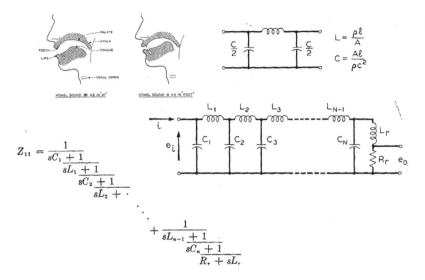

Excerpt from the E. E. David paper showing the analog model of vocal tract.

found in current-day digital filters (but using slightly less elegant notation). That's 60 years ago! And there is reference to an earlier paper in the same journal concerning z-transforms.

It is also not a surprise to see a paper in the same (December 1956) issue on a related matter: *"System Theory as an Extension of Circuit Theory,"* by W. K. Linvill, pp. 217-223. This new notion of *"system theory"* was beginning to spring forth, to ultimately become a part of our Society's name – but (as shown above) that would take another dozen years. However, even in the following year, 1957, the September issue of the IRE Transactions on Circuit Theory gave us an early introduction to something that would soon be known as *"state equations."* There we find this paper: *"The A matrix, new network description,"* by T. H. Bashkow, pp. 117-119. The same issue also contains a short correspondence from H. Freeman, (anticipating DSP): *"A simplified procedure for the long-division inversion of z-transform expressions."* By the following year, March 1958, we find, in *"Synthesis of sampled-signal networks"* by P. M. Lewis II, pp. 74-77, figures that illustrate RLC circuits being replaced by *"short-circuited lines and open-circuited lines."*

By 1963, another matter was arising. There were some in the IEEE Circuit Theory Group who evidently were questioning the *"relevance"* of circuit theory. The March 1963 issue of the IEEE Transactions on Circuit Theory had a lead article by D. O. Pederson, entitled *"Circuit Theory in Orbit?"* Its first paragraph ends with: *"It might be said that we have launched circuit theory and circuit theorists into orbit, for all the contact the real world has with them."* And in the very next (June 1963) issue of the Transactions, the lead article is entitled *"What is system theory and where is it going? – a panel discussion."* Contributors to this discussion were: L. A. Zadeh, W. K. Linvill, R. E. Kalman,

$$y_{k+1} = y_k - \frac{1}{10} y_k^2$$

$$y_0 = 1.12.$$

K	y_K	y_K^2	y_K^2 (rounded)	$y_{K+1} = y_K - \frac{1}{10} y_K^2$ (rounded)
0	1.12	1.2544	1.3	$y_1 = 1.12 - 0.13 = 0.99$
1	0.99	0.9801	1.0	$y_2 = 0.99 - 0.10 = 0.89$
2	0.89	0.7921	0.8	$y_3 = 0.89 - 0.08 = 0.81$
3	0.81	0.6561	0.7	$y_4 = 0.81 - 0.07 = 0.74$
4	0.74	0.5476	0.5	$y_5 = 0.73 - 0.05 = 0.68$

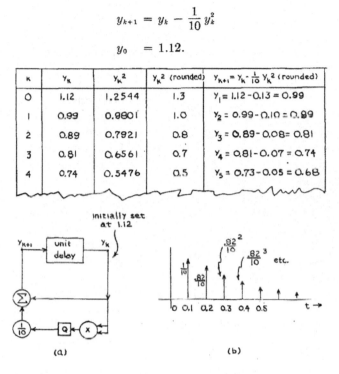

(a)   (b)

Excerpt from the Widrow's paper showing a simple example of discrete-time signal processing.

G. F. Franklin, S. Seshu and Y. C. Ho. The concluding remarks by Kalman contained this sage advice: "Stick to working and stop philosophizing." I think most in the Circuit Theory Group took this advice, but there was still a good bit of introspection afoot. It was clear at that time that many favored "system theory" being incorporated into the name of the Circuit Theory Group and its Transactions - many, but not all. I can recall a presentation at an evening session of the Allerton Conference in the late 1960s at which Mac VanValkenburg spoke on this topic. But one attendee was getting a lot of attention by proposing that *"system theory is just circuit theory without examples!"*

The December 1968 Transactions was a Special Issue on Modern Filter Design. It led off with an editorial introduction by Gabor Temes, entitled *"The present and future of filter theory,"* which was followed by a discussion entitled *"What, if any, are the important unsolved questions facing filter theorists today?"* This was addressed by Sidney Darlington (of Bell Telephone Labs.), Nai-Ta Ming (of Standard Electrik Lorenz AG), H. J. Orchard (of Lenkurt Electric Co., Inc.) and Hitoshi Watanabe (of Nippon Electric Co., Ltd.). Sid Darlington specified work needed in CAD as well as in digital filters. He says:

"Some other aspects of modern filter theory relate to digital filters, which are important now and will become increasingly important in the future. Digital

filters may be applied to analog, as well as to digital signals. The relative costs of logic circuits and linear components are such that digital filters, combined with analog-to-digital and digital-to-analog converters, may compete with analog filters of linear components. Progress is needed in both theoretical and practical aspects of this art."

A year earlier, in 1967, I had been offered a job by Sid Darlington in the Research Division of Bell Labs, and even though I had intended to take a teaching position, I couldn't say no to the offer. (I was totally free to work on whatever research topics I wanted; how could I not take such a job!) By 1968 I was deep into nonlinear circuit theory. But I was well aware of research that was under way at Bell Labs in the digital filter area, much of it being published

Alan Oppenheim lecturing at MIT. (MIT-OpenCourseWare)

in the Bell System Technical Journal. In my own small group there were researchers who are well known for their digital filter research: Jim Kaiser, Lee Jackson, Dave Goodman, Leon Harmon, Hank McDonald, and Irwin Sandberg, for example. And "right down the hall" were the likes of Larry Rabiner, Ron Schafer, and Dick Hamming, not to mention Sid Darlington. There couldn't have been a better place for me to *"grow up"* – technically speaking. Before long I too began working on some interesting digital filter research problems. E.g., my paper (sort of bridging the gap between nonlinear and DSP): *"Limit cycles due to adder overflow in digital filters,"* IEEE Transactions on Circuit Theory, vol. CT-19, pp. 342-346, July 1972.

Incidentally, if the name Leon Harmon (among the above-listed names of my former Bell Labs colleagues) is unfamiliar to you, perhaps you are familiar with his two-dimensional signal processing research, which led to a well-known quantized image of Abraham Lincoln. Leon did this at Bell Labs in 1971 and, with his colleague, Bela Julesz, published it in the journal Science, in their 1973 paper *"Masking in visual recognition: effects of two-dimensional filtered noise,"* pp. 1194-1197. Interestingly, just three years later the Lincoln picture became art: Salvador Dalí used it in his famous 1976 portrait *"Gala contemplating the Mediterranean Sea."* (By then Leon had become the Department Head at the Department of Biomedical Engineering at Case Western Reserve University.)

I was quite aware that significant work on digital signal processing was also being pursued in Europe in the mid-1960s to early 1970s (e.g., W. Schüssler, and O. Herrmann) and at MIT (Oppenheim, Gold, Rader) and at nearby Princeton University (Bede Liu). Gold and Rader published what I believe to be the first book on DSP: Digital Processing of Signals, McGraw Hill, 1969.

I've mentioned Larry Rabiner. Much of his early FIR filter research was published in The Bell System Technical Journal or in the IEEE Transactions on Audio and Electroacoustics. The famous "Parks-McClellan algorithm" which employed the Remez algorithm was, however, introduced in a March 1972 IEEE Transactions on Circuit Theory publication by T.W. Parks and J.H. McClellan, entitled "Chebyshev approximation for nonrecursive digital filters with linear phase," pp. 189-194. The following year (balancing the name sequencing?) J.H. McClellan and T.W. Parks published "A unified approach to the design of optimum FIR linear phase digital filters," IEEE Transactions on Circuit Theory, vol. pp. 697-701. Larry Rabiner became involved with this work too, and eventually teamed up with them. This is mentioned in "A personal history of the Parks-McClellan algorithm," by Parks and McClellan, which appeared in the March 2005 issue of the IEEE Signal Processing Magazine, where they write: "much of the credit for the algorithm's robustness goes to Larry Rabiner. Larry was the author of two competing filter design techniques: frequency sampling and a linear programming implementation. But he jumped on the Remez bandwagon and became a strong advocate for our method once he was convinced that it would live up to our billing."

Parks and McClellan go on to explain that, in comparison to university computing facilities, "in the 1970s Bell Labs had extensive computer resources" so Larry Rabiner was able to run hundreds of filter simulations per day and could thereby expeditiously improve the level of capability of the Remez approach.

Having mentioned the IEEE Transactions on Audio and Electroacoustics, it is only right to point out that this journal could be viewed as a "competitor" to the IEEE Trans. on Circuit Theory for publishing papers on the topic of signal processing. The journal is sponsored by what was in the early 1970s called the IEEE Group on Acoustics, Speech, and Signal Processing. Here's its evolution over time:

1948 IRE Professional Group on Audio started
1965 Name-change to IEEE Group on Audio and Electroacoustics
1974 Name-change to IEEE Group on Acoustics, Speech, and Signal Processing
1976 Status change: Group on ASSP became IEEE ASSP Society
1990 Society name change to IEEE Signal Processing Society

Today the group is, of course, known as the IEEE Signal Processing Society. Interestingly, in the early years both the CAS and ASSP Groups/Societies published some early DSP papers and both had "other interests." For CAS it was the ongoing research in analog circuits, analog filter theory and systems theory, while for ASSP it was audio, electroacoustics and speech processing. As shown above, ASSP didn't change its name to the "Signal Processing Society" until 1990, well after the "Circuits and Systems Society" reached "steady state" in 1973. For both Societies, times were changing in many ways, but technology was certainly moving toward digital systems. Rather than thinking of the two Societies as competitors, it might be better to acknowledge that they each have their own areas of interest which happen to overlap a bit, particularly in the area of digital filters and more generally *DSP.*

It is interesting how even in a fast-paced blossoming time, and even for work that exudes novelty in an important new field, getting one's results published can sometimes be difficult. (For those readers who have ever had a paper rejected, the following may make you feel a little better.) In the IEEE oral history files (Center for the History of Electrical Engineering, The Institute of Electrical and Electronics Engineers, Inc.) there are many interesting stories. One of our recently departed long-time CAS colleagues, Alfred Fettweis, was extremely well known for his creation of *"Wave Digital Filters."* Many CAS colleagues have studied his work on this topic and have made their own contributions to this field of Signal Processing. Alfred was quite involved in the activities of the IEEE Circuits and Systems Society. But where do you think his 1971 breakthrough on this new class of filters was published? It was in the German journal AEÜ (Arch. Elek. Übertragung) which is now operated by Springer publishing company using the English title: *"International Journal of Electronics and Communications."* Alfred wanted to publish his original Wave Digital Filter paper in our IEEE CAS Transactions. But it was turned down! Here are Alfred's own words (from the IEEE oral history files):

Fettweis: *"They said it needed experimental verification before they could publish it. It's one of the classical examples of what can happen to a truly fundamental paper. It's certainly the most fundamental paper I have written, no question about it. It was published here in Germany, but not by the IEEE, and the reason was that they indeed rejected it."*

Nebeker [Interviewer]: *"Do you think that it was because it was regarded as too fundamental?"*

Fettweis: *"Yes. If you write a paper in line with the present way of thinking, in which you enhance the work of some others, people appreciate it. If you come up, like I did, with a completely different way of looking at digital signal processing by going back to classical circuits and showing how you can carry ideas from classical circuits over to digital signal processing, then you go against the trend. People were largely thinking exclusively in terms of digital signal processing. They felt they could forget the classical area. In the new field they were doing things completely differently. This paper went against their line."*

I might add that The CAS Transactions didn't completely lose out on this breakthrough. It happens that the Feb. 1971 AEÜ paper was followed, just a month later, by this (a condensed version): A. Fettweis, *"Some principals of designing digital filters imitating classical filter structures,"* IEEE Trans. Circuit Theory, vol. CT-18, pp. 314-316, Mar. 1971.

The Nov. 1971 issue of the IEEE Trans. on Circuit Theory was a special issue dedicated to two topics of importance to the Circuits and Systems Society:

active RC networks and digital filters. Irwin Sandberg was the Guest Editor for the issue and this is what he wrote in his Editorial in regard to digital filters:

"The digital-filtering area warrants our serious attention not only because it is now economically feasible to consider the associated special-purpose computer-type implementations, but also because it has given rise to many interesting and fundamental problems of a type not ordinarily encountered by the network theoretician. While it is certainly true that some significant work has been done, and is being done, in the area of digital filtering, it is my opinion that the field is still very much in an embryonic state in the sense that several basic questions concerning, for example, approaches to design, the effects of roundoff, and the extent to which a given implementation is optimal have not yet been answered. On the one hand, I have the feeling that digital-filtering approaches will be of considerable value in the future if the developments concerning the cost of digital integrated circuits turn out to be favorable, and that these favorable developments will probably occur; on the other hand, it seems to me that so much can be done, and needs to be done, in this promising area that in the not too distant future we will view almost all of the digital-filtering-synthesis approaches and techniques of today as unsophisticated, inefficient, and obsolete. Good hunting!   *– Irwin W. Sandberg, Guest Editor"*

Irwin's predictions were certainly correct! Both active RC networks and digital filters have been, and continue to be, very productive and very valuable research areas which have received ever increasing recognition as more and more engineers have focused on them.

In 1973 I left Bell Labs and began what would be a 40+ year career at UCLA. My Bell Labs work had optimally positioned me to create the first digital signal processing courses at UCLA and they did, indeed, catch on. My third Ph.D. student at UCLA was Henry Samueli (who went on to first become a UCLA professor, then to found Broadcom Corp.) and, based on his Ph.D. research, we published the paper *"Almost period P sequences and the analysis of forced overflow oscillations in digital filters,"* in the IEEE Transactions on CAS, pp. 510-515 in August 1982. Henry had actually started as my advisee when he was still a UCLA undergraduate. His classmate, John Adams, became my fourth Ph.D. student. Based on John's Ph.D. research, we published *"A new approach to FIR digital filters with fewer multipliers and reduced sensitivity,"* in the IEEE Transactions on CAS, vol. 30, pp. 277-283, in May 1983. This paper, and its surprising new *"prefilter and amplitude-equalizer"* cascade FIR design, was very well received and has been referenced and used by many subsequent researchers. In fact, it was the winner of the 1985 IEEE W.R.G. Baker Prize Paper Award (given for the most outstanding paper reporting original work published in all Transactions, Journals, and Magazines of the IEEE Societies or in the Proceedings of the IEEE)–Not bad for a student's first journal paper! (And presumably a credit to the CAS Transactions.)

Like me, others who have chosen academic careers have become mentors to a whole new generation of researchers in signal processing. For instance L. B. Jackson moved to the Electrical Engineering Department at the University of

Rhode Island in 1974 where he still serves as a faculty member. Ron Schafer also left Bell Labs that same year and went to Georgia Tech where he, during a 30-year academic career, introduced thousands of students to DSP, co-authored six widely used textbooks and supervised graduate research in speech, image, biomedical and communication signal processing. Dave Goodman left to become a professor at Rutgers, the State University of New Jersey, and later moved to Brooklyn Poly. These and other signal processing researchers have trained vast numbers of students in their field specialties and several generations of new signal processing experts have been produced.

Examples of more recent CAS Society publications in signal processing are abundant; for instance, to name just a few:

- C. M. Rader and L. B. Jackson, *"Approximating noncausal IIR digital filters having arbitrary poles, including new Hilbert transformer designs, via forward/backward block recursion,"* IEEE Trans. CAS–I, vol. 53, pp. 2779-2787, Dec. 2006.
- R. Bregovic, Ya Jun Yu, T. Saramäki, and Y. C. Lim, *"Implementation of linear-phase FIR filters for a rational sampling-rate conversion utilizing the coefficient symmetry,"* IEEE Trans. CAS–I, vol. 58, pp. 548-560, Mar. 2011.

And a very new contribution is to be found in a very recent issue of the CAS Society Magazine:

- A. Mehrnia and A. N. Willson, Jr, *"FIR filter design using optimal factoring: A walkthrough and summary of benefits,"* CAS Society Magazine, vol. 16, No. 1, pp. 8-21, First Quarter 2016.

So, it's obvious that, thanks in large measure to the support of the CAS Society, the future of digital signal processing is in good hands!

Alan N. Willson, Jr.
Charles P. Reames Research Professor
Electrical Engineering Department
Henry Samueli School of Engineering, UCLA

CAS Image Processing

The technical field of image processing began during the latter half of the nineteenth century in the early days of chemical photography. Image processing then branched into the evolving fields of radio wave transmission and X-ray technology during the early twentieth century. Based on this evolutionary trend image processing became highly interdisciplinary during the twentieth century and has continued to quickly move forward during the twenty-first century. When the IEEE Circuits Theory Group began in the 1950's and then later became the IEEE Circuits and Systems (CAS) Society in the 1970's, it has long been an organization that supports many image processing technologies, some of which remained in the domain of the CAS Society such as video circuits and video technology, and others that quickly moved forward into newly emerging areas of image processing technology such as medical imaging and radar imaging.

Throughout the last 50 years the CAS Society has remained heavily involved with underlying theories and practical methodologies of signal processing across interdisciplinary fields of synthetic aperture radar (SAR) imaging, biomedical medical imaging (CAT, MRI, Ultrasound, etc.), high definition television (HDTV), and non-destructive testing (NDT) The following paragraphs highlight the interdisciplinary nature of the signal processing required in each of these technical areas and they emphasize how multi-dimensional system theory and 2-D digital signal processing have facilitated rapid growth in these CAS-related areas.

When technology of the post-WWII era began rapidly advancing in the 1950's analog TV appeared in the commercial marketplace and entered the homes of U.S. families, as well as homes throughout many parts of the world. It is noteworthy that many decades later, i.e. in approximately 2010, HDTV was the

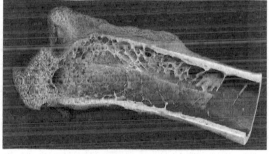

CAT scanned image of a a femur bone.

digital imaging technology that most recently invaded homes throughout the entire world because commercial broadcasting companies such as Comcast stopped transmitting analog TV and replaced all broadcasting with digital TV (HDTV). Since radar research had been aggressively pursued during World War II synthetic aperture radar (SAR) underwent rapid development for military applications, imaging the earth's surface from space, and commercial applications in airports and weather prediction [1]. Due to the invention of the transistor in 1947, electronic computers started down the road of rapid advancement. Although internal imaging of objects began early with the discovery of X-rays, the advent of computers strongly motivated the development of more sophisticated internal imaging techniques. X-rays were first discovered in 1895 by W. K. Roentgen, but X-ray imaging did not come forward until digital computers became available to implement the required digital image processing. Two of the most widely used types of medical imaging are Computer-Aided Tomography (CAT) [2] and Magnetic Resonance Imaging (MRI) [3].

Computer-aided tomography (CAT), magnetic resonance imaging (MRI) and Synthetic Aperture Radar (SAR) are well-known techniques for constructing high-resolution images by processing data obtained from many different perspective views of a target area. The CAT scan is an X-ray technique that enables the imaging of two-dimensional cross sections of solid objects. The basis of MRI is a directional magnetic field associated with charged particles in motion. Nuclei containing an odd number of protons and/or neutrons have a characteristic motion, or precession. Because nuclei are charged particles, this precession produces small magnetic moments that lead to the creation of high resolution images using digital image reconstruction techniques.

SAR image that was taken on the Magellan mission to Venus.

In particular, tomography is used extensively for noninvasive medical examination of internal organs and in nondestructive testing of manufactured items. Although SAR is well known to a more exclusive community (DoD) than the CAS Society, it too is a well-developed technique for producing high-resolution images. In typical SAR systems the desired image is a terrain map. The data are collected by means of airborne or spaceborne microwave radars that illuminate the target area from different perspectives. An early form of SAR, known as unfocused strip mapping, was demonstrated experimentally at the University of Illinois as far back as the early 1950's. In strip-mapping SAR (both focused and unfocused), the position of the antenna remains fixed relative to the aircraft, thereby illuminating a strip of terrain as the aircraft flies. Proper processing of the returned signals allows the effective synthesis of a very large antenna, providing high resolution. Extensive developmental work on optical processing of data collected in strip-mapping SAR was subsequently carried out at the Willow Run Research Laboratory at the University of Michigan, that later became the Environmental Research Institute of Michigan (ERIM). At ERIM researchers made a breakthrough in characterizing the requirements for optically processing coherent radar data collected from targets placed on a rotating platform, an experimental setup designed to simulate an airborne radar flying around a stationary ground patch, which was the creation of the first spotlight-mode SAR.

The microwave frequencies used in SAR permit successful imaging through cloud cover and rain, thereby providing an excellent ground mapping modality in all types of weather conditions. SAR has proved to be a very effective microwave imaging technique for high resolution ground mapping, remote sensing, and surveillance applications. In the 1980's, as part of the U.S. NASA Magellan mission, a SAR was placed into orbit around Venus to provide mankind with the first complete unobstructed view of the surface of this cloud-covered planet.

Throughout the literature it had been shown numerous times that the underlying mathematical structure of SAR image reconstruction is similar to the image reconstruction theory encountered in computer-aided tomography CAT [4]. The similarity in these two otherwise seemingly different imaging systems suggested that reconstruction algorithms used in CAT could also be used in SAR, and vice versa. In fact, it was known that the direct Fourier method was indeed a state-of-the-art imaging algorithm in

Geometry for data collection in spotlight-mode SAR (Fig. 4, Ref. [4])

both SAR and CAT signal processing communities. However, one of the most popular CAT signal processing algorithms, the convolution back-projection (CBP) algorithm, had not previously been used in SAR. Although various aspects of using the CBP algorithm in SAR were discussed throughout the literature, a complete treatment of how the CBP algorithm performs in SAR had not previously appeared. In the 1980's work in the Coordinated Science Laboratory at the University of Illinois, the Naval Weapons Center in China Lake, CA, and the MIT Lincoln Laboratory in Lexington, MA, demonstrated that the CBP algorithm can be modified for SAR processing and that interesting trade-offs in computational complexity, image quality, and algorithmic parallelism result [5].

Another area of imaging technology that evolved during the latter half of the twentieth century was based on the transmission of ultrasound waveforms. Medical ultrasound imaging is a diagnostic technique that uses ultrasound to reveal details of internal body structures such as tendons, muscles, joints, vessels and internal organs. One of the strengths of ultrasound imaging is its ability to measure blood and tissue velocities with high precision and high frame rates. The knowledge of volumetric blood flow rate is an important quantity in the diagnosis of various diseases and trauma, as well as in cardiovascular research. Volume blood flow is one of the best indicators of available oxygen and also of the ability of the heart to maintain normal body processes. Many of the classical methods of blood flow mea-

Cover page of the book Nonlinear Image Processing

surement, dye dilution, and angiographic techniques were invasive and could be potentially harmful to the patient.

One problem with time-domain image processing methods is that they tend to be very computationally intensive, and in the early days of ultrasound imaging measurements were generally restricted to off-line analysis. In order to be clinically useful, an ultrasonic blood flowmeter should be capable of producing results in real-time. In the 1980's interdisciplinary research that bridged the gap between biomedical research and image processing research resulted in a real-time ultrasound time domain correlation (UTDC) blood flowmeter that incorporated a high-speed residue-number system (RNS) correlator to replace hardware multipliers with high-speed lookup tables stored in ROMs. The result was a UTDC flowmeter capable of producing a flow velocity versus range profile every 0.34 seconds (real-time), whereas previous UTDC blood flowmeters required up to 90 seconds (off-line) in order to compute the same result [6].

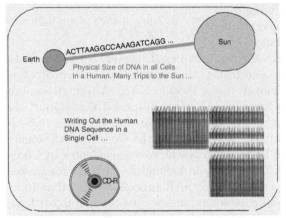

The lenght of DNA in all the (5 trillion) cells in an average human covers the distance from earth to the Sun (93 million miles), about 50 times [12].

Since over the past fifty years many CAS members have been involved in various areas of image processing technology it is not possible to provide a comprehensive review of the CAS Society's contributions to the various fields of imaging processing. However, it is useful to sample some image processing contributions of a few CAS members, several of who served as President of the CAS Society. For example, Sanjit K. Mitra (CAS President in 1986) made significant contributions in the areas of digital signal processing, image and video processing, data compression, image and video enhancement, image analysis, and mixed analog-digital signal processing. He was the coauthor of one book entitled Nonlinear Image Processing [7] and another entitled Multidimensional Processing of Video Signals [8]. M.N.S. Swamy (CAS President in 2004) is a well known author and researcher in the general area of circuits and systems, and in the specific areas of multidimensional digital signal processing, distributed parameter networks and image processing. He was the coauthor of an interdisciplinary paper that coupled RNS arithmetic with digital image processing [9]. Nirmal Bose, a world-renowned expert and CAS member in multidimensional signals and systems theory, conducted research on high resolution reconstructions of blurred and noisy images, and processing of noisy images. In 1983 he published a pioneering book entitled Applied Multidimensional Systems Theory [10]. George Moschytz (CAS President in 1999) has authored and co-authored many books on analog, digital, switched-capacitor, and adaptive circuit and filter design, and has published many papers in the field of network theory and design, signal processing, and circuit sensitivity. One of his publications that relates to imaging processing proposed a Fingerprint Verification System [11]. W. Kenneth Jenkins, (CAS President in 1985) has conducted multidisciplinary research in the areas of adaptive signal processing, multidimensional array processing, and bio-inspired optimization algorithms for intelligent signal processing. During the 1980's he was engaged in interdisciplinary work involving SAR image processing, CAT and MRI medical imaging, and ultrasonic blood flow detection [4,5,6].

P.P Vaidyanathan's research activities lie in the areas of digital signal processing, compressive sensing and sparse reconstruction, spectrum sensing and applications, multi-rate systems and filter banks, wavelet transforms, signal processing for digital communications, and genomic signal processing [12]. Rui de Figueiredo was best known for his pioneering contributions to the mathematical foundations of linear and nonlinear problems in pattern recognition,
signal and image processing, and neural networks. His work that often connected with the CAS image processing community was internationally recognized [13].

Although the contributions of the CAS members mentioned above provide only a few examples of how the CAS Society continued to have significant impact on various interdisciplinary fields of image processing, there is no doubt that in general CAS Society members have nurtured the development of image processing technologies over many years. It is very likely that the CAS Society will continue these trends dur-

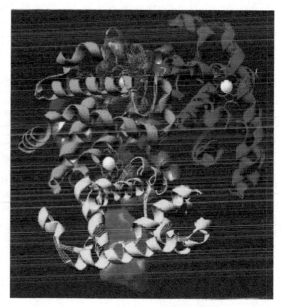

An example of a protein (Hemoglobin, human). [12].

ing the entire twenty-first century as human-machine interactions and biomedical technologies become more centralized in the lives of people throughout the entire world.

References

[1] K. Tomiyasu, *"Tutorial Review of Synthetic-Aperture Radar (SAR) with Applications to Imaging of the Ocean Surface,"* Proc. IEEE, vol. 66, pp. 563-583, May 1978.

[2] H. J. Scudder, *"Introduction to Computer Aided Tomography,"* Proc.IEEE, vol. 66, pp. 628-637, June 1978.

[3] Brown, Mark A. and Richard C. Semelka, *"MRI: Basic Principles and Applications,"* New York: Wiley-Liss, 1999

[4] D. C. Munson, Jr., J. D. O'Brien, and W. K. Jenkins, *"A Tomographic Formulation for Synthetic aperture radar,"* IEEE Proceedings, vol. 71, no. 6, pp. 917-925, August 1983.

[5] M. D. Desai and W. K. Jenkins, *"Convolution Back-projection Image Processing for Synthetic Aperture Radar,"* IEEE Transactions on Signal Processing, vol. 1, no. 4, pp. 505-517, October 1992.

[6] I. A. Hein, J. T. Chen, W. K. Jenkins, and W. D. O'Brien, Jr., *"A Real-time Ultrasound Time-domain Correlation Blood Flowmeter: Part I - theory and design,"* Proceedings of the IEEE Transactions on Ultrasonics, Ferroelectrics, and Frequency Control, vol. 40, no. 6, pp. 766-775, November 1993.

[7] S. K. Mitra and G. L Sicuranza, *"Nonlinear Image Processing,"* Academic Press, New York, NY, September 2000.

[8] S. K. Mitra and G. L Sicuranza, *"Multidimensional Processing of Video Signals,"* Kluwer Academic Publishers, Boston, MA, 1992.

[9] Wei Wang, M.N.S. Swamy, and M. O. Ahmad, *"RNS Application for Digital Image Processing,"* Proceedings of the 4th IEEE International Workshop on System-on-Chip for Real-Time Applications, August 2004.

[10] Nirmal K. Bose, *"Applied Multidimensional Systems Theory,"* Van Nostrand Reinhold, 1982.

[11] Q. Gao and G. Moschytz, *"A CNN-Based Fingerprint Verification System in Complex Computing Networks: A Link between Brain-like and Wave-oriented Electrodynamics Algorithms,"* Springer Proceedings in Physics, vol. 1094, March 2006.

[12] P. P. Vaidyanathan, *"Genomics and Proteomics: a Signal Processor's Tour,"* IEEE Circuits and Systems Magazine, Jan. 2005.

[13] Rui de Figueiredo and T. Eltoft, *"A New Neural Network for Cluster Detection and Labeling,"* IEEE Trans. on Neural Networks, pp. 10211035, vol. 9, no. 5, 1998.

W. Kenneth Jenkins

IEEE Life Fellow

The Pennsylvania State University

Data Converters

The history of modern data converters is a fascinating topic, and reaches back much further in time than one might imagine [1]. It's surprising to find that many of the basic principles of operation that are taught and used today were known in antiquity, but are now scaled to the point where an entire A/D converter (ADC) can no longer be seen by the naked eye. There is virtually no aspect of modern technology that would be possible without data converters. One example is the modern cell-phone; it is estimated that an iPhone 6 has at least 60 data converters, most of which are deeply embedded in custom IC's that are implemented in aggressive process nodes and surrounded by DSP blocks operating at high clock-rates.

The evolution of data converters since 1970 is very closely tied to advances in semiconductor processes and Moore's law. As the cost and size of transistors has dramatically shrunk, signal processing techniques that were previously either impractical or were done using analog circuits are now implemented digitally, and that has driven the data converter industry to develop algorithms and circuits that are suitable for the signals that must be funneled in and out of those processing blocks. In addition, the requirement that converters and DSP cores must be integrated on a single chip has driven the converter industry towards designs that survive well when implemented in process nodes that are optimized for digital circuitry. In some cases this only impacts the details of the circuit implementation, but in general there is a deeper and more profound impact on the choice of the fundamental conversion architecture and algorithmic approach.

Early History

It is difficult to search for very early examples of the principles of data conversion, as they were generally mechanical or hydraulic in nature. One very early example of a D/A converter (DAC) is a hydraulic system used in Turkey in the 1700's, designed to provide precise quantized control of water flow from a dam. It used a tank of water (kept completely full by using a spillway, providing a precision "*reference*") and a series of gated nozzles that were sized proportionally to provide a controlled rate of flow. The ratio of the nozzle widths was not binary,

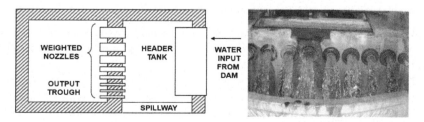

Nozzle outputs from the header tank in the Mahmud II Dam, Reproduced with the permission of Istanbul Technical University, Istanbul, Turkey [2]

but nevertheless a particular pattern of nozzle on/off states would result in a repeatable flow rate.

This example may just be the oldest D/A converter (DAC) in the world, and it provided a degree of precision control over water flow rates that must have been deemed important at the time.

The first example of an electronic DAC was shown by Lord Kelvin in the late 1800's, and employed a cascaded variable-tap-position resistor-string topology that today would be called a "*string DAC*". The switching elements were mechanical switches or relays, so this DAC would have been quite noisy in operation! This topology is still in use today.

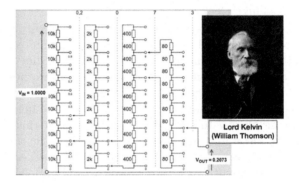

String DAC built by Lord Kelvin, mid 1800's.

Early data converters were primarily used for measuring (or producing) *dc* levels, and therefore there was no thought given to converting *ac* signals.

Conversion of *ac* signals using *PCM* was used before the Nyquist sampling theorem was developed, but there was at least an intuitive understanding that the sample rate should be at least somewhat larger than the highest frequency of interest. One example of this can be found in the early telephone industry, where there was a desire to multiplex multiple analog signals onto a single telephone line. In a patent by Williard Miner in 1903 [3],

My present invention proceeds upon the theory that for the successful transmission of speech over any one of the branch or sub circuits successively connected to the line, the closures of connection for that branch must be repeated with a frequency or rapidity approximating the frequency or average frequency of the finer or more complex vibrations which are characteristic of the voice.

This intuition was formalized in the 1920's by Harry Nyquist, resulting in the famous Nyquist theorem stating that a signal with bandwidth W must be sampled at a minimum rate of 2W.

The first use of PCM was demonstrated in a patent by Paul M. Rainey of Western Electric in 1921 [4]. In this patent he describes a system for transmitting facsimiles over telegraph lines using serial pulses where the PCM code represents a quantized photocell current. The current is converted to digital form using a galvanometer, which steered a light beam onto 1 of 32 photocells, which was then converted to PCM form through a series of relays.

The Tube Era

The widespread introduction of tubes made all-electronic data converters practical. Alec Reeves disclosed the first record of an all-electronic data conversion and transmission system in a patent in 1939 [5]. In this patent, an A/D converter is described that uses an analog sampler driving a voltage-controlled one-shot, followed by a counter that is enabled by the one-shot. The counter value, at the end of the one-shot period, is proportional to the value of the sampled analog signal. The DAC circuit uses the inverse of the ADC technique, converting the digital input to an analog pulse-width, followed by a lowpass filter.

World-War 2 accelerated the development of PCM systems, at least partly due to the need for secure communications. During this period of time, Bell Labs became extensively involved in PCM conversion and communications, and developed the *"Sigsaly"* system [6] that was used to provide secure communications among the allied powers. An example of this system can be found at the *"Crypto Museum"* in Washington DC and it it occupies an entire room.

Another technique that arose during this era was the use of cathode ray tubes for A/D conversion. A shadow mask was inserted into the tube with a succession of binary codes arranged from top to bottom. The sampled analog input signal was used to control Y deflection of the scan, and therefore a unique code can be read out

1954 "DATRAC" 11 bit, 50-kSPS Vacuum Tube ADC Designed by Bernard M. Gordon at EPSCO. (With the permission of Analogic Corporation, Peabody, MA.)

directly during a horizontal scan. Somewhat surprisingly, the performance of this technique was good enough that it survived into the 1960's, due in part to the fact that it was the only high-speed conversion technique that gave reasonable performance. Bernie Gordon showed the first example of a commercial-use A/D converter in 1953 [7]. This converter achieved 11 bit accuracy at a sample-rate of 50KS/s, and included both a sample-and-hold as well as a shift-programmable successive-approximation architecture, features that are both common to today's data converters. With a hefty power consumption of 500 watts, it would not fare well on a modern figure-of-merit graph!

The Semiconductor Era

The advent of transistors in the 1950's brought rapid change to the fledgling data converter industry. Bell Labs was again in the forefront, driven by the needs of the defense industry to provide converters with resolutions of 8-10 bits at speeds up to 10MHz. These converters were required to implement phased-array radar systems, which were part of the missile defense system spurred by the Cold War, and were based on discrete transistor designs housed in rack-mounted boxes drawing hundreds of watts.

As more highly integrated semiconductors became available in the mid-1960's, the cost, size and power of data converters began to shrink dramatically, and a large number of companies rushed in and began producing products. Some of the many names include Analogic (founded by Bernard M. Gordon), Pastoriza Electronics (later acquired by Analog Devices), Computer Labs (also acquired by Analog Devices), Adage, Burr Brown, General Instrument, Radiation, Inc., Redcor Corporation, Beckman Instruments, Reeves Instruments, Texas Instruments, Raytheon Computer, Preston Scientific, and Zeltex, Inc. By the end of the 1960's, a 12-bit 100KHz ADC was available on a circuit board the size of an index card and sold for about $800.00.

Until the late 1960's, data converters often used discrete matched resistors to achieve the accuracy that was required. These designs were gradually replaced by more integrated designs, and discrete resistors were replaced by resistor arrays and eventually the arrays themselves were integrated onto a single chip. However, the inherent matching accuracy was often not sufficient to meet the requirements of, say, a 14-bit converter. This brought about the need for on-chip trimming, and several companies developed proprietary laser-trimming processes to actively adjust the values of on-chip resistors, pushing the limits of accuracy to 16 bits or even higher. The downside of laser-trimming is that test times could often be counted in minutes, resulting in high production costs and limited ability to scale to the high volumes demanded in the consumer market.

In the early days of data converters, accuracy was measured on a per-sample basis, but in the 1970's many new markets arose that required excellent AC performance, typically measured in the frequency domain. An example might be a DAC reproducing an audio signal; if each DAC output sample is highly accurate, but the way that the DAC transitions from one sample to the next is non-linear and code-dependent, then the AC performance will likely be very poor. Today we commonly divide data converters into those that excel at data-acquisition, and those that excel at *"signal acquisition,"* where signal-acquisition implies that frequency-domain characteristics such as harmonic distortion are the most important characteristic.

During the early 1970's, most designs were either hybrids (at the higher-end of the performance scale) or based on bipolar process technology. CMOS designs began to become available in the late 1970's, and one of their chief advantages is that CMOS transistors make much better analog switches than their bipolar counterparts. The use of CMOS also dramatically improves the ability to mix converters with logic circuitry, an advantage that was to become increasingly crucial as Moore's law fed an explosion in logic density.

During the 1970's, the primary application of data converters was still solidly rooted in the industrial and military markets, and as a result data converters continued to demand high prices. However, starting in the late 1970's, Sony and Phillips jointly introduced the CD player, and this proved to be a watershed moment in data converter history. Suddenly there was a market with volumes in

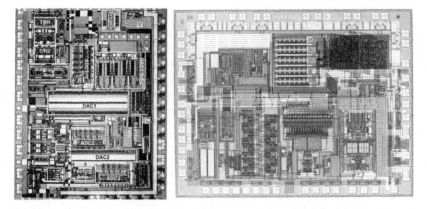

Left: 2000 state-of-the art BiCMOS pipeline converter (26 mm^2) [8]. Right: 2009 CMOS implementation of pipeline converter (10 mm^2) [9]

the 10's of millions, demanding high performance (16 bit accuracy). The initial products were based on variants of traditional laser-trimmed DAC architectures, and cost around $12 in high volume. A decade later, virtually 100% of consumer audio converters were based on delta-sigma modulation techniques, and sold for less than 30 cents (stereo version!). Later still, the DACs would become integrated with one of the major system-on-chip designs and largely disappear from view. A detailed look at delta sigma conversion will be given in another chapter, but the chief advantage when applied to the consumer market is that the technique yields converters that are inherently linear with no trimming (even at very small signal levels), and can easily be executed in digital-friendly CMOS process nodes.

Architectures

Space does not permit a detailed look at the various architectures in use today, but suffice it to say that the successive approximation algorithm that first arose 60 years ago is now just one of many approaches. A brief list of other approaches includes delta-sigma modulation (for low cost and excellent ac performance, many variants), pipeline converters (for high-speed conversion), integrating converters (slow, with high dc accuracy, often dual or quad-slope), flash converters (super-high-speed, medium resolution), and algorithmic converters (a recursive pipeline, low-cost). In addition, there are many converters on the market today that do not have an input sample-and-hold circuit, but rather have a continuous-time feedback system wrapped around a low-resolution converter, which makes these devices very easy to drive, while defying 50 years of accepted ADC dogma.

Conclusion

Today the data converter industry is a multi-billion dollar industry supported by thousands of engineers who specialize in various aspects of converter design.

Since so many data converters are now deeply embedded in complex application-specific integrated circuits, it's not easy to accurately estimate the total value of the data-converter market, or the total number of converter instances. But it's a safe bet to assume that the demand for data converters will continue to increase, and performance levels will continue to rise in response to market demands. The challenge of how to convert from analog to digital and back again has been the focus of attention for many of the brightest minds of the last 60 years, and it undoubtedly will continue to challenge every new generation of engineers.

References

[1] W. Kester, *The Data Conversion Handbook*. Burlington, VT: Analog Devices and Newnes-Elsevier, 2005.

[2] Kâzim Çeçen, *Sinan's Water Supply System in Istanbul*, Istanbul Technical University, Istanbul Water and Sewage Administration, Istanbul Turkey, 1992-1993, pp. 165-167.

[3] W. M. Miner, *Multiplex telephony*, U.S. patent 745,734, Dec. 1, 1903.

[4] P. M. Rainey, *Facsimile telegraph system*, U.S. Patent 1,608,527, Nov. 30, 1926.

[5] A. H. Reeves, *Electric signalling system*, U.S. patent 2,272,070, Feb. 3, 1942.

[6] https://www.nsa.gov/about/cryptologic_heritage/museum/virtual_tour/museum_tour_text.shtml#sigsaly

[7] B. M. Gordon and E. T. Colton, *Signal conversion apparatus*, U.S. patent 2,997,704, Aug. 22, 1961.

[8] C. Moreland, M. Elliott, F. Murden, J. Young, M. Hensley, R. Stop*A 14b 100MSample/s 3-Stage A/D Converter*, IEEE JSSC,pp. 1791-1798, 2000.

[9] S. Devarajan, L. Singer, D. Kelly, S. Decker, A. Kamath, and P. Wilkins, *A 16-bit, 125 MS/s, 385 mW, 78.7 dB SNRcCMOS Pipeline ADC*, IEEE JSSC,pp. 3305-3313, 2009.

Robert Adams

Analog Devices, Boston, MA

Delta Sigma Converters

The Need for Oversampling Converters

Computational and signal processing tasks are now performed predominantly by digital means, since digital circuits are robust and can be realized by extremely small and simple structures which can in turn be combined to obtain very complex, accurate and fast systems. Every year, the speed and density of digital integrated circuits (ICs) is increased, enhancing the dominance of digital methods in almost all areas of communications and consumer products. Since the physical world nevertheless remains stubbornly analog, data converters are needed to interface with the digital signal processing (DSP) core. As the speed and capability of DSP cores increases, so too must the speed and accuracy of the converters associated with them. This presents a continued challenge to the lucky few engineers dedicated to the design of data converters!

Data converters (both ADCs and DACs) can be classified into two main categories: Nyquist-rate and oversampled converters. In the former category, there exists a one-to-one correspondence between the input and output samples. Each input sample is separately processed, regardless of the earlier input samples; the converter has no memory. As the name implies, the sampling rate f_s of Nyquist-rate converters can be ideally as low as Nyquist's criterion requires, i.e., twice the bandwidth f_B of the input signal. (For practical reasons, the actual rate is usually somewhat higher than this minimum value.)

In most cases, the linearity and accuracy of Nyquist-rate converters is determined by the matching accuracy of the analog components (resistors, current sources or capacitors). Practical conditions restrict the matching accuracy to about 0.02%, and hence the *effective number of bits* (ENOB) to about 12, for such converters.

In many applications (such as digital audio), higher resolution and linearity is required, perhaps as much as 18 or even 20 bits. The only Nyquist-rate converters capable of such accuracy are the integrating or counting ones. These, however, require at least 2^N clock periods to convert a single sample with N-bit accuracy, and hence are too slow for most signal-processing applications.

Oversampling data converters are able to achieve over 20 *ENOB* resolution at reasonably high conversion speeds by relying on a trade-off. They use sampling rates much higher than the Nyquist rate, typically higher by a factor between 16 and 512, and generate each output utilizing all preceding input values. Thus, the converter incorporates memory elements in its structure. This property destroys the one-to-one relation between input and output samples. Now only a comparison of the complete input and output waveforms can be used to evaluate the converter's accuracy, either in the time or in the frequency domain.

The implementation of oversampling converters requires a considerable amount of digital circuitry, in addition to some analog stages. Both need to be operated faster than the Nyquist rate. However, the accuracy requirements on the analog components are relaxed compared to those associated with Nyquist-rate converters. The cost paid for high accuracy thus includes faster operation and added digital circuitry; both of these are getting cheaper as digital IC technology advances. Hence, the trade-off offered by $\Delta\Sigma$ converters is gradually improving. As a result, they are gradually taking over many applications previously dominated by Nyquist-rate converters.

Delta and Delta-Sigma Modulation

Next, oversampling analog-to-digital converters processing *baseband signals* (i.e., signals with spectra centered around dc) will be discussed. Such data converters contain several stages. Analog and digital filter stages may be used before and after the stage (called the *modulator*, or *converter loop*) which performs the actual analog-to-digital conversion. The two main types of oversampling modulators are the *delta* modulator and the *delta-sigma* modulator. Fig. 1 shows a basic delta modulator used as an ADC. It is a feedback loop, containing an internal low-resolution ADC and DAC, as well as a loop filter (here, an integrator).

The name *delta modulator* is derived from the fact that the output contains the difference (*delta*) between the current sample $u(n)$ of the input and a predicted value $u(n-1)$ of that sample. In the general case, the loop filter may be a higher-order circuit, which generates a more accurate prediction of the input sample $u(n)$ than $u(n-1)$, to subtract from the actual $u(n)$. This type of modulator is sometimes called a *predictive encoder*.

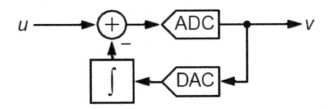

Fig. 1 - A delta modulator used as an ADC

The advantage of this structure is that for oversampled signals the difference $(u(n) - u(n-1))$ is much smaller than $u(n)$ itself, on average, and hence larger input signals can be allowed. There are, however, several disadvantages. The loop filter (integrator for the first-order loop shown) is in the feedback path, and hence its non-idealities limit the achievable linearity and accuracy. Also, in the demodulator, a DAC and a demodulation filter (for first-order modulators, an integrator) are needed. The filter has a high gain in the signal band, and hence it will amplify the nonlinear distortion of the DAC as well as any noise picked up by the signal between the modulator and demodulator.

The delta modulation (Δ modulation) scheme of Fig.1 is also called an *error feedback structure*. It was proposed in 1952 by de Jager [1], and in a different form in 1954 by Cutler [2].

An alternative oversampling structure which avoids the shortcomings of the

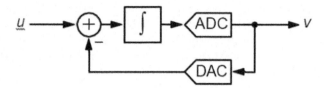

Fig. 2 - A delta-sigma modulator used as an ADC

predictive modulator is shown in Fig. 2. It is again a feedback loop, containing
a loop filter as well as an internal low-resolution ADC and DAC, but the loop
filter is now in the forward path of the loop. Analysis shows that the digital
output contains a delayed, but otherwise unchanged replica of the analog input
signal u, and a differentiated version of the quantization error ϵ_Q. Since the
signal is not changed by the modulation process, the demodulation operation
does not need an integrator as was the case for the delta modulator. Hence, the
amplification of in-band noise and distortion at the receiver does not take place.
Furthermore, the differentiation of the error ϵ_Q suppresses it at frequencies
which are small compared to the sampling rate f_s. In general, if the loop filter
has a high gain in the signal band, the in-band quantization "noise" is strongly
attenuated, a process now commonly called noise shaping.

Any nonlinearity of the ADC is simply combined with the quantization error
ϵ_Q, and is thus suppressed in-band along with ϵ_Q. Non-linear distortion in the
DAC, however, affects the output signal without any shaping, and hence it
represents a major limitation on the attainable performance. This effect can be
handled in various ways. The simplest, and historically earliest, method is to use
single-bit quantization. In this case, the input/output characteristic of the DAC
consists of only two points, and hence the DAC's operation is inherently linear.
For multi bit (typically, 2-5 bit) quantization, digital correction or dynamic
matching techniques may be used.

Any nonlinearity of the ADC is simply combined with the quantization error
ϵ_Q, and is thus suppressed in-band along with ϵ_Q. Non-linear distortion in the
DAC, however, affects the output signal without any shaping, and hence it
represents a major limitation on the attainable performance. This effect can be
handled in various ways. The simplest, and historically earliest, method is to use
single-bit quantization. In this case, the input/output characteristic of the DAC
consists of only two points, and hence the DAC's operation is inherently linear.
For multi-bit (typically, 2-5 bit) quantization, digital correction or dynamic
matching techniques may be used.

It can be shown that the system of Fig. 2 can be obtained from that of
Fig. 1 by cascading an integrator or summing block with the delta modulator.
Hence, the structure of Fig. 2 came to be called a *sigma-delta* ($\Sigma\Delta$) *modulator*.
Alternatively, one can observe the differencing at the input, followed by the
summation in the loop filter, and hence call the structure a *delta-sigma* ($\Delta\Sigma$)
modulator. This was the name used by the inventors Inose et al. [3]. Both

terms have been used in the past to denote the first-order system of Fig. 2 with a single-bit quantizer. Other systems with higher-order loop filters, multi-bit quantizers, etc. are most properly called *noise-shaping modulators*, but it is common to extend the term $\Delta\Sigma$ modulator (or $\Sigma\Delta$ modulator) to these systems as well.

Delta-Sigma Digital-to-Analog Converters

The motivation for using $\Delta\Sigma$ modulation to realize high-performance DACs is the same as for ADCs: it is difficult if not impossible to achieve a linearity and accuracy better than about 14 bits for DACs operated at Nyquist rate. Using $\Delta\Sigma$ modulation, this task becomes feasible. A $\Delta\Sigma$ DAC system is illustrated in Fig. 4. By operating a fully digital $\Delta\Sigma$ modulator loop at an oversampled clock rate, a data stream with (say) 18-bit word length may be changed into a single-bit digital signal such that the baseband spectrum is preserved. The large amount of truncation noise generated in the loop is shaped in order to make the in-band noise negligible. The single-bit digital output signal can then be converted with high (ideally, perfect) linearity into an analog signal using a simple two-level DAC circuit. The out-of-band truncation noise can be subsequently removed using analog low-pass filters.

Early History; Performance and Architectural Trends

Although the basic idea of using feedback to improve the accuracy of data conversion has been around for about 50 years, the concept of noise shaping

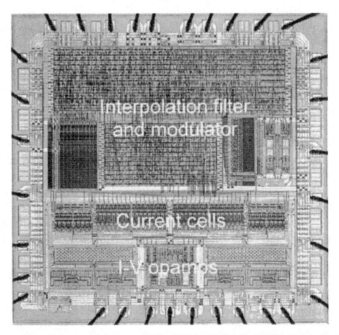

Fig. 3 - Chip microphotograph of a 113-dB SNR $\Delta\Sigma$ DAC. IEEE Journal of Solid State Circuits, Dec. 1998, p.1877.

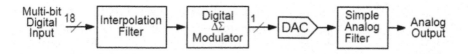

Fig. 4 - A $\Delta\Sigma$ DAC system

was probably first proposed (along with the name delta-sigma modulation) in 1962 by Inose et al. [3]. They described a system containing a continuous-time integrator as the loop filter, and a Schmitt trigger as the quantizer, which achieved (nearly) 40 dB SNR, and had a signal bandwidth of about 5 kHz. Since the trade-off between analog accuracy and higher speed plus additional digital hardware was not particularly attractive at the time, further research on this topic was relatively sparse for a while.

Twelve years later, Ritchie proposed the use of higher-order loop filters [4]. Useful theory, as well as analysis and design techniques were developed by Candy and his collaborators at Bell Laboratories [5]-[9]. Candy and Huynh also proposed the MASH concept for the digital modulators used in $\Delta\Sigma$ DACs [10]. In 1986, Adams described an 18-bit $\Delta\Sigma$ ADC which used a third-order continuous-time loop filter, and a 4-bit quantizer with trimmed resistors performing as the DAC [11]. The MASH configuration was first applied to $\Delta\Sigma$ ADCs by Hayashi et al. [12] in 1986.

Using a multi-bit internal quantizer in a $\Delta\Sigma$ loop with digital linearity correction was proposed by Larson et al. [13] in 1988; the use of dynamic matching (randomization) was also introduced for the internal DAC of a $\Delta\Sigma$ ADC by Carley and

Fig. 5 - James C. Candy, pioneer of understanding and analyzing the operation of $\Delta\Sigma$ modulators.

Kenney in 1988 [14]. Various mismatch-shaping algorithms were suggested subsequently by Leung and Sutarja [15], Story [16], Redman-White and Bourner [17], Jackson [18], Adams and Kwan [19], Baird and Fiez [20], Schreier and Zhang [21], and Galton [22].

Bandpass $\Delta\Sigma$ modulators were motivated for their potential applications in wireless communications, and emerged in the late 1980s [23]-[25]. Current design trends in $\Delta\Sigma$ converters are aimed at extending the signal frequency range without any reduction in SNR. This will open up new applications in digital video, wireless and wired communications, radar, etc. Higher speed can often be achieved by using high-resolution (typically, 5-bit) internal quantizers, and a multi-stage (2- or 3-stage) MASH architecture. To correct for the nonlinearity of the internal DAC and for quantization noise leakage, improved digital correction algorithms have been proposed [26] for $\Delta\Sigma$ ADCs. A great deal of effort is also being applied to improving the performance of bandpass $\Delta\Sigma$ ADCs [27]-[30].

Fig. 6 - Robert Adams, pioneer of implementing $\Delta\Sigma$ DAC modulators.

Technological trends (finer line widths, accompanied with lower breakdown voltages) stimulated research into $\Delta\Sigma$ modulators needing only low supply voltages. Also, applications opening up in portable devices motivated the development of low-power design techniques for $\Delta\Sigma$ data converters. Finally, applications in the instrumentation and measurements area, including biomedical sensor interfaces motivated the development of low-frequency and very-high-accuracy ADCs, often realized by periodically reset $\Delta\Sigma$ modulators, called *incremental data converters* [31]-[33].

A recent trend is to realize $\Delta\Sigma$ ADCs with *continuous-time* loop filters. This enables faster operation, and allows the modulator to perform anti-aliasing filtering [34]-[37].

As technology develops, and noise-shaping theory and practice continue to mature, $\Delta\Sigma$ data converters can be expected to expand their range of application even further.

References

[1] F. de Jager, "Delta modulation - a method of PCM transmission using the one unit code," Philips Res. Rept.,vol. 7, pp. 442-466, 1952.

[2] C. C. Cutler, "Transmission system employing quantization," U.S. Patent # 2,927,962, March 8, 1960.

[3] H. Inose, Y. Yasuda and J. Murakami, "A telemetering system by code modulation $\Delta\Sigma$ modulation," IRE Trans. Space Electron. Telemetry, vol. 8, pp. 204-209, Sept. 1962.

[4] G. R. Ritchie, J. C. Candy and W. H. Ninke, "Interpolative digital to analog converters," IEEE Transactions on Communications, vol. 22, pp. 1797-1806, Nov. 1974.

[5] J. C. Candy, "A use of limit cycle oscillations to obtain robust analog-to-digital converters,"IEEE Transactions on Communications, vol. 22, no. 3, pp. 298-305, March 1974.

[6] J. C. Candy, B. A. Wooley and O. J. Benjamin, "A voiceband codec with digital filtering," IEEE Transactions on Communications, vol. 29, no. 6, pp. 815-830, June 1981.

[7] J. C. Candy and O. J. Benjamin, "The structure of quantization noise from sigma-delta modulation," IEEE Transactions on Communications, vol. 29, no. 9, pp. 1316-1323, Sept. 1981.

[8] J. C. Candy, "A use of double integration in sigma-delta modulation," IEEE Transactions on Communications, vol. 33, no. 3, pp. 249-258, March 1985.

[9] J. C. Candy, "Decimation for sigma-delta modulation," IEEE Transactions on Communications, vol. 34, no. 1, pp. 72-76, Jan. 1986.

[10] J. C. Candy and A. Huynh, "Double integration for digital-to-analog conversion," IEEE Transactions on Communications, vol. 34, no. 1, pp. 77-81, Jan. 1986.

[11] R. W. Adams, "Design and implementation of an audio 18-bit analog-to-digital converter using oversampling techniques," Journal of the Audio Engineering Society, vol. 34, pp. 153- 166, March 1986.

[12] T. Hayashi, Y. Inabe, K. Uchimura and A. Iwata, "A multistage delta-sigma modulator without double integration loop," ISSCC Digest of Technical Papers, pp. 182-183, Feb. 1986.

[13] L. E. Larson, T. Cataltepe and G. C. Temes, "Multi-bit oversampled $\Sigma\Delta$ A/D converter with digital error correction," Electronics Letters, vol. 24, pp. 1051-1052, Aug. 1988.

[14] R. Carley and J. Kenney, "A 16-bit 4th order noise-shaping D/A converter," IEEE Proceedings of the Custom Integrated Circuits Conference, pp. 21.7.1-21.7.4, 1988.

[15] B. H. Leung and S. Sutarja, "Multi-bit $\Sigma\Delta$ A/D converter incorporating a novel class of dynamic element matching," IEEE Transactions on Circuits and Systems II, vol. 39, pp. 35- 51, Jan. 1992.

[16] M. J. Story, "Digital to analogue converter adapted to select input sources based on a preselected algorithm once per cycle of a sampling signal," U.S. patent number 5138317, Aug. 11 1992 (filed Feb. 10 1989).

[17] W. Redman-White and D. J. L. Bourner, "Improved dynamic linearity in multi-level $\Delta\Sigma$ converters by spectral dispersion of D/A distortion products," IEE Conference Publication European Conference on Circuit Theory and Design, pp. 205-208, Sept. 5-8 1989.

[18] H. S. Jackson, "Circuit and method for cancelling nonlinearity error associated with component value mismatches in a data converter," U.S. patent number 5221926, June 22 1993 (filed July 1 1992).

[19] R. W. Adams and T. W. Kwan, "Data-directed scrambler for multi-bit noise-shaping D/A converters," U.S. patent number 5404142, April 4 1995 (filed Aug. 1993).

[20] R. T. Baird and T. S. Fiez, "Linearity enhancement of multibit $\Delta\Sigma$ A/D and D/A converters using data weighted averaging," IEEE Transactions on Circuits and Systems II, vol. 42, no. 12, pp. 753-762, Dec. 1995.

[21] R. Schreier and B. Zhang, "Noise-shaped multibit D/A convertor employing unit elements," Electronics Letters, vol. 31, no. 20, pp. 1712-1713, Sept. 28 1995.

[22] I. Galton, "Noise-shaping D/A converters for $\Delta\Sigma$ modulation," IEEE Transactions on Circuits and Systems II, Proceedings of the 1996 IEEE International Symposium on Circuits and Systems, vol. 1, pp. 441-444, May 1996.

[23] T. H. Pearce and A. C. Baker, "Analogue to digital conversion requirements for HF radio receivers," Proceedings of the IEE Colloquium on system aspects and applications of ADCs for radar, sonar and communications, London, Nov. 1987, Digest No 1987/92.

[24] P. H. Gailus, W. J. Turney and F. R. Yester, Jr., "Method and arrangement for a sigma delta convertor for bandpass signals," US Patent number 4,857,928, Aug. 1989 (filed Jan. 1988).

[25] R. Schreier and W. M. Snelgrove, "Bandpass sigma-delta modulation," Electronics Letters, vol. 25, no. 23, pp. 1560-1561, Nov. 9 1989.

[26] X. Wang, U. Moon, M. Liu and G. C.Temes, "Digital correlation technique for the estimation and correction of DAC errors in multibit MASH $\Delta\Sigma$ ADCs," 2002 IEEE International Symposium on Circuits and Systems, vol. 4, pp. 691-694, May 2002.

[27] W. Gao and W. M. Snelgrove, "A 950-MHz IF second-order integrated LC bandpass delta- sigma modulator," IEEE Journal of Solid-State Circuits, vol. 33, no. 5, pp. 723-732, May 1998.

[28] G. Raghavan, J.F. Jensen, J. Laskowski, M. Kardos, M. G. Case, M. Sokolich and S. Thomas III, "Architecture, design, and test of continuous-time tunable intermediate-frequency bandpass delta-sigma modulators," IEEE Journal of Solid-State Circuits, vol. 36, no. 1, pp. 5-13, Jan. 2001.

[29] P. Cusinato, D. Tonietto, F. Stefani and A. Baschirotto, "A 3.3-V CMOS 10.7-MHz sixth- order bandpass $\Delta\Sigma$ modulator with 74-dB dynamic range," IEEE Journal of Solid-State Circuits, vol. 36, no. 4, pp. 629-638, April 2001.

[30] R. Schreier, J. Lloyd, L. Singer, D. Paterson, M. Timko, M. Hensley, G. Patterson, K. Behel, J. Zhou and W. J, Martin, "A 50 mW Bandpass $\Delta\Sigma$ ADC with 333 kHz BW and

90 dB DR," International Solid-State Circuits Conference Digest of Technical Papers, pp. 216-217, Feb. 2002.

[31] J. Markus, J. Silva, and G. C. Temes, "Theory and applications of incremental delta sigma converters," IEEE Trans. Circuits Syst. I, vol.51, no. 4, pp. 678690, Apr. 2004.

[32] V. Quiquempoix, P. Deval, A. Barreto, G. Bellini, J. Markus, J. Silva, and G. C. Temes, "A low-power 22-bit incremental ADC," IEEE J. Solid-State Circuits, vol. 41, no. 7, pp. 1562-1571, Jul. 2006.

[33] C.-H. Chen, Y. Zhang, T. He, P. Chiang and G. C. Temes, "A micro-power two-step incremental analog-to-digital converter," IEEE J. of Solid-State Circuits, vol. 50, no.8, pp. 1796-1808, 2015.

[34] P. Shettigar and S. Pavan, "Design technique for wideband single-bit continuous-time $\Delta\Sigma$ modulators with FIR feedback DACs," IEEE J. Solid-State Circuits, vol. 47, no. 12, pp. 2865-2879, Dec. 2012.

[35] A. Jain, M. Venkatesan, and S. Pavan, "Analysis and design of a high speed continuous-time $\Delta\Sigma$ modulator using the assisted opamp technique," IEEE J. Solid-State Circuits, vol. 47, no. 7, pp. 1615-1625, Jul. 2012.

[36] Y. Zhang, C. H. Chen, T. He, G. C. Temes, "A continuous-time delta-sigma modulator for biomedical ultrasound beamformer using digital ELD compensation and FIR feedback," IEEE Transactions on Circuits and Systems I, vol. 62, no. 7, pp. 1689-1698, 2015.

[37] Y. Dong, W. Yang, R. Schreier, A. Sheikholeslami, S. Korrapati, "A continuous-time 0-3 MASH ADC achieving 88 dB DR with 53 MHz BW in 28 nm CMOS," IEEE J. Solid-State Circuits, vol. 49, no. 12, pp. 2868-2877, 2014.

Gabor Temes

Oregon State University

Sensor and Sensors Systems

Since the antiquity, it was quite a challenge for men to measure physical quantities. One element of great interest was temperature. Is not know if ancient Greeks or Chinese had ways to measure temperature, but certainly during the Renaissance many studies for defining standards and for obtaining quantitative temperature measurements flourished.

In 1664, Robert Hooke proposed the freezing point of water as zero point. In the same years, Ole Røemer suggested to use the boiling point of water as second relevant point and proposed to use a linear interpolation between the two points for measuring intermediate temperatures. In 1742, the Swedish astronomer Anders Celsius (1701-1744) divided that range into 100 equal parts but the method was in a reversed manner with respect to the one used today: 0°C represented the boiling point of water, 100°C represented the freezing point of water. During the 19th century, Gay-Lussac observed that the volume of a gas at constant pressure augments by the fraction of 1/267 per °C, (later revised to 1/273.15). The extrapolation of that effect leads to the zero absolute temperature (-273.15 °C) and the absolute temperature definition [1].

Around 1592 Galileo build a device for measuring temperature. It was based on the contraction of air in a vessel capable of moving a column of water. Santorio Santorii in 1612 sealed a liquid inside a glass tube and observed that the level moves up because of the expansion caused by temperature. The was the invention of the thermometer [2] used in many forms for centuries.

Many discoveries were inspired by the demand for control. An example, using again temperature, is the thermostat, patented in 1883 by Warren S. Johnson. It was a bimetal-actuated electric thermostat (Fig. 1) *"relates to devices adapted to indicate, at any convenient point, the relative temperature in rooms, conservatories, cellars, etc., situated at a point remote from the indica-*

Anders Celsius [Wikipedia]).

tors of my device; and it consists of certain peculiarities of construction, as will be more fully set forth hereinafter.," as the text of the patent says [3].

Temperature is just an example but there are many other examples for various quantities like distance, speed, pressure, weight, ... In all cases the sensors invented in the past centuries furnished the output in the form of mechanical displacement and not as electrical signal. Having the result in the electrical domain is the key that leads to the present sensor systems. An electrical signal permits analog processing, data conversion and storing or transmission of results. Moreover, an electrical system can operate in the

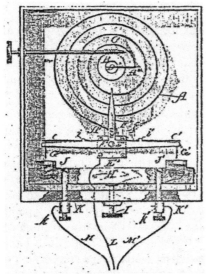

Fig. 1 - Picture of the thermostat. (W. S. Johnson's patent. Source:1980, 2007 ASME Milwaukee Section.)

reverse direction and generate with an electrical control a physical quantity. Devices named *actuators* allow this.

Many different types of sensors, both analogue and digital have been developed in the years. Generally, all of them can be classified as passive or active. A good example of passive sensor is tha strain gauge which is basically a pressure-sensitive resistance arranged in a bridge. In order to obtain the value of the pressure a current through it measure its electrical resistance. The obtained value is linearly proportional to the amount of strain or force being applied.

The oxygen sensor used to control the emission of cars is chemical: it determines the gasoline/oxygen ratio and generates a voltage. After the conversion from analog to digital a sophisticated processing engine determines if the mixture is not optimal and adapts the balance. The processing accounts for non-linearity and dependence on the environmental conditions. The result, a combination of sensing and processing, constitutes a sensor system.

Sensors Systems

Sensor systems has evolved in the history of Circuits and Systems till proposing very complex systems with applications in several fields, including but not limited to robotic, biomedicine, automotive, just to mention few. Furthermore, the word *"sensor"* usually refers to very different devices that may sense completely different signals. In general, a sensor is a device that transduces into an electrical signal a specific signal form the environment where it is located. into an electrical signal that is usually measured by means of an electrical system. The electrical system that enables the readability of the transduced signal is called readout circuit.

Good examples of sensory systems are the biosensors. Biosensors are sensors that transduce biological signals into electrical ones. A biosensor is a device typically incorporating a biological material intimately integrated with a transducer. The biological material may be a macromolecule (e.g., enzymes, antibodies, nucleic acids, aptamers, etc.), a biological cell, a microorganism or a bio-mimicking material (e.g., mutant proteins, synthetic receptors, biomimetic catalysts, imprinted polymers, etc.). The transducer typically is a semiconductor device that transduces a chemical signal (e.g., the product of a redox reaction or the pairing of two biomolecules) into an electrical signal. The trans-

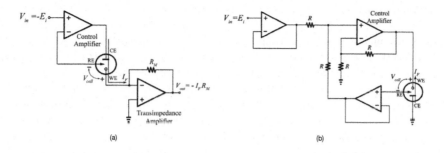

Fig. 2 - Typical sensor readout connected in grounded (a) or in grounded counter (b) configuration [4].

ducer may be optical, electrochemical, thermometric, piezoelectric, magnetic, micromechanical, etc. The generated electrical signal is always quantitatively related to the concentration of chemical species that have generated the original biological signal.

Typical examples of readout circuits for sensors are those related to electrochemical biosensors. An electrochemical biosensor is a sensing device where the transduction is related to a redox reaction involving biochemical species (e.g., human metabolites) and the transduction is toward signals in current or in potential. Typical architectures used as readout circuit for electrochemical biosensors are shown in Fig 2, where two basic configurations for the electrical reading of an electrochemical sensor (represented by the cell that includes the three electrodes named CE, RE, and WE) are shown. In the figure, the cell that includes the three electrodes named CE, RE, and WE represents the electrochemical sensor. Of course, modern challenges typically consequence of new needs emerged in the field of personalized medicine have required an effort in developing new circuits and new architectures that could not only measure just one biological signal but many in parallel. More that that, multi-panel systems are

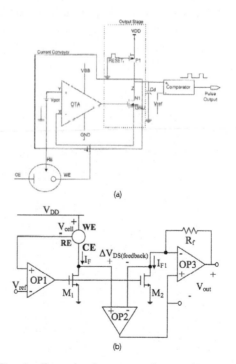

Fig. 3 - Example of sensor readout architectures. (a) With current-to-frequency converter [5], (b) in ground counter configurations [6].

now required to be fully autonomous, remotely powered or self-sufficient in term of energy demand, easy-to-use, fully-connected. All these new requirements and specifications for more sophisticated sensory systems have generated a lot of new and innovative architectures that appeared in literature and new and innovative devices that appeared in the market or that are under development right now to enter in market the nearest future. For example, new and more precise current readers have been developed taking into account the possibility of a direct conversion from current to frequency, which is extremely useful to directly convert in digital what is indeed an analog signal, or fault tolerant readers have been proposed too with the aim to implement the cancellation of errors due to component mismatch in CMOS design. For example, Fig. 3 shows a couple of such an innovative architectures.

On the other hand, the needs for multi-panel devices offered the possibility to design and test more complex readouts that apply more precise current measure, e.g. the current-to-frequency conversion method, to a panel of diversified sensor. For example, in Fig. 4(a) is shown the readout of a CMOS design that assures the reading of five different working electrodes (WE1 – WE5), which may be used to electrochemically detect five different human metabolites. With such a circuit, endogenous human metabolites like glucose, lactate, cholesterol, bilirubin, and glutamate are simultaneously and continuously detectable just by integrating the right biological macromolecule (typically the right enzyme) in the corresponding working electrode. The cell realizing the five electrochemical-sensors may have a single counter electrode (CE) and a single reference one (RE)

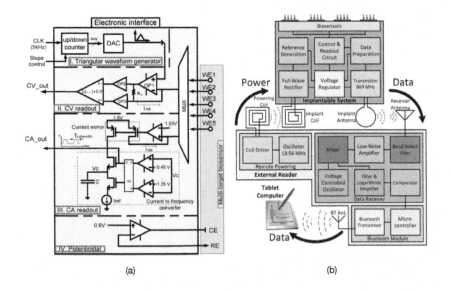

(a)   (b)

Fig. 4 - Examples of readout for more precise current measure by current-to-frequency conversion (A) or by error-cancellation (reprinted from [7] and from [8], respectively)

meanwhile offering the possibility of several working electrodes to implement the multi-panel sensing. Similar kinds of multi-panel devices have been also considered to develop fully implantable biosensors that can offer continuous monitoring of several disease biomarkers as well as therapeutic compounds. In case of fully-implanted biosensor, the further need to assure minimal invasive surgeries pushed out cleaver solutions for the remote transmission of data and power.

For example, Fig. 4(b) shows the readout of a CMOS design that assures the transmission of power (on the top of the figure) to a multi-panel biosensor, which autonomously transmits the data (on the right of the figure) once acquired, and finally send those acquired data to a tablet (on the bottom of the figure). By implementing such kinds of circuits and systems architectures and then embedding the so-obtained integrated circuits in fully functional multi-panel devices that assure so small sizes thanks to the more advanced scale of integration in CMOS technology, fully implantation under-the-skin may be conceived. In fact, the remote powering allows avoiding bulky and not-biocompatible batteries meanwhile remote data transmission allows communication through the skin barrier. The possibility to have multi-panel sensory systems allows indeed the continuous monitoring of patients by through the measure of bio-signals related to the unbalance of human metabolites associated to several diseases.

Of course, the need to assure minimally invasive devices in order to avoid complex surgeries requires extremely integrated heterogeneous systems, which includes sensors, readout, microcontrollers, memories, power managers, and communication modules. Although the complexity of such kind of systems is quite challenging and it requires a multi disciplinary approach already in the design phases, such kind of extremely heterogeneous systems has been recently demonstrated, as shown by the two photographs of Fig. 5. The applicability of these multi-panel circuits and systems for distributed diagnostics and personalized medicine has been demonstrated too. In fact, the device

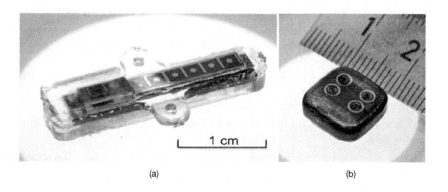

(a)   (b)

Fig. 5 - Examples of fully–implantable multi–panel sensor systems. (a) With multiple sensors [7], (b) fully independent system [9].

shown in Fig 5(b) have been successfully implanted in mice and tested for several endogenous metabolites and therapeutic compounds as well [10]. Of course, the multi-disciplinary approach usually required by the design phases in these kind of sensor systems asks for a co-design of the different layers of the architectures as well as of the integration of the bio and nano materials [11]. Such new approach in addressing all the required design layers assure the best performing and fully adapted system architectures to address the new needs for the breakthrough that is expected in the next ten years from now in the field of distributed diagnostics. The new concept of distributed diagnostics is expected to become fully embedded also in our personal electronics. In the last ten years or so, similar and parallel developments in co-design of circuits and systems for multi-panel sensing have been obtained as well, in the last ten years or so, in other fields of applications of sensor systems, such as robotics, avionics, and automotive, just to mention some.

References

[1] History of Sensors http://www.slideshare.net/AbdallaAli7/history-of-sensors

[2] Sensor History https://www.allsensors.com/engineering-resources/sensor-history

[3] http://controltrends.org/building-automation-and-integration/06/stattalk -warren-s-johnson-patented-the-thermostat-in-1883-thanks-asme-milwaukee-for -the-history-heritage/

[4] M. M. Ahmadi, G. A. Jullien, Current-mirror-based potentiostats for three-electrode amperometric electrochemical sensors. Circuits and Systems I: Regular Papers, IEEE Transactions on 56, 1339 (2009).

[5] H. S. Narula, J. G. Harris, A time-based VLSI potentiostat for ion current measurements. Sensors Journal, IEEE 6, 239 (2006).

[6] M. Razzaghpour, S. Rodriguez, E. Alarcón, A. Rusu, in Biomedical Circuits and Systems Conference (BioCAS), 2011 IEEE. (IEEE, 2011), pp. 5-8.

[7] S. S. Ghoreishizadeh, C. Baj-Rossi, A. Cavallini, S. Carrara, G. De Micheli, An integrated control and readout circuit for implantable multi-target electrochemical biosensing. Biomedical Circuits and Systems, IEEE Transactions on 8, 891 (2014).

[8] E. G. Kilinc et al., A System for Wireless Power Transfer and Data Communication of Long Term Bio-Monitoring. Sensors Journal, IEEE 15, 6559 (2015).

[9] C. Baj-Rossi et al., Full Fabrication and Packaging of an Implantable Multi-Panel Device for Monitoring of Metabolites in Small Animals. Biomedical Circuits and Systems, IEEE Transactions on 8, 636 (2014).

[10] C. Baj-Rossi, G. De Micheli, S. Carrara, in Engineering in Medicine and Biology Society (EMBC), 2014 36th Annual International Conference of the IEEE. (IEEE, 2014), pp. 2020-2023.

[11] S. Carrara, Bio/CMOS interfaces and co-design. (Springer Science & Business Media, 2012).

Sandro Carrara

École Polytechnique Fédérale de Lausanne, EPFL, Lausanne

Biological Sensor Systems

What is a biosensor?

Biosensors systems in their broadest definition are referred to analytical devices which convert biological responses into quantifiable and processable signal [1]. Under this definition, a clinical thermometer, a blood pressure sensor, and a pregnancy test kit are all considered biosensor systems. More specifically, however, biosensors are devices that combine biological and chemical materials and transducers for the detection or measurement of samples, such as drugs, metabolites, pollutants, and controlled parameters by converting a biochemical signal into a measurable physiochemical signal which in turn quantify the amount of that sample [2]. Under this definition, only the last example, a pregnancy test kit qualifies as a biosensor.

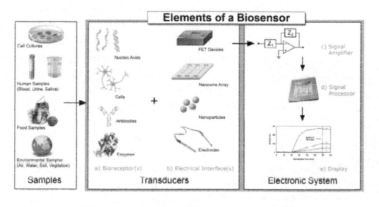

Elements of a typical biosensor system [3].

Biosensor systems have broad applications, which can be categorized into: 1) Industrial: in monitoring and control of various processes for manufacturing foods, drinks, and drugs. They are also used in monitoring materials and microorganisms in packaged and transported food products and pharmaceuticals to indicate whether they are safe to be used. 2) Medical: there are various types of biosensors that are used for measuring chemical analytes in body fluids, particularly in blood and urine samples, both acutely and chronically. 3) Domestic and environmental: monitoring pollutants in air, quality of water resources, contaminants in land fields, anywhere from large cities to remote sites and from factories to home environment. 4) Military and homeland security: battlefield monitoring of poisonous gases, nerve agents, and other harmful chemicals to civilians and military personnel, as well as bioterrorism.

A biosensing system often has four major components, as shown in the above figure. First part is dedicated to handling the sample, which would range from a simple small reservoir to a sophisticated network of microfluidic channels and chambers that would not only deliver the sample to different parts of the biosensor but also apply pre-treatment steps, such as filtering, mixing, temperature adjustment, dilution, etc. Second part is a sensitive bio-recognition element such as enzymes, antibodies, receptor tissue, microorganisms, organelles, cell receptors, nucleic acids, etc. a biologically derived material or biomimetic component that directly interacts with, i.e. binds to or recognizes, the sample under study. Utilizing the selectivity of the biological element is the main driving force behind the strength of biosensors. The third part is a transducer or detector element that uses an electrochemical, optical (absorbance, fluorescence and chemiluminense), piezoelectric (acoustic and ultrasonic), thermometric, magnetic, etc. phenomenon in proximity of the reaction to transform the signal resulting from the interaction between the sample and bio-recognition element into a different signal that is often easier to detect and quantify, such as a change in current, voltage, impedance, transparency, color, reflectance, spectrum, temperature, resonance frequency, etc. Some of the most common electrochemical methods that are currently used in biosensing systems are potentiometry (potential difference at zero current level), voltammetry (current measurement by sweeping voltage), amperometry (current measurement at constant voltage), and conductimetry (measuring the solution resistance) [4].

A physician using his senses of sight, touch, smell, and taste to make a diagnosis [5].

It is often possible and recommended to include a reference element in parallel with the main reaction to produce a reference signal without the analyte/sample to serve as a control of the abovementioned experiment and account for other sources of inaccuracy and interference, such as aging, temperature change, electromagnetic noise, etc. In this case the difference between the two signals analyzed, which would, for instance, be proportional to the concentration of the material being measured. It is worthwhile noting that the environment in steps before the transducer is often wet, while it is a dry environment afterwards. Finally the fourth part is a reader device with its associated signal conditioning and processing elements that are primarily responsible for displaying of the results in a user-friendly way. Desired characteristics in biosensor systems are high sensitivity and specificity, small size, mass manufacturability at low cost, high speed in determining the results within the timescale of the process or diagnostic test, economical in terms of cost of ownership, consumables, and maintenance, self-calibrating, requiring minimal action by the user, and clear means for visualizing the results in a user-friendly environment [3].

Early History

During mediaeval times, attempts were made by physicians to identify various diseases by examining urine samples for appearance, color, sediment, and even taste [5]. Another example of early biosensors were bright yellow canary birds, which were considered the coal miners' life insurance and the root of the *"canary in a coal mine"* idiom. They were carried below ground in small cages, and because of their highly sensitive and active metabolism detected methane and carbon monoxide gasses that gave an early warning for potential explosions or poisonous air. Canaries were normally chirp and sing all day long. However, if the carbon monoxide levels were too high, the canaries would have trouble breathing, and may even die. When the canaries were no longer singing, miners would know that they should leave the mine as quickly as possible to avoid being caught in an explosion or poisoned [6].

A coal miner holding a canary in a small cage that would warn him about high levels of potentially dangerous gasses.

The first reference to bioelectronics in the broader definition of biosensors, was published in 1912, and focused on measurement of electrical signals generated by the body, which became the basis of the electrocardiogram (ECG) [7]. In the 1960s two new trends in bioelectronics began to appear. One trend, enabled by the invention of the transistor, centered on the development of implantable electronic devices and systems to stimulate organs, particularly the pacemaker for the heart [8]. In the same time frame, fundamental studies were reported on electron transfer in electrochemical reactions [3]. In this regard, within the more specific definition of biosensors and considering the prevalence of diabetes in developed and developing countries, the biosensor systems for glucose monitoring in diabetes patients have historically been the market driver for expansion and advancement in this field both in academia and industry. For instance, it is estimated that 29.1 million Americans (9.3% of the population) have type-I or type-II diabetes at the total costs of diagnosed cases being $245 billion in 2012 [9].

Modern Biosensor Systems, the History of Blood Glucose Meters

Professor Leland C. Clark is known as the father of biosensor concept and inventor of the first modern biosensor for measuring the concentration of glucose in a solution, like urine or blood. In 1956, Clark published his seminal paper on the oxygen electrode [10]. The concept was illustrated by an experiment in which enzyme glucose oxidase was entrapped over a Clark oxygen electrode using dialysis membrane [11]. The amount of glucose was proportional to the reduction in the dissolved oxygen concentration in the solution. The term *"enzyme electrode"* was coined in a published paper where the first biosensor was described [12]. In 1963

Leland C. Clark [Wikipedia]).

G.A. Rechnitz and S. Katz introduced one of the first papers in the field of biosensors with the direct potentiometric determination of urea after urease hydrolysis. Updike and Hicks then used the same term in 1967 to describe a device in which they immobilized glucose oxidase in a polyacrylamide gel onto the surface of an oxygen electrode for rapid and quantitative determination of glucose [13].

In 1970s, many other researchers started to couple various enzymes and electrochemical sensors to either develop new biosensors or further improve the existing ones. Researchers also tried to extend the range of sensors to non-electrochemical active compounds and non-ionic compounds, while research on ion selective electrodes (ISE) became very popular in that period. In 1969 G. Guilbault introduced the potentiometric urea electrode, in 1973 P. Racinee and W. Mindt developed a lactate electrode, in 1976 the first microbe-based biosensor was invented, and in 1977 K. Cammann coined the term *"biosensor"* [3]. Clark's ideas were finally commercialized in 1975 with the successful launch of the glucose analyser based on the amperometric detection of hydrogen peroxide by Yellow Springs Instrument Company in Ohio [11].

In 1965 the Ames Research Division of Miles Laboratories in Elkhart, Indiana, USA, developed the first blood glucose test strip, called the Dextrostix [5]. A paper reagent strip, which used the glucose oxidase/peroxidise reaction with an outer semipermeable membrane that trapped red blood cells but allowed glucose to pass through and react with the dry reagents. A drop of blood (50-100 μL) was applied to the test strip and was gently washed away, then the strip color was assessed against a color chart to indicate the blood glucose value. Unfortunately the colours were difficult to visualise as the color blocks were affected by the ambient lighting, and variations in the user's visual acuity, resulting in difficulty in obtaining accurate and precise readings.

Ames Reflectance Meter.

A. Clemens, also at Ames, developed the first instrument to produce a quantitative blood glucose output in the late 1960s by capturing the light reflexed from the surface of Dextrostix test strips and applying it to a photoelectric cell to produce a signal that was displayed by an analog scales, to indicate the concentration of blood glucose [5]. This device, shown in the figure, which was called the Ames Reflectance Meter (ARM), weighed 1.2 kg due to its casing and lead acid rechargeable batteries, and cost around $495, became commercially available in 1970 [14]. The sample application, wash and blot technique to remove the red blood cells, and timing were all critical to the ARM precision, and considerable nonlinearity due to saturation was observed compared to laboratory methods [5].

The pioneering work of Clemens sat the stage for further developments to improve blood glucose meter systems (BGMS), inviting other companies to diversify this device in appearance, technology, and performance. In 1974, Boehringer Mannheim produced the Reflomat, a reflectance meter using a modified reagent strip, which required a much smaller volume of blood (20-30 μL), which was removed more simply by wiping with a cotton wool ball. Up to this point, the BGMS were designed only for testing in doctors' offices, but in mid-1970s the idea of diabetics self-testing was contemplated using smaller, lower cost, and easier to use devices. The Dextrometer, launched in 1980, was the first BGMS with a digital display and could be operated either by battery or mains. Glucochek, the first of a series of BGMS produced by Lifescan, became available in 1980, and later known as Glucoscan, a battery driven digital reflectance meter manufactured by Medistron [5].

The 1980s was an active phase in the evolution of meters, by becoming easier to use, smaller in size, often with memory to store and retrieve results. Reagent strips were also changing to accept smaller volumes of blood, and some were barcoded for auto-calibration and quality assurance. The OneTouch meter was introduced in 1987 and was regarded as a second generation BGMS because it utilised a modified sampling procedure with a small volume of blood applied to the reagent strip that was already inserted in the meter, timing began automatically, and results displayed after 45 s. Moreover, the strip required no washing, wiping or blotting, and reduced operator variation. Towards the end of 80s, the first enzyme electrode strips were introduced (ExacTech launched in 1987 by MediSense), providing a choice of either using reflectance or electrochemical principles to measure blood glucose, which designated the third-generation BGMS. Furthermore, the new BGMS were available in two new forms, a slim pen or a thin card the size of a credit card [5].

In the 90's, glucose became one of the most frequently measured analytes both in clinical units and at patients' homes, thanks to user friendly biosensors. In a rapidly growing market, major pharmaceutical companies; Bayer, Abbott, and Roche purchased pioneer firms; Ames, MediSense, and Boehringer Mannheim, respectively. However, there were still operator-dependent difficulties, such as obtaining a sufficient volume of blood, timing the application and removal of blood from the test strip, calibration and coding errors, lack of maintenance, and quality control procedures. Therefore, many manufacturers started developing systems that minimised those operator-dependent steps and

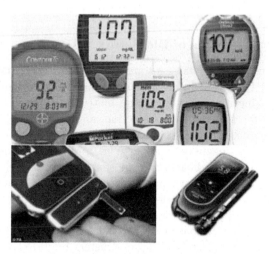

Today's blood glucose meters.

made the system easier to use, such as Accuchek Advantage by Roche and Medisense Precision by Abbott [5].

The latest trend in diabetes care and management is continuous glucose monitoring systems (CGMS), which give considerably greater insight into the direction, magnitude, duration, frequency and possible causes of glucose fluctuations in response to meals, insulin injections, hypoglycaemic episodes and exercise throughout the day. The CGMS systems work by inserting a small catheter or implant containing the sensor subcutaneously. The sensor measures the glucose in the interstitial fluid and results are transmitted to a monitor or smartphone for storage or immediate display. Glucowatch Biographer (Cyngnus, CA), which was worn as a wristwatch and used reverse iontophoresis to stimulate the secretion of subcutaneous fluid and measure glucose content using an electrode-biosensor unit, became available in 1999 and received FDA approval in 2001. This device, however, still required frequent calibration with finger pricks, and did not do well in subsequent clinical trials, and it was discontinued [5]. In 2007 an implanted glucose biosensor (freestyle Navigator system) operated for five days [15]. With the latest implantable CGMS, results are often provided every 10 minutes for up to 90 days [16], [17].

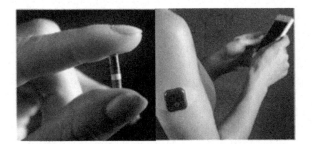

An implantable continuous glucose monitoring systems from Senseonics Inc. [16].

Other Types of Biosensors

Custom nanomaterials are being designed and fabricated with unique properties for biosensor applications, thanks to integration of material science, molecular engineering, chemistry, and biotechnology, to improve the sensitivity and specificity of biomolecule detection, atomic and molecular manipulation, pathogenic diagnosis, and environmental monitoring. Gold nanoparticles, carbon nanotubes, magnetic nanoparticles, and quantum dots are being heavily researched for their applications in biosensor systems [18]. Acoustic wave biosensors use quartz crystal or silicon micromachined resonators that change their resonance frequency due to deposition of mass of any material on their surface, which is directly proportional to the concentration of the sample [3]. Some of the nanoparticle-based sensors, including the optical biosensors, use resonance enhancement of metal nanoclusters bound to a surface by bio-recognitive interactions for enhancing bio-optical sensing devices. The phenomena of surface plasma resonance (SPR), which occurs during optical illumination of a metal surface, has shown good biosensing potential, as demonstrated by BIAcore^{TM}, enabling real time detection and monitoring of biomolecular binding events and paving the way for better understanding of biochemical mechanisms [19]. Gold nanoparticles have also been used as a new class of universal fluorescence quenchers to develop an optical biosensor for recognizing and detecting specific DNA sequences [3].

Integrated biosensor arrays, also known as gene chips or DNA-chips, use the same excitation source and measurement process for a large number of elements. Most of them are based on the use of nucleic acids as sensing elements but antibodies, enzymes, and cellular components can also be used. These biosensors offer an exciting alternative to traditional methods, allowing rapid real time, and multiple analyses for detection, diagnosis, and estimation of any sample. A series of new biosensors are under development to be placed in laboratory animals for physiological and neurochemical in vivo measurements or implanted in the human body for health check purposes and metabolite monitoring. The main challenges in the field of implantable biosensors are the stability, selectivity, and biocompatibility of the sensor [15], [17].

For medical applications, nano-biosensors, integrated biosensors, and implantable biosensors will reduce the cost and detection time thereby increasing the efficiency of the tests. Also disposable biochips offer an added advantage of rapid point-of-care (PoC) medical diagnostics of diseases without the need for sending samples to a laboratory for analysis. New biosensor systems present a great opportunity for development of robust, low cost, rapid, and specific detection and analysis [1], [3], [5].

References

[1] C. R. Lowe, "*Biosensors*," Trends in Biotechnology, vol. 2, no. 3, pp. 5965, 1984.

[2] N. Arora, "*Recent advances in biosensors technology: a review*," Octa Journal of Biosciences, vol. 1, no. 2, pp. 147-150, Dec. 2013.

[3] D. Grieshaber, R. MacKenzie, J. Voros, and E. Reimhult, "*Electrochemical biosensors - sensor principles and architectures*," Sensors, vol. 8, pp. 1400-1458, Mar. 2008.

[4] M. Mascini, "*A brief story of biosensor technology*," Biotechnological Applications of Photosynthetic Proteins: Biochips, Biosensors and Biodevices, eds. M.T. Giardi and E.V. Piletska., Landes Bioscience, 2006.

[5] S.F. Clarke and J.R. Foster, "*A history of blood glucose meters and their role in self-monitoring of diabetes mellitus*," British Journal of Biomedical Science, vol. 69, no. 2, pp. 83-93, 2012.

[6] Wikipedia, [Online] Available: https://en.wikipedia.org/wiki/Biosensor

[7] M.C. Potter, "*Electrical effects accompanying the decomposition of organic compounds*," Proceedings of the Royal Society, vol. 84, pp. 290-276, 1912.

[8] M.A. Wood and K.A. Ellenbogen, "*Cardiac pacemakers from the patient's perspective*," Circulation, vol. 105, pp. 2136-2138, 2002.

[9] American Diabetes Association, [Online] Available: http://www.diabetes.org/diabetes-basics/statistics/

[10] L.C. Clark, "*Monitor and control of blood and tissue oxygenation*," Trans. Am. Soc. Artif. Intern. Organs, vol. 2, pp. 41-48, 1956.

[11] A.P.F. Turner, I. Karube, and G.S. Wilson, "*Biosensors Fundamentals and Applications*," Oxford University Press, Oxford, 1987.

[12] L.C. Clark and C. Lyons, "*Electrode systems for continuous monitoring cardiovascular surgery*," Ann NY Acad Sci, vol. 102, pp. 29-45, 1962.

[13] S.J. Updike and G.P. Hicks "*The enzyme electrode*," Nature, vol. 214, pp. 986-988, 1967.

[14] E.L. Mazzaferri, R.G. Skillman, R.R. Lanese, and M. Keller, "*Use of test strips with a colour meter to measure blood-glucose*," Lancet, vol. 7642, pp. 331-333, 1970.

[15] S. Borgmann, A. Schulte, S. Neugebauer, and W. Schuhmann, "*Amperometric biosensors*," Advances in Electrochemical Science and Engineering, 2011.

[16] Senseonics Inc., [Online] Available: http://www.senseonics.com/products.

[17] A. DeHennis, S. Getzlaff, D. Grice, and M. Mailand, "*An NFC-enabled CMOS IC for a wireless fully implantable glucose sensor*," IEEE J. Biomedical and Health Informatics, vol. 20, no. 1, Jan. 2016.

[18] X. Zhang, Q. Guo, and D. Cui, "*Recent advances in nanotechnology applied to biosensors*," Sensors, vol. 9, pp. 1033-1053, 2009.

[19] P. Leonard, S. Hearty, J. Brennan, L. Dunne, J. Quinn, T. Chakraborty, and R. O'Kennedy, "*Advances in biosensors for detection of pathogens in food and water*," Enzyme and Microbial Technology. vol. 32, pp. 313, 2003.

Maysam Ghovanloo

Georgia Institute of Technology, Atlanta, GA

Biomedical Circuits and Systems

This section summarizes the origin and exponentially increasing biomedical circuits and systems activities with CAS Society. It includes early activities within ISCAS, then the starting of BioCAS TC, then the BioCAS conference. A summary of the adopted topic at early stage, then main research directions including the bioelectronics intended for wearable body and brain interfaces (recording and treatment), the wireless bidirectional data links and power harvesting mechanism will be described which are followed by a conclusion.

More than a century ago electricity was used to develop solutions to help people struggle the various dysfunctions; for example Franklin in 1752 was using electrical stimulation for pain relief, and to treat a person with convulsions, and various palsies, and Galvani in 1791 stimulated inactive frog nerve-muscle tissue. In 1892, Einthoven recorded the first human electrocardiogram in Europe using the Lippmann capillary electrometer.

Nowadays, medical devices is currently used for a very large number of medical and biological applications such as Pacemaker, Cochlear Implant, Deep brain stimulator, etc. More recently, IEEE members were motivated to develop modern solutions to enhance medical organ functions.

Early after year 2000, Circuits and Systems Society (CASS) enlarged activities in this direction. Then members of the CASS launched the International conference on biomedical circuits and systems (BioCAS) in Singapore chaired by Yong Lian in December 2004. The following edition was organized by Chris Toumazou in London UK in 2006, then Mohamad Sawan hosted in Montreal the 2007 edition. The following editions were held in Baltimore, Beijing, Cyprus, San Diego, Hsinchu, Rotterdam, Lausanne, and Atlanta hosted respectively by Ralph Etienne-Cummings, Zhihua Wang, Julius Georgiou, Gert Cauwenberghs and Andrew Masson, Wai-Chi Fang, Wouter Serdijn, Sandro Carrara and Maysam Ghovanloo.

The above is a picture took to celebrate the 10th Anniversary of BioCAS in November 2014. It is grouping the respectives general chairs of the conference,which now rotates around the main continents (Europe, Americas, Asia).

BioCAS brought a strong collaboration with biotechnologies by facilitating interdisciplinary collaborations among scientists, engineers and medical researchers and practitioners to solve complex problems and innovating in rapidly growing area of research.

The multidisciplinary approach of BioCAS enabled the CAS community to cover a wide range of topics: biofeedback and electrical stimulation, bioinspired circuits and systems, biomedical imaging technologies and image processing, BioMEMS, biomedical instrumentations, biosensors, bioactuators, bio-signal processing, body area networks/body sensor networks, electronics for brain science and brain machine interfaces, implantable electronics, innovative circuits and systems for medical applications, lab-on-chip, medical information systems and wireless and energy harvesting /scavenging technology in medicine. BioCAS events attracted keynotes of world's renowned experts from medical and biological arenas, which motivated more the CASS's members to find niches for their contributions to innovate. As a result, BioCAS has received an increasing number of high quality contributions from the CASS community, and has become considerably more selective by maintaining its single-track program.

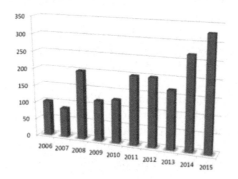

The figure at the left shows the submissions variations over the 10 past years.

Almost in parallel to the first edition of BioCAS Conference, a technical track named BioCAS within ISCAS was launched too. Five CASS members were the initiators: Chris Toumazou, Tor Sverre (Bassen) Lande, Yong Lian, Mohamad Sawan and Wael Badawy. Biocas activities within CASS illustrates a new wave of circuits and systems inspired by biology and healthcare, life sciences, physical sciences and engineering with application to medical problems.This rapidly growing community motivated Bassen and past chairs of Biocas conference to launch the IEEE Transactions on Biomedical Circuits and systems (TBio-CAS). Tor Sverre (Bassen) Lande was the first Editor in Chief, followed by Gert Cauwenberghs, then Mohamad Sawan took over the task for the window 2016-18. The first volume of TBioCAS was delivred on March 2007. In addition to CASS, TBioCAS was supported by the IEEE Engineering in Medicine and Biology Society (EMBS). The status of the TBioCAS is great. The impact factor is ranked among the top journals in the field, it went up to 3.371. The Article Influence Score (AIS) is ranked 31 over 274 by Journal Citation Report (JCR) in Electrical and Electronic Engineering, and 7 over 77 in Biomedical Engineering (BE). The current number of regular submissions is more than 300/year following the BioCAS community grows.

Mohamad Sawan
Polytechnique Montreal, QC, Canada

Switching Power Converter Circuits

Power converter circuits, which root in various fundamental circuit concepts, witnessed a revolution with the advent of switch-mode power converter circuits and in turn modern power electronics, which flourished in the early 70s mainly around the Caltech group of R. D. Middlebrook, with former research in device modelling and circuit theory, and Slobodan Ćuk, with a strong circuit theory and magnetic component background. This revolution was in particular catalyzed mainly by aerospace applications, and later by those which would benefit the

most from the reduced volume and weight and increased efficiency of these power processing circuits. Emerging in the fertile overlap of power devices, circuit theory, the control discipline and signal processing, the very research field of switch-mode power converters, circuits which are nowadays at the backbone of any energy distribution architecture ranging from mW up to kW, spun off various fundamental contributions within the CAS Society, mainly in topics closer to other adjacent areas such as analog processing and circuits as well as RF electronics, pre-

David Middlebrook (www.caltec.edu)

dominantly with a circuit-centric design-oriented modeling approach. Indeed, the manifold contributions emerging out of the CASS community as mothership include switched capacitor power converters, class-E power amplifiers and converters, modelling and control of DC DC power converters, and miniaturized and integrated power supplies.

Switched-capacitor power converters

Switched-capacitor converters has been one of the topics of power electronics circuits for which the IEEE CAS Society played an overwhelming role in their development. Probably, this was due to the previous research of the circuit theorists on switched-capacitor filters which inspired the power electronics community in CASS to look for means to replace the magnetic components in the controlled energy processing by switched-capacitors. Today it is unconceivable to think of all the portable electronic equipment without the light weight, small-size, high power density switched-capacitor electronic power supplies.

Following old ideas of Greinacher (1919), Cockcroft-Walton (1932), Midgley-Sigger (1974), Dickson (1976) and Kunzinger-Sohn (1978), the group of Prof. Ueno proposed at the end of the 80s and beginning of 90s the first switched-capacitor power converters. The world became aware of this new subject due to their publications in CASS conferences [1-2]. If these first power supplies were suffering from a lack of regulation for changes in the input voltage and from a fixed input-to-output voltage ratio, soon the group of A. Ioinovici and his students published in ISCAS and in CAS journals the first papers presenting

Prof. Funio Ueno (T. IEE Japan, 120 C, 2000)

a fully regulated switched-capacitor converter [3,4]. Unlike the charge-pump capacitor-diode circuits able of increasing the input voltage by a number of times, the SC converters can provide an output voltage lower or higher than the input one, or even of opposite polarity, the load voltage being constant despite variation in the input and/or load. The theory of the new circuits was studied in many CASS published papers, including [5]. Implementation of switched-capacitor converters in integrated circuit technology was initialized in another ISCAS paper [6]. The number of publications and the extension of the results spread in the CAS publications about switched-capacitor converters was so vast, that in 2001 the writing of a tutorial [7] became compulsory, as proved by more than 130 citations (Scopus, 2016). The research in this topic continued intensively in our century, leading to the development of switched-capacitor power supplies with enhanced features, better adapted to the new challenges. AC-DC converters, DC-AC inverters and bidirectional power electronics containing only capacitors and switches were also presented. Advanced control methods like frequency modelling or mixed adaptive control, deep-green mode operation for a system-on-a-chip application or interleaving for better shaping of the input current have made the object of recent studies [8-11]. Interesting new applications of these converters operating with a very high efficiency like powering micro-satellites, energy supply of liquid crystal displays, power management of sub-mW energy harvesting, ultra low power chip for a self-sustainable oceanic sensing platform and many more others have been recently published by researchers of CASS [12-14]. The power electronics researchers of the CASS initiated the answer to another challenge imposed by the development of the environmental friendly sources of energy: the solar or fuel cells provide energy of a too low level and variable voltage to be used as such in the front end of an electrical supply system. In addition, in order to prolong the life of the green energy sources, the power electronics circuit has to absorb a non-pulsating input current, making the use of the pure switched-capacitor converters unsuitable. A new idea was developed for realizing a very large DC gain converter able to efficiently increase and control the voltage of the new energy cells. Starting from an old idea of Middlebrook (1988) about a voltage divider inserted in a buck converter, Ioinovici and his group proposed basic switched-capacitor-inductor structures to be integrated in classical boost converters [15]. The idea of this paper published in Transactions on CAS was taken by power electronics researchers from CASS [16] but also from other IEEE societies in a large number (the paper was quoted in over 325 publications, 2016).

Ultra high gain converters have been obtained by combining classical converters with a coupled-inductor and a switched-capacitor circuit, answering thus requests from the telecommunication and data industries [17].

Invention, Development, and Applications of Class E High-Efficiency Power Amplifier

The class-E high-efficiency switching-mode power resonant amplifier was invented in 1975 by the Sokals [18]. The transistor is operated in this amplifier as a switch. The switching loss can be reduced to zero because the transistor turns on at zero voltage. This type of operation is called zero-voltage switching (ZVS) or soft-switching. For nominal operation, the switch also turns on at zero slope of the voltage. This type of operation is termed zero-derivative switching (ZDS). The first theoretical analysis of this amplifier was published by Raab in the IEEE Transactions on Circuits and Systems [19] in 1977. Subsequently, various aspects of the Class-E amplifier were studied and many contributions were published by Kazimierczuk, Ebert, Puczko, Suetsugu, Kessler, Sekiya, and Hayati in the IEEE Transactions on Circuits and Systems [20]-[29], such as analysis of the amplifier, amplitude modulation, matching circuits, maximum operating frequency, design procedures, off-nominal operation, effects on nonlinear MOSFET capacitances, Class-E rectifiers, and DC-DC resonant converters. Class-E amplifier has the highest operating frequency among all high-frequency power amplifiers. In addition, many papers were presented at the IEEE ISCAS and the IEEE Midwest Symposium on CAS. There are many applications of the Class-E amplifiers, such as RF and microwave transmitters, DC-DC power converters, electronic ballasts, wireless power transfer, MRI, and battery chargers of biomedical implants.

Modeling and control of pulse-width modulated DC-DC power converters

There was a need to characterize dynamic performance of switching-mode power supplies (*SMPS*). Prof. Middlebrook had developed a state-space description of the power stages of switching dc-dc power converters. Contributions at CASS in this field encompass pioneer works by S. Sanders and G. Verghese [30] on enhanced average circuit models, S. Singer and R.W. Erickson [31] on power conservative models for converter topologies, A. Kislovski and R. Redl [32] on injected-current dynamic models and F. Guinjoan, A. Poveda and L. Martinez-Salamero on small-signal models based upon separation of converter dynamics [33]. Later, small-signal models of pulse-width modulated dc-dc power converters have been developed using circuit-average techniques and the principle of conservation of energy by Kazimierczuk and Czarkowski. Subsequently, small-signal models of various converters were developed. These models were used to derive small-signal transfer functions and transient responses of *PWM* converters for continuous, discontinuous, and boundary conduction modes of operation by Kazimierczuk, Czarkowski, Reatti, and Ayachit. The dynamic characterization of power stages allowed for the development and design of feedback control loops, using various control techniques. Current-mode control of PWM converters were studied by Kazimierczuk, Bryant, and Kondrath. A Z-source converter for steady-state operation was studied by Galigekere and Kazimierczuk. Many papers were published on these subjects in IEEE Transactions on Circuits and Systems and presented at IEEE ISCAS and the IEEE CAS Midwest Symposium. These converters have very broad range of applications such as efficient power supplies, battery charges, and active power factor correctors. Many of the concepts published in the IEEE TCAS were extended in other IEEE societies, such Power Electronics, Industrial Electronics, Industry Applications, and Microwave Societies.

Switching power converter miniaturization and on-chip integration

Consistent with the integration trends that drove research activity in CASS aligned to the SoC approach and RF CMOS in the 90s, CASS researchers pioneered work in increased miniaturization, increased power density and eventually on-chip compatibility of switch-mode power converters in the past decade. Special sessions [34, 35] organized by Eduard Alarcón, Luigi Fortina, Amit Patra and Dariusz Czarkowski crystallized emergent activities within CASS and engaged the community, thereby subsequently yielding full integrated switch-mode power management circuits, as in those contributions from Philip KT Mok [36], Wing-Hung Ki [37], Gabriel A Rincon-Mora [38] and Ke-Horng Chen [39]. The thrust for miniaturization of power supplies stands from the outstanding overall impact of this power processing subsystem within a whole system in terms of volume/area, weight and efficiency, being thus power management the limiting factor in the portability and operating lifetime, particularly in wireless and mobile systems such as cellular phone terminals and wireless sensor network nodes. The ultimate step consequently consists in the fully monolithic integration of the power regulators together with the same circuits which constitute their load within either the same substrate or chip package, yielding a complete Powered System on a Chip (PSOC). Some of the later results encompass extensions to wideband power supplies aiming adaptive power management of RF power amplifiers [40], and energy management architectures for ultra-low power regimes in the context of Energy Harvesting [41].

References

[1] T. Umeno, K. Takahashi, I. Oota, F. Ueno, and T. Inoue ,"New switched-capacitor DC-DC converter with low input current ripple and its hybridization", in Proc. 33rd IEEE Midwest Symp. on Circuits and Systems, Calgary, Canada, Aug.1990, pp.1091-1094.

[2] F. Ueno, T. Inoue, and I. Oota, "Realization of a switched-capacitor AC-DC converter using a new phase controller", in Proc. ISCAS, June 199, pp. 1057-1060.

[3] S. V. Cheong, H. Chung, and A. Ioinovici, "Development of power electronics based on switched-capacitor circuits ", in Proc. ISCAS, 1992, pp.1907-1911.

[4] S. V. Cheong, S. H. Chung, and A. Ioinovici, "Duty-cycle control boosts dc-dc converters," IEEE Circuits and Devices Magazine, pp.36-37, 1993.

[5] M. S. Makowski, "Realizability conditions and bounds on synthesis of switched-capacitor DC-DC voltage multiplier circuits," IEEE Transactions on CAS, pp. 684-692, 1997.

[6] W .Chen, W. H. Ki, P. K. T. Mok, and M. Chan, "Switched-capacitor power converters with integrated low dropout regulator", in Proc. ISCAS, 2001, pp. III 293-296.

[7] A. Ioinovici, "Switched-capacitor power electronics circuits," IEEE Circuits and Systems Magazine, pp. 37-42, 2001.

[8] F. Su and W. H. Ki, "Component-efficient multiphase switched-capacitor DC-DC converter with configurable conversion ratios for LCD driver applications," IEEE Trans. on Circuits and Systems II, pp.753-757, 2008.

[9] Y. H. Chang, "Design and analysis of multistage multiphase switched-capacitor boost DC-AC inverter," IEEE Trans. on Circuits and Systems I, pp. 205-208, 2011.

[10] W. C. Chen et al. "A wide load range and high efficiency switched-capacitor DC-DC converter with pseudo-clock controlled load-dependent frequency," IEEE Trans. on Circuits and Systems I, pp. 911-921, 2014.

[11] T. Souvignet, R Allard and X. Lin-Shi, "Sampled-data modeling of switched- capacitor voltage regulator with frequency-modulation control," IEEE Trans. on Circuits and Systems I, pp. 957-966, 2015.

[12] Y. C. Lin et al. " Liquid Crystal Display (LCD) Supplied by Highly Integrated Dual-Side Dual-Output Switched-Capacitor DC-DC Converter With Only Two Flying Capacitors," IEEE Trans. on Circuits and Systems I, pp. 439-446, 2012.

[13] I. Lee et al., " System-On-Mud: Ultra-Low Power Oceanic Sensing Platform Powered by Small-Scale Benthic Microbial Fuel Cells," IEEE Trans. CAS-I, pp. 1126-1135, 2015.

[14] I. Vaisband, M. Saadat and B. Murmann, "A closed-loop reconfigurable switched-capacitor DC-DC Converter for sub-mW energy harvesting applications," IEEE Trans. on Circuits and Systems I, pp. 385-394, 2015.

[15] B. Axelrod, Y. Berkovich and A. Ioinovici, "Switched capacitor/switched inductor structures for getting transformerless hybrid DC-DC PWM converters," IEEE Trans. on Circuits and Systems I, pp. 687-696, 2008.

[16] S. Xiong, S. C. Tan and S. C. Wong, "Analysis and Design of a High-Voltage-Gain Hybrid Switched-Capacitor Buck Converter," IEEE Trans. on Circuits and Systems I, pp. 1132-1141, 2012.

[17] T. J. Liang et al. "Ultra large gain step-up switched-capacitor DC-DC converter with coupled inductor for alternative sources of energy," IEEE TCAS-I, pp.864-874, 2012.

[18] N. O. Sokal and A. D. Sokal, "Class E a new class of high-efficiency tuned single-ended switching power amplifiers," IEEE J. of Solid-State Circuits, pp. 168-176, 1975.

[19] F. H. Raab, "Idealized operation of Class E tuned power amplifier," IEEE Trans. Circuits and Systems, pp. 725-735, 1977.

[20] J. Ebert and M. K. Kazimierczuk, "Class E high-efficiency tuned oscillator," IEEE Journal of Solid-State Circuits, 62-66, 1981.

[21] M. K. Kazimierczuk, "Collector amplitude modulation of the Class E tuned power amplifier," IEEE Trans. Circuits and Systems, pp. 543-549, 1984.

[22] M. K. Kazimierczuk and K. Puczko, "Exact analysis of Class E tuned power amplifier at any Q and switch duty cycle," IEEE Trans. Circuits and Systems, pp. 149-159, 1987.

[23] M. K. Kazimierczuk and W. Tabisz, " Class C-E high-efficiency tuned power amplifier," IEEE Trans. Circuits and Systems, pp. 421-428, 1989.

[24] T. Suetsugu and M. K. Kazimierczuk, "Comparison of Class E amplifier with nonlinear and linear shunt capacitance," IEEE Trans. Circuits and Systems- I, pp. 1089-1097, 2003.

[25]] T. Suetsugu and M. K. Kazimierczuk, "Analysis and design of Class E amplifier with shunt capacitance composed of linear and nonlinear capacitances," IEEE Trans. Circuits and Systems- I, pp. 1261-1268, 2004.

[26] D. Kesser and M. K. Kazimierczuk, "Power losses of Class E amplifier with any duty cycle," IEEE Trans. Circuits and Systems-I, pp. 1675-1689, 2004.

[27] T. Suetsugu and M. K. Kazimierczuk, "Design procedure for Class E amplifier for off-nominal operation at 50% duty cycle," IEEE Trans. Circuits and Systems-I, pp. 1468-1476, 2006.

[28] T. Suetsugu and M. K. Kazimierczuk, "Off-nominal operation of Class E amplifier at any duty cycle," IEEE Trans. Circuits and Systems-I, pp. 1389-1397, 2007.

[29] H. Sekiya, N. Sagawa, and M. K. Kazimierczuk, "Analysis of Class-DE amplifier with linear and nonlinear shunt capacitances at 25% duty cycle," IEEE Trans. Circuits and Systems-I, pp. 2334-2342, 2010.

[30] Sanders, S.R.; Verghese, George C., "Synthesis of averaged circuit models for switched power converters" Circuits and Systems, IEEE Transactions, pp. 905 - 915, 1991.

[31] Singer, S.; Erickson, R.W., "Control implied input/output buffering of power conservative two port networks" IEEE ISCAS, Vol. 4 pp. 1911 - 1913 vol.4, 1992.

[32] Kislovski, A.S.; Redl, R., "Generalization of the injected-absorbed-current dynamic analysis method of DC-DC switching power cells", IEEE ISCAS, Vol. 4 Pages: 1895 - 1898, 1992.

[33] F. Guinjoan ; A. Poveda ; L. Martinez ; L. G. Vicuna, "An accurate small-signal modelling approach for switching DC-DC converters", IEEE ISCAS, 1993.

[34] Special session "new trends in switching power converters towards circuit integration", IEEE ISCAS (Bangkok, Thailand) - (E. Alarcón, UPC and L. Fortina, Università di Catania), 2003.

[35] Special session "Monolithic integration of switching power regulators: trends in On-chip power management techniques" IEEE ISCAS (E. Alarcón, UPC, D. Czarkowski, Polytechnic University, Brooklyn, US and A. Patra, Indian Institute of Technology, Kharagpur, India), 2006.

[36] Chi Yat Leung, Philip KT Mok, Ka Nang Leung, Mansun Chan, "An integrated CMOS current-sensing circuit for low-voltage current-mode buck regulator", IEEE Circuits and Systems II, pages 394-397, 2005.

[37] Suet-Chui Koon, Yat-Hei Lam, Wing-Hung Ki, "Integrated charge-control single-inductor dual-output step-up/step-down converter", IEEE ISCAS pp 3071, 2005.

[38] Biranchinath Sahu, Gabriel A Rincon-Mora, "An accurate, low-voltage, CMOS switching power supply with adaptive on-time pulse-frequency modulation (PFM) control" IEEE Transactions on Circuits and Systems-I, pp. 312-321, 2007.

[39] K. H. Chen, H. W. Huang and S. Y. Kuo, "Fast-Transient DCDC Converter With On-Chip Compensated Error Amplifier", IEEE Transactions on Circuits and Systems-IIpp 1150 1154, 2007.

[40] Vahid Yousefzadeh, Eduard Alarcón, Dragan Maksimović, "Efficiency optimization in linear-assisted switching power converters for envelope tracking in RF power amplifiers", IEEE ISCAS, pp 1302-1305, 2005.

[41] G.A. Rincon-Mora, "Introduction to the Special Section on Energy-Harvesting/Scavenging Circuits and Systems", IEEE Transactions on Circuits and Systems-II, pp. 785-786, 2011.

Eduard Alarcón
Technical University Catalunya, UPC BarcelonaTech, Spain

Marian K. Kazimierczuk
Wright State University, OH, USA

Adrian Ioinovici
Holon Institute of Technology, Israel

Control Theory

Control theory studies the fundamental mechanisms that regulate and ensure the correct function of almost every system and device surrounding us. It addresses one of the oldest dreams of mankind – that of making systems and devices behave in a desired and reliable manner without any direct human intervention. Its development is intertwined with the technological and social progress of humanity and its roots can be traced back to the ancient civilizations of Greece and Rome.

Indeed one of the first example of control is Ktesibios' water clock (or clepsydra) that appeared in the ancient city of Alexandria (Egypt) around 250 BCE as a functional device to measure the passage of time (see Fig. 1). In particular, Ktesibios (285-222 BCE), who is thought to be the first director of the Museum of Alexandria, was faced with the problem of ensuring a reliable outflow of water droplets to measure time; the problem being that the rate of flow increased when there was more water in the device while, as it emptied, the decrease in pressure slowed the dripping. To solve this problem, Ktesibios proposed "a three tier system in which a large body of water emptied into the clepsydra to insure it remained full" [1] establishing a feedback between the water level in the device, as sensed by a floating device, and the valve regulating the inflow. His solution to this problem was so effective that water clocks remained the best way to measure time until the invention of the pendulum clock by the Dutch physicist and polymath Christiaan Huygens in the 17th century.

After such an initial success, control went into a long hybernation that lasted till the 17th century, apart from some applications of control for entertainment purposes with the construction of the mechanical automata showcased around

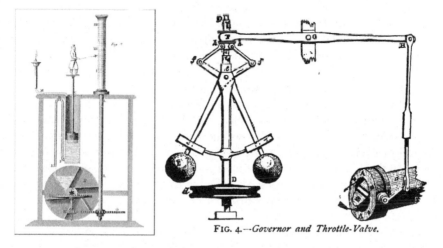

FIG. 4.—*Governor and Throttle-Valve.*

Fig. 1 – Ktesibios' clock (left panel) and Watts' flyball governor (right panel). [Wikipedia]

Europe's royal courts in the Reinassance and Baroque age. Then, in 1620 Cornelius Drebbel (1527-1633), a Dutch immigrant to England, invented the Thermostat, a device able to maintain temperature inside a box at a desired value. Later in the 18th century, with the onset of the industrial revolution in England, new technological challenges pushed the invention of new control devices. Most notably, the first industrial application of control arose in 1788 with James Watt's invention of the Flyball governor to control a steam engine.

In his ingenious design (see Fig. 1), James Watt (1736-1819) connected the rotating shaft of the steam engine via a belt or chain to a spindle with two masses mounted on lever arms. As the rotating speed of the engine increases, the spindle rotates faster so that the two masses are subject to a higher centrifugal force. As the masses move apart, they move a link to reduce the aperture of a throttle valve feeding the engine with steam so that the engine shafts slows down. Vice versa, slowing the motion of the engine causes the masses to rotate slower and hence move towards each other causing the steam valve to open thereby pushing up the engine rotational speed. In so do-ing, the rotational speed of the engine

Portrait of James Watt (1736 -1819) by Carl Frederik von Breda

shaft is controlled preventing unwanted speed fluctuations and overspeeding.

Later in the 19th century it was the same principle Ktesibios had used back in ancient Greece that led the British plumber Thomas Crapper (1836-1910), later knighted by Queen Victoria, to revolutionize the design and reliability of flushing toilets making them the common household device they are today [2]. His invention of the ballcock (or float valve) consists of a simple mechanical feedback control mechanism whereas a float mounted at the end of a lever controls the opening or closing of a valve providing the required inflow to fill a water tank to a desired level. With this invention, Control Technology entered our daily lives and home automation was officially born.

In all the applications described so far, control design had been entirely a matter of intuition and practical skills. In 1840 the British Astronomer Royal, George Biddell Airy (1801-1892), was the first to use differential equations to solve the problem of making a telescope at Greenwich move in a controlled way in order to track a celestial object of interest. Mathematical Control Theory was born and the use of mathematics, also thanks to the later work of James Clerk Maxwell (1831-1879) on governors in 1868, became from then on a fundamental part of any control design exercise.

With the advent of electricity, the need for power supply to be reliable and for it to be kept constant motivated further applications of control theory and technology. Also new devices for meaurement, signal transmission and actuation were invented, which control engineers started to utilize.

The first decades of the 20th century saw the development of feedback control strategies for voltage, current and frequency regulation, speed control of electric motors, and the invention of flight itself depended upon mechanisms for steering and auto-stabilization. As the breadth and scope of control applications increased, the need became apparent for better analysis and design methods. This led to the pioneering work of Nicholas Minorksy (1885-1970) who was the first to present a clear analysis of the classical proportional-integral-derivative (PID) controllers as applied to position control systems. Control solutions were also fundamental to solve the amplification problem that was a serious obstacle to long-distance communication via telephone. The solution proposed by Harold Stephen Black (1898-1983) in 1927 was a negative feedback amplifier where feedback was used to reduce distortion due to noise and component drift. It was the work on the negative feedback amplifier that led Harry Nyquist (1889-1976) to set the foundations in the 1930s to the analysis and design tools in the frequency domain that now bear his name and did not require any explicit use of differential equations.

Later extensions of the methodology by Hendrik Bode (1905-1982) in the 1940s established it as a fundamental tool for control analysis and design, which is still used today. At around the same time, work on PID controllers by John G. Ziegler (1909-1997) and Nathaniel B. Nichols (1914-1997) gave the world their famous empirical rules for tuning the gains of such controllers, which have been used in industry since their first appearance as a practical, yet powerful, control solution for a number of different applications. Other relevant work of this period includes the pioneering work on relay systems by Aleksandr Andronov (1901-1952) and his school in Nizhny Novgorod (then

Hendrik Wade Bode (Wikipedia)

known as Gorkii), and later Leonid Tsypkin in Moscow, which established the foundations of what was to become nonlinear control theory [3].

During the Second World War, with the invention of the radar, automatic control became essential for the detection and tracking of aircraft and for the design of automatic servo-mechanism for fast positioning of anti-aircraft guns. Work in this direction saw the interdisciplinary collaboration between mechanical, electrical and electronic engineers that brought about new developments in Control Theory and the understanding of how to integrate different systems and devices to improve control systems design. By this time both frequency domain tools based on the use of Nyquist and Bode diagrams and Nichols charts as well as tools based on the use of Laplace transforms (the s-domain) had become well established. Also Russian work in the 1930s started to appear in translated books and articles in the West.

Then came the Space Race and the Cold War of the 1950s and 1960s and the need to develop controllers for missile control systems to address questions such as how to control nonlinear systems and how to control something in the "best" possible way. The use of Lyapunov-based methods was introduced and the pioneering work of Lev Pontryagin (1908-1988), Richard Bellman (1920-1984) and others laid the foundations of Optimal Control. With the pioneering work by Rudolf Emil Kalman in the late 1950s on multivariable control, controllability, observability, and his development together with Bucy of what we now call the Kalman filter, the theory of Modern Control Systems was born. State Space methods became quickly more popular than those in the frequency domain, although these latter techniques returned back into fashion in the late 1960s with their extension to multivariable control and their relevance in industrial applications.

The introduction of digital systems in the 1950s also brought about major changes in Control Systems, that continued through the 1960s and flourished in the 1970s with the introduction of microprocessor-based digital controllers. Also in the mid 1960s programmable logic controllers (PLCs) appeared to replace electromechanical relay systems. In the 1970s and the 1980s, the introduction of digital computers allowed the development and implementation of more sophisticated control approaches such as stochastic, robust, adaptive, switching and optimal control algorithms.

Sojourner (rover), Wikipedia, www.nasa. gov/mission_pages/mars-pathfinder/

The 1990s saw the emergence of intelligent control strategies based on the use of neural networks and fuzzy logic and the launch of the first ever autonomous rover, Sojourner, which arrived on Mars on 4th July 1997 and autonomously operated for 12 days (longer than the 7 days that were originally planned). Also, switched and hybrid control systems became popular as extensions of the early work on variable structure systems by Alksei Fedorovich Filippov (1923-2006) and Vadim I. Utkin in the late 1960s and 1970s, and a lot of interest emerged particularly within the IEEE Circuits and Systems Society as well as the Control Systems Society on the analysis and control of nonlinear and chaotic systems.

With the beginning of the new millenium and the availability of cheap

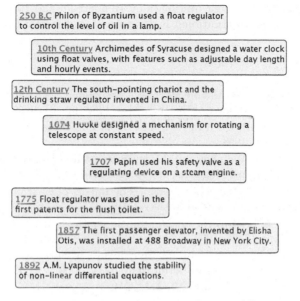

Notable steps in the history of control [7].

and reliable microcontrollers, control strategies became embedded everywhere from washing machines to digital cameras. As control systems turn out to be increasingly crucial in diverse areas such transportation systems (airplanes, automatic cars), robotics, power electronics, and even medicine and the life sciences, new applications are arising that call for new control solutions.

Today the paradigm has quickly shifted from the control of single systems or devices to the control of networks of interacting agents. Examples from applications include control of congestion in the Internet, flocking and swarming robots, traffic control, power distribution systems, and cell populatons. In all of these examples, control is distributed and decentralized. Agents in the network need to communicate with each other in order to steer their collective behaviour in a desired manner. New control problems have emerged in this context such as that of achieving consensus or synchronization of all agents in the network to some desired common target state, or that of investigating the controllability or observability of a given network. Many challenges and open problems remain that make Control Theory a fascinating and stimulating area of research which never ends to surprise.

Control Theory and Technology and its applications have been an important area of research for members of the IEEE Circuits and Systems Society over the years, particularly on interdisciplinary topics in the area of nonlinear circuits and systems, and complex networks [4]. So it is particularly relevant to celebrate here Control and its History with an outlook to its promising applications along with its challenging interdisciplinary open problems.

This brief historical overview is loosely based on a number of different sources including the excellent overviews by Bennett [5] and Bissell [6], and the history website of the International Federation of Automatic Control [7].

References

[1]   L. Nocks, "The robot: the life story of a technology", Greenwood Publishing Group, 2007." Proceedings of the IRE, May 1962, pp. 872-878.

[2]   W. Reyburn, "Flushed with Pride: The Story of Thomas Crapper", Trafalgar Square Publishing, 1998.

[3]   C.C. Bissell, "A.A. Andronov and the development of Soviet control engineering", IEEE Control Systems Magazine, 18, pp. 56–62, 1998.

[4]   T. Samad and A.M. Annaswamy (eds.), The Impact of Control Technology, 2nd edition, IEEE Control Systems Society, 2014

[5]   S. Bennett, "A brief history of automatic control", IEEE Control Systems Magazine, 16, pp.17–25, 1996.

[6]   C. C. Bissell, "A history of automatic control", in Springer Handbook of Automation, pp. 53-69. Springer Berlin Heidelberg, 2009.

[7]   http://controlrc.ifac-control.org/history/history-of-control.

Mario di Bernardo

University of Naples Federico II, Italy

University of Bristol, U.K.

SPICE and Compact Modeling

The first Silicon Integrated Circuits (IC) developed at the beginning of the 1960's were requiring a new design approach compared to the printed-circuit-board (PCB) electronic circuits built with discreet components. While the latter could be verified by breadboards and corrected through trial and error, the former needed to go through many process steps to be corrected if not fulfilling the initial design specification. The non-working Silicon was thrown away adding to the cost and lengthening the time-to-market of an IC.

A new design approach was needed to get to first-working Silicon. It was soon realized that the complexity of accurately solving the equations representing the electronic circuit could be addressed only by using the aid of a computer. About the same time in the 1960s powerful scientific computers such as the Control-Data Corporation (CDC) 6600 were available to solve the equations capturing the behavior of ICs. These factors led to efforts to develop computer programs to solve the electric problem in more than one research laboratory around the world.

At the University of California at Berkeley, Don Pederson who in 1962 recognized the overwhelming impact that ICs were going to have and had the first semiconductor fabrication facility being built in a university saw the need for computer-aided design. With Don's support a young professor, Ron Rohrer, organized a class project on Computer-Aided Design (CAD) as part of a graduate class in 1969-70. The goal of this class project was to "produce the best available computer program for the simulation of practical ICs." The main need that the program was going to fulfill was to provide quick access to the operating point of ICs; in the early days of the semiconductor industry linear bipolar ICs such as opamps were holding a key role and their performance was wholly dependent on correct biasing.

The different solution codes developed by the students in this class were assembled by Larry Nagel into a single program called CANCER (Computer-Aided Nonlinear Circuit analysis Excluding Radiation) [1], which performed DC, AC and time-domain analysis. CANCER had a couple of issues stacked against it; first, it was its unappealing name but more importantly it was a proprietary program. This was contrary to Don's philosophy that university research should be in the public domain and contribute to the advancement of science. Additional research, improvements and a new name resulted in the simulator SPICE first released in the public domain as SPICE1 in the fall of 1971. Today, more than four decades and one billion transistors per chip later, SPICE, in its various commercial or public-domain releases, has evolved into an IC design standard used for every single transistor-level circuit design.

SPICE1 introduced a slew of innovations starting with the circuit description language, which today is the de-facto inter-change format among different IC CAD tools; it was complemented by algorithmic innovations such as the incidence matrix circuit representation, node-to-datum voltage solution, sparse-matrix arithmetic, Newton-Raphson iterations and implicit numerical integration. The sequence of these algorithms is commonly referred to as direct-methods solution [2].

SPICE1 was used by students in all circuit classes at UC Berkeley as well as by "friendly" designers in industry. Valuable feedback pointed out shortcomings of this first version of SPICE. The most important ones concerned the component number limitation declared at compile time, the constraint of voltage sources to be grounded (node-to-datum voltage equations), and fixed-time-step integration.

The above shortcomings motivated research leading to SPICE2 first released in 1975. The early versions of SPICE 2A and 2B addressed the latter two issues, node-to-datum voltage equations were replaced by the Modified-Nodal Analysis (MNA), and a variable time-step algorithm was added [3]; the component number limitations was only resolved by Ellis Cohen in SPICE 2D in 1976; Ellis restructured the data structures and wrote a memory management package in Fortran to allocate all available computer memory to the circuit independent of the number of each type of component (transistor, resistor, capacitor, etc.).

At this point SPICE2 was ready for wider use and a procedure was set in place at the Electronics Research Laboratory (ERL), UC Berkeley, to release the program including source code and documentation to any interested party.

The acceptance of the program in industry was however not unanimous; there were many doubters lining up behind Bob Pease who had a column in the Electronic Design magazine entitled "What's All This SPICEY Stuff, Anyhow?" and stated that "there is nothing SPICE can do that I cannot do by hand." It is only human that a disruptive technology meets with some resistance; however, with the number of adopters growing rapidly with the increasing complexity of ICs, most semiconductor companies were using SPICE by the end of the 1970s.

With industry-wide use of the program more improvements and additions became necessary; Don assigned this challenge to Andrei Vladimirescu. The key issue to be addressed was the robustness of the solution determined by the Newton algorithm's convergence, or the lack of it. SPICE versions 2F and 2G added algorithmic refinements such as circuit matrix reordering based on actual conductance values (pivoting), continuation methods for operating point solution and device-specific current-limiting algorithms from Newton iteration to iteration. This made SPICE 2G6 [4] released in 1980, and distributed freely by UC Berkeley, the most robust software program ever produced by a university. The industry-strength SPICE 2G6 served as basis for the initial HSPICE program, which today is the market leader among the commercial SPICE simulators.

Big companies such as Hewlett Packard, Analog Devices, Tektronix, Texas Instruments and others formed their internal CAD groups with in-house SPICE experts Dick Dowell, Steve Hamm, Graham Boyle, Burt Epler who adapted the Berkeley program to their specific needs. During the almost one decade from the latter part of the 1970s to the beginning of the 1980s, communication between Berkeley and the industry SPICE experts was open and continuous where the latter group provided feedback on bug fixes and improvements, which Berkeley introduced in future releases. This was a unique example of collaboration between university research and industry for the benefit of the entire semiconductor community.

Algorithmic innovation in circuit simulation continued with the conception of the first parallel simulator called CLASSIE [5] based on the solution algorithms of SPICE designed for the first supercomputer CRAY-1, a vector (Single-Instruction-Multiple Date, SIMD) architecture, introduced in 1976. This architecture is similar to that of a modern Nvidia graphics processor. The techniques introduced in CLASSIE can be found today in the leading simulators, HSPICE, Spectre and Eldo when running on multi-cores, and FASTSPICE simulators, which take advantage of decoupled Bordered-Block Diagonal Format (BBDF) circuit matrices.

At the end of the 1970's UC Berkeley released the Berkeley Standard Distribution (bsd) Unix operating system, an extension to Unix, first introduced by Bell Laboratories. The C language, intimately connected to Unix, offered among other facilities interactive execution of an application program. Tom Quarles undertook re-architecting SPICE2 and its data structures from Fortran to C; he released SPICE3, an interactive program with graphic capabilities in 1983. Improvements and additions to SPICE3 continued until 1993 when the last version, SPICE 3f5, was released by UC Berkeley as no further research projects could be envisaged since the solution algorithms were mature.

In 1998 Don Pederson was awarded the IEEE Gold Medal for his determinant role in

Don Pederson

the SPICE project, and its adoption as the de facto standard for the design of ICs; he was officially named the "Father of SPICE" on this occasion.

Advancement of the state-of-the-art in the last two decades was carried out by the major EDA companies Synopsys, Cadence and Mentor; their simulators HSPICE, Spectre and Eldo added such features as steady-state and harmonic analysis for RF, hardware description languages, Verilog-AMS and VHDL-AMS, usability features and a slew of new semiconductor device models. The tradition of public-domain open-source SPICE software is continued by such programs as ngspice, QUCS (Quite Universal Circuit Simulator) and Xyce of Sandia Labs.

Good algorithms are only one part of the accuracy equation in circuit simulation; the representation of semiconductor devices in the circuit description is another part. This observation led to the edict that one's simulation results are only as accurate as the models used. In order to introduce the branch-constitutive equations (BCE) of transistors and diodes an analytical formulation of the dependence between terminal voltages and currents of the device was needed; this representation is referred to as the device model.

A semiconductor device model can assume different levels of detail. The most accurate representation is based on the detailed 2- or 3-dimensional (2D, 3D) geometrical and processing characteristics of the device structure; these are models for *device-level* simulators also referred to as *Technology* CAD (TCAD). These simulators solve Poisson's, transport and continuity equations to find the electron and hole concentrations and derive the currents flowing in the device under given external bias conditions.

The above kind of model is impractical for circuit simulation due to the very complex and time-consuming solution needed for every single device. Mathematical models intended for circuit-level simulation that describe the current and charge behavior of semiconductor devices as a function of voltage, process, electrical, environment and geometry parameters are referred to as *compact models.* *Circuit-level* models used in SPICE describe a component relative to its terminals; they are based on the physics of the device but are more abstract and include also some empirical approximations in order to keep evaluation time for each device reasonable.

An important decision in the SPICE project was to make semiconductor device models an integral part of the simulator. This was a very inspired engineering decision as IC designers are not device physicists and they were not going to develop their own device equations. This decision set apart SPICE from a circuit simulation program developed at IBM, ASTAP, which relegated device models to external libraries rather than integrating them into the simulator.

Spice Commemorative Plaque

The first models in SPICE were very simple such as the Ebers-Moll model for the bipolar transistor with only 16 model parameters. As SPICE started being used in industry for the design of actual ICs accurate semiconductor device models were required. Bipolar transistors, on which the majority of ICs were based in the late 1960s and early 1970s, benefited from the introduction of the Gummel-Poon model.

The big merit of this model is that it captured most of the second-order effects such as collector-current dependency of the gain and transit time in a simple and unique formulation as a function of the normalized base charge. The current equations are valid in all regions of operation and have continuous derivatives; this is a very important requirement of device equations for circuit simulation for both the convergence of the Newton algorithm and accuracy of the small-signal analysis based on the first derivative, and computation of linearity and distortion terms based on higher-order derivatives.

With shrinking dimensions and increased requirements for simulation accuracy more geometry and physics detail were added in the MEXTRAM BJT model, also referred to as the 503/504 Philips model. This model considered the three-dimensional aspect of a BJT, segmenting it into several regions such as intrinsic, the vertical slab below the emitter, the extrinsic, base and collector beyond the emitter and pnp transistor to the substrate. This approach allowed one to accurately predict physical effects such as quasi-saturation, base push-out, current crowding and to accurately compute diffusion and depletion charges. A similar approach was taken by the VBIC95 model, which builds the three-dimensional detail of the BJT structure on top of the Gummel-Poon model. VBIC falls back on Gummel-Poon when the parameters representing the additional detail are not present. Another model preferred for power BJTs is HICUM.

In the early 1970's fabrication of MOS ICs became mature first as pMOS-only, which culminated with the first microprocessor, the Intel 4004, and was then followed by nMOS Large-Scale-Integrated (LSI) circuits by the mid 1970's. At this time the SPICE MOSFET model was limited to an augmented Shichman-Hodges model, also referred to as Frohman-Grove, which was inadequate. This model covered only operation in strong-inversion.

Simulation model development has evolved in step with the progress of the fabrication process and device-size scaling as first described by Dennard [6]. Thus, MOS models are customarily classified in five generations. A second classification is based on the physical quantity taken as reference in deriving the current equation; there are three types of models according to this criterion, models based on *threshold voltage* (VT) covering the first three generations, models based on *surface potential*, the fourth generation and *charge*-based models, the fifth generation.

The first generation of models were implemented in the 1970s; they encompass models referred to as LEVEL=1, 2 and 3, developed at UC Berkeley [7]. The LEVEL=1 model is the quadratic model of Shichman and Hodges introduced in 1968, which is applicable to long-channel devices with L of a few micrometers. Today this model is mostly used in textbooks for describing the operation of analog and digital MOS circuits. The LEVEL=2 model is a physics-based model that includes second-order effects such as subthreshold conduction, velocity saturation and drain-induced barrier lowering (DIBL) also referred to as *short-channel* effects, apparent in MOS devices when the channel length is reduced to 1 μm and below. The LEVEL=3 model represents a semi-empirical abstraction of the LEVEL=2 model obtained by reducing the complexity of the I-V equations, eliminating time-consuming floating-point operations such as power and transcendental functions, and defining only one or two model parameters for each physical effect present in LEVEL=2, thus speeding up evaluation and facilitating model parameter extraction. While the LEVEL=2 is rarely used today, the LEVEL=3 model is of interest for a quick first-order estimation of circuit performance due to its simplicity, reduced number of parameters, and ability to reasonably match the I-V characteristics of submicron devices.

The second-generation models represented by BSIM1 (LEVEL=4) (1985) and BSIM2 (LEVEL=5) (1990), acronyms for Berkeley Short-channel Insulated-gate field-effect transistor Model, started out to correct two shortcomings of the first models: improve the robustness of device equations and add the impact of geometry aspect ratios and scaling on the characteristics of MOSFETs. They were inspired by a semi-empirical model developed at ATT Bell Laboratories, CSIM (Compact Short-channel Igfet Model), which was similar in concept to LEVEL=3.

About this time with an important effort being dedicated to the completion and release of SPICE3 and other projects like the mixed-mode simulator SPLICE and layout editor KIC going on in the CAD group, Don Pederson realized that compact model development required a dedicated effort and skill set and asked a young faculty member, Chenming Hu, to oversee this activity; the different generations of BSIM models were developed by his research students.

The mathematical conditioning of the device equations led to the introduction of many new parameters and gave an empirical character to these models making parameter extraction more difficult due to the disconnect between parameters and underlying physics. The first objective, robustness, was not attained in the public domain formulations leading to the development of the proprietary HSPICE LEVEL=28 model, the only solid implementation of BSIM1. This development set the precedent for proprietary device models that locked designers into a specific commercial simulator; fortunately for the SPICE user community, this trend was reversed in the third-generation MOS models. The second-generation models are not used anymore except perhaps for old designs, where the only available process parameters are for these models.

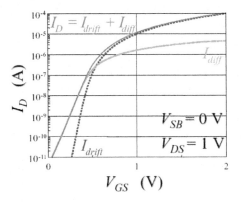

Drain current of MOSFET in weak and strong inversion. (Source: Philips).

The biggest challenge for capturing the behavior of a MOS transistor in a single analytical expression covering all regions of operation is the different physical nature of the drain current in weak inversion: a diffusion current with exponential voltage dependency and in strong inversion, a drift current with linear or super-linear voltage dependency, see the figure on the left. It had been recognized ever since the weak inversion region was added to LEVEL=2,3 that the only way to achieve a single expression with continuous first- and higher-order derivatives was to base the model on the surface potential. The difficulty arose from the fact that the surface potential cannot be expressed directly as a function of terminal voltages and a second level of iterations would have been required to solve the implicit equation. A practical solution to this challenge did not come until the PSP model was introduced in the fourth generation.

The shortcomings of the above models led to the development of two new third-generation models in the public domain, BSIM3 [8] at UC Berkeley, and MOS9 at Philips Research. The purpose of these new models was to unify the MOSFET current equation into a single analytical expression over all regions of operation for improved accuracy and simulation robustness. The unified current expression was obtained by mathematical means rather than physical fundamentals. BSIM3 [8] was released as Version 2 and 3, with BSIM3v3 being the de-facto standard for process characterization by all foundries over a period of more than ten years, covering technologies from 0.6μm down to 65 nm. As BSIM3 covered several technology generations, upgrades in terms of supported physical effects such as gate current, became necessary leading to the introduction of BSIM4 in 2000. BSIM4 is a superset of BSIM3.

Another third-generation model based on the time of its introduction is the EKV MOS model,, named after its three authors, Enz [12], Krummenacher and Vittoz. Derived from the charge-sheet formulation it is the only third-generation model, which is not VT based but charge-based; the EKV model was adopted by many analog designers due to its physical derivation and relative simplicity compared to BSIM3.

These three generations of MOS models are also referred to as *threshold-voltage based* models; the name derives from a very elaborate analytical formulation of VTH that gathers all first- and second-order effects for getting an accurate estimation of the drain current. With the reduction in dimensions of the MOSFET structure below $0.1\mu m$, accurate modeling needed to get closer to device physics, which led to the fourth generation, or *surface-potential based models* such as PSP and HiSIM. Expressing the MOSFET current as a function of the surface potential is a sure way to have a single, physics-based, analytical formulation for all regions of operation.

A number of shortcomings and inaccuracies of the BSIM3/4 have been addressed with the introduction of the BSIM6 model [13] in 2011; this is a charge-based fifth generation model resulting from the collaboration between the device research groups at UC Berkeley and at the Ecole Polytechnique Federale de Lausanne (EPFL). BSIM6 builds upon the ideas of the EKV and corrects the shortcomings of its BSIM predecessors by its symmetry, continuity of high-order derivatives and physics-based derivation. Fifth generation models address also the new Multi-Gate (MG) device structures, FinFET and Fully-Depleted Silicon-On-Insulator (FDSOI), introduced for process nodes of 28nm and below. A model, which covers both structures is BSIM-MG with its two versions, BSIM-CMG (Common MG) for FinFETs and BSIM IMG (Independent MG) for FDSOI. BSIM-MG is a surface-potential-based model similar to the ST-Microelectronics model for FDSOI available in commercial SPICE simulators. The chronology of the BSIM family of MOSFET models is shown in the below figure.

It is important to note that the main semiconductor device models in use today are in the public domain and are available in the main commercial SPICE versions. A user is supposed to get the same results for the same set of model parameters of a given model no matter which simulator supporting that model is used.

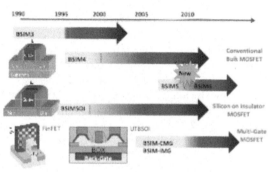

BSIM Chronology [www-device.eecs.berkeley.edu/bsim].

An important role in promoting the work of research groups engaged in developing compact models for circuit simulation has had, for more than a decade, the Compact Modeling Council (www.cmc.org) and the MOS-AK Compact Modeling (CM) Group driven by its animator Wladek Grabinski. This is a forum of volunteers as well as representatives from IC companies and universities, where new models and improvements to existing ones are discussed, evaluated and recommendations are issued to the fabrication and design communities.

It is also important to acknowledge the contributions of other people and groups to semiconductor device modeling. An important contribution to MOS-FET modeling was brought by the device group at the University of Santa Caterina in Florianopolis, Brasil, which developed the ACM compact model. A leading text dedicated to MOSFET operation and modeling was contributed by Yannis Tsividis [9] from Columbia University. Last but not least, the contribution to a mixed circuit-device modeling approach was made by the CODECS project at the University of California Berkeley [10] foreseeing the importance that a mixed device-circuit simulator will acquire for designing increasingly larger semiconductor memories.

References

[1] L. W. Nagel and R. Rohrer. "Computer Analysis of Nonlinear Circuits, Excluding Radiation (CANCER)." IEEE J. Solid-State Circuits, Vol. SC-6. Aug. 1971.

[2] A. Vladimirescu, The SPICE Book, J. Wiley and Sons, New York, NY, 1994.

[3] L.W. Nagel. "SPICE - A Computer Program to Simulate Semiconductor Circuits", ERL Memo No. ERL-M520. University of California. Berkeley, May 1975.

[4] A. Vladimirescu, et al., SPICE Version 2G User's Guide, University of California, Berkeley, August 1981.

[5] A. Vladimirescu, "LSI Circuit Simulation on Vector Computers," UCB/ERL Memo M82/75, Univ. of California, Berkeley, Oct. 1982.

[6] R.H. Dennard, et al., "Design of ion-implanted MOSFET's with very small physical dimensions", IEEE Journal of Solid State Circuits, Vol. SC9, Oct. 1974.

[7] A. Vladimirescu and S. Liu. The Simulation of MOS Integrated Circuits Using SPICE2, UCB/ERL Memo M80/7, Univ. of California, Berkeley, Feb. 1980.

[8] Y. Cheng and C. Hu, MOSFET Modeling and BSIM3 User's Guide, Kluwer Academic Publishers, 1999.

[9] Y. Tsividis and C. McAndrew, Operation and Modelling of the MOS Transistor, 3rd Ed., Oxford University Press, 2010.

[10] K. Mayaram, CODECS : A Mixed-Level Circuit and Device Simulator, UC Berkeley, Techinical Report No. UCB/ERL M88/71, 1988.

[11] IEEE Solid-State Circuits Magazine, SPICE Commemorative Issue, 2nd Printing, Vol. 3, No. 2, 2011.

[12] C.C. Enz and E.A. Vittoz, Charge-Based MOS Transistor Modeling: The EKV Model for Low-Power and RF IC Design, J. Wiley, 2006.

[13] Y.S. Chauhan et al., BSIM6: Analog and RF Compact Model for Bulk MOSFET, IEEE Trans. ED, vol. 61, nr. 2, 2014.

Andrei Vladimirescu

Department of EECS, University of California, Berkeley

Corsi e Ricorsi: The EDA Story

I dedicate this paper to the memory of Don Pederson and Richard Newton, two leaders of the field, who are no longer with us and to the many friends/colleagues and students with whom I had the great pleasure to work.

Preamble

The paper is a reflection on the development of the EDA field from its early days to its explosive growth and to its present maturity. It is based on "The Tides of EDA", a paper that was published in 2003 by the IEEE Design and Test Magazine after a keynote given in 2003 on occasion of the 40th anniversary of the Design Automation Conference.

This paper is complementary to the next one written by Lucio Lanza on the business of EDA industry. In this paper, I give some insights also on the technology transfer aspects from research to industry trying to focus on parts which will not be covered by the Lanza paper.

The Ages of EDA

In this period, the foundations of EDA were laid out. While browsing the early-day EDA papers, it is possible to find seminal papers that have still a strong impact today. I classified them by relevant topics and I clustered fundamental contributions in six areas: circuit simulation, logic simulation and testing, MOS timing simulation, PCB layout systems, wire routing, and regular arrays.

The Early Ages (1964-1978)

In this period, the foundations of EDA were laid out. I was surprised while browsing the early-day EDA papers, to find seminal papers that have still a strong impact today. I classified them by relevant topics and I clustered fundamental contributions in six areas: circuit simulation, logic simulation and testing, MOS timing simulation, PCB layout systems, wire routing, and regular arrays

Relevant Technologies

Circuit Simulation

Circuit simulation has been an important topic for EDA especially in the area of IC design. I argue that the great success of circuit simulation in important IC designs has been a dominant reason for the birth of EDA as an industry. IBM researchers were the most prominent force in these years: Frank Branin was a pioneer in determining the architecture of a circuit simulation. In my opinion, Brayton, Hachtel, and colleagues at IBM T. J. Watson Research Center in Yorktown, who introduced all the algorithms that are behind circuit simulation, as we know it today, from sparse matrices to backward differentiation formulae, made the fundamental revolutionary contributions to this field. They enabled the development of two important programs like ASTAP at IBM and SPICE at Berkeley, where the early contributions of Ron Rohrer and Don Pederson were essential in making this program the work-horse for circuit simulation for years and years.

Logic Simulation and Testing

Logic simulation was in use already for some time when DAC started. Computer companies, such as IBM and CDC, were relying on this technology to debug their logic design. The contributions to this field were countless. The founding fathers of the field had also a fundamental role in developing automatic test pattern generation and fault simulation. The roles of Ulrich, Hayes at Michigan, Breuer at USC, and Szygenda at UT Austin, to name a few, were invaluable. The invention of the LSSD methodology by Tom Williams and colleagues and the consequent development of the D-algorithm by Paul Roth of IBM changed the way in which computer hardware was designed and test patterns generated.

MOS Timing Simulation

Exploiting the quasi-unidirectional characteristic of a MOS transistor to speed up circuit simulation first appeared in the work of Hermann Gummel and colleagues of Bell Labs in 1975. Gummel's intuition was to approximate the solution of the ordinary differential equations describing the circuit equations with a fast relaxation-based heuristic algorithm. The insight was to recognize that when analyzing the timing of a digital circuit, it is not important to have accurate waveforms as long as the switching events could be correctly placed in time. Fast circuit simulation is still based today on this idea.

Wire Routing

Most of the techniques used today in tools are based on the results that were obtained in this period. The infamous Lee maze router developed in 1961, earlier than the first Design Automation Conference, is still the basis for most of the routers that are in use today. The Hightower line extension algorithm dates back to 1969, and the idea of channel routing came in 1971 with the work by Hashimoto and Stevens.

Regular Arrays

With the advent of large-scale integrated circuits, the appeal of regular layout patterns for reducing design time was strong. Gate-arrays (called master slices at IBM) and standard cells (master images at IBM) were developed as alternatives to custom layouts. At IBM and Bell Labs, integrated tools for the automatic layout of circuits using these design styles were extensively used in this period. I remember the beginning of the idea when a gate array integrated circuit had four gates and a researcher from Schlumberger discussed the use of automated tools to optimize the use of the circuit. Look at the distance we have covered since then: in forty years, from four to more than one billion gates on a single chip.

The Importance of CAD in Industry

CAD was considered to be of strategic value to system industry, and in particular, to the computer industry. IBM and the other large companies dealing with computer design considered it so important to warrant a sizeable investment in funding and resources. Internal CAD groups were powerful organizations in engineering. Looking at the DAC proceedings, we also notice a strong presence of the Japanese computer and communication system industry in papers describing unified approaches to system design using tools, thus showing the strategic value of the technology worldwide.

During this period, the first generation CAD companies were founded. Applicon was established in 1969, Calma in 1970, Computer Vision in 1972. They all supported pretty much the same design activity: artwork editing on customized workstations. Their business model was centered on the sale of the workstations; the software was considered an add-on to the hardware. Of these first generation CAD companies, none is alive today. We can identify several causes for their demise:

- The architecture of their products: customized hardware with complex software written mostly in assembly code. It was very difficult for them to keep abreast of the technology advances because of the investments needed to design novel workstations with severe software porting problems and limited sale volume to support the investments.

- Limited customer loyalty due to the perceived small value added of the products.

- Limited understanding of market evolutions and customer needs.

These factors contributed to complete lack of innovation capabilities and eventual obsolescence.

The Golden Age (1979-1993)

Between 1979 and 1993 the EDA field exploded in all its aspects. So many essential contributions were made in all aspects of EDA from physical verification to layout synthesis, from logic synthesis to formal verification, from system level design to hardware acceleration. The technical community expanded to reach areas of expertise in nonlinear and combinatorial optimization, control, artificial intelligence, and logic. In these years, the most successful EDA companies were founded. The most prominent research groups hired a number of PhDs in EDA; there was strong pressure from students to enter this area.

Relevant Contributions

Verification and Testing

In this field, there were two main lines of work: one was focused on making circuit simulation orders of magnitude faster than with SPICE, the other on formal techniques to prove that the circuit would perform the correct function. In the first domain, the work on relaxation-based techniques and mixed-mode simulation by, among others, Lelarasmee Newton, Ruehli and me around 1980, posed the basis for the fast MOS simulators in use today. Following this period, Jacob White, fully characterized the waveform-relaxation algorithm and studied acceleration techniques that made this approach one of the most studied even in numerical analysis circles.

Simulation for digital circuits with accuracy falling in between the ones that could be achieved by circuit and logic simulation, but with two orders of magnitude or more speedup versus SPICE yielded the work on MOSSIM by Randy Bryant in 1980. Interconnects started showing their impact as geometries scaled down. The interconnect delay model introduced by Penfield and Rubinstein in 1981 was extensively used. Interconnect simulation was tackled by Pileggi and Rohrer at CMU with their results on asymptotic waveform evaluation (AWE) in 1988.

Formal techniques first aimed at answering the question whether two different networks of gates computed the same Boolean function. The first use of formal verification was again at IBM by Bahnsen during the early ages. The seminal work by Randy Bryant on Binary Decision Diagrams in 1986 revolutionized the field by introducing a canonical form for Boolean functions and very fast manipulation algorithms. The work by Coudert and Madre on finite-state machine equivalence using BDDs, and the one by Ed Clarke, Joseph Sifakis, Ken McMillan, Dave Dill and Bob Kurshan on model checking in the early 1990s, elevated formal verification to higher levels of abstraction. They tackled the problem of verifying whether a sequential system represented by an FSM would satisfy a property described with a logic proposition defined on the states and transitions. In the testing area, the famous PODEM-X program based on the D-algorithm saw the light at IBM under the direction of Prabhu Goel in 1981.

Layout

At the representation and basic manipulation level, artwork editing saw a fundamental change. In the previous period, designers used different data repositories for each of the design steps and went through a great deal of tedious translation work from one data format to another. The work at Berkeley in the early 1980s by Newton with Squid, Oct, VEM and Ousterhout with the Magic systems revolutionized the field by showing that it was possible to have a unified database and graphical user interface. The impact of this work cannot be overemphasized. The SI2 Open Access initiative is based on the seminal work of Newton; some of the data representation for advanced physical design is based on corner stitching introduced by Ousterhout.

Another revolutionary idea was brought forward by the work on silicon compilation and layout languages around the same period at MIT (John Battali), CalTech (Carver Mead's group with the Bristle Block silicon compiler) and Bell Labs (Hermann Gummel and colleagues, Misha Buric and colleagues with the L layout language). The idea was to inject some parts of computer science culture into IC design. While the approach was intellectually elegant and powerful, there is not much left in present tools.

Kirkpatrick and Gelatt, two physicists who used their knowledge of spin systems to develop this technique, introduced *simulated annealing* at IBM to solve a placement problem for gate array layout, in 1980. Once the approach was made public, a great deal research on efficient implementation of the algorithm and on its theoretical properties started. Customization of the algorithm took place for standard cell and macro cell layout as well as for global routing. Most of the major companies, from Intel to DEC, from Motorola to TI used the TimberWolf system written by Sechen for the layout of standard cells followed by a version for macro-cell layout. In these years, Fabio Romeo, Debasis Mitra, I, and most notably Greg Sorkin with his seminal thesis contributed to the understanding of the mathematical properties of the simulated annealing algorithm going beyond the analogy to spin systems and laying the foundations for statistical optimization techniques.

In these years, we also saw a great deal of interest from theoretical computer scientists towards the combinatorial aspects of layout design. The work of Rivest and Pinter on routing at MIT, of Karp and students on placement and routing at Berkeley were examples of this involvement; a clear sign of the success that the EDA field was having in attracting other communities to contribute. In these years, the work of Kuh at Berkeley yielded integrated layout systems for macro cell design and point tools for placement and routing of great impact in the field.

Logic Synthesis

Logic synthesis was introduced in the EDA vocabulary in 1979 with a seminal paper by Darringer, Joyner and Trevillyan who used peephole rule-based optimization to generate efficient gate-level representations of a design. Immediately following this work and somewhat independently, another approach to Boolean optimization began at IBM in 1979 with Brayton and Hachtel in collaboration with Newton and me at Berkeley with the two-level logic optimizer ESPRESSO, and the multi-level logic optimizers Yorktown Silicon Compiler and MIS system. This work, supported by DARPA, spanned a period of over ten years. During the early stage of development of MIS, adopted by all major companies including Intel, ST, TI, Motorola, Honeywell, DEC, and Philips, a technology independent phase where Boolean functions were manipulated and optimized, was followed by a technology-mapping step that would map the optimized Boolean function to a library of gates. For this second phase, the common approach was to use rule-based techniques like in the SOCRATES system developed at GE by Aart DeGeus and colleagues. When I met Kurt Keutzer during a talk at Princeton in 1982, he showed me the use of the work by Aho and Ullman on compilation to obtain a very efficient technology mapping. The idea was to formulate the problem as a tree-covering problem and use dynamic programming to solve it; a great idea that made it in most of the logic synthesis systems in use today. The work in Japan at Fujitsu, NTT and NEC was outstanding in producing working logic-synthesis systems that were mostly based on the original work of Darringer at IBM.

Logic synthesis was a great achievement of our community and as such, it originated a large number of papers and interests in other connected fields. The work by Keutzer, Devadas, Malik, McGeer and Saldanha in collaboration with Brayton, Newton and me showed a number of results in redundancy removal and delay testing using logic synthesis techniques. New testing algorithms came out with V. Agrawal, Tim Cheng and others that showed the contamination of logic synthesis and testing two fields.

In the mature period of logic synthesis, Coudert and Madre were able to speed up considerably logic synthesis algorithms based on CNF representations of Boolean function by using BDDs.

Hardware Description Languages

The technical work that has been characterized as logic synthesis should have been better classified as logic optimization since the algorithms were changing a Boolean representation of a digital circuit into an optimized, equivalent one. Synthesis implies a bridge between two layers of abstractions. Hardware description languages were born to represent digital systems more efficiently and compactly than Boolean functions. Then the real synthesis job was to map an HDL description into a netlist of gates. Unfortunately, the development of HDLs at the beginning was independent from the work on logic optimization. This implied that not all constructs of HDLs such as Verilog, proposed by Moorby and colleagues, and VHDL could be tackled by logic synthesis algorithms. We had then to restrict the use of these languages to what was called their *synthesizable subset*. While HDLs were indeed a great advance for reducing design time by introducing verification early in the design cycle, the need for sub-setting showed that they did have problems on the semantic side.

The HDL battle was a very interesting one during these years: Verilog was a proprietary language (Gateway Design was selling a Verilog simulator and licensed the language) while VHDL was born as an open standard supported by DARPA within the VHSIC program. We had countless debates about the superiority of one language over the other at DAC. When Verilog was made public, there were cosmetic differences between the two even though you would find people strongly attached to one versus the other according to their personal tastes. Joe Costello in his key note talk at the Design Automation Conference in 1991 is very eloquent in arguing that the adoption of two standards for a single task is in general a bad idea:

"...Adoption of VHDL was one of the biggest mistakes in the history of design automation, causing users and EDA vendors to waste hundreds of millions of dollars..."

Hardware Acceleration

All EDA approaches require a massive amount of compute time to execute complex algorithms on very large sets of data. In a great expansion period of the technology, the appeal of customized hardware to speed up the execution of EDA algorithms was great. As usual, IBM, which I believe is the single institution that had the most impact on the field, proposed a special-purpose architecture for logic simulation, the Yorktown Simulation Engine (YSE) by

Pfister and colleagues. The great advantages in performance originated a strong activity in the industrial domain where a number of new companies were formed to serve this market.

The idea of hardware acceleration was extended to other EDA field, for example, wire routing (Ravi Nair and colleagues at IBM) but it did not have enough appeal for the limitations of the algorithms implemented to create the interest of the YSE.

In parallel to this work, the alternative idea was pursued of using general-purpose parallel computers to achieve similar performance advantages but with a lower development cost. In the late 1980s there was a great deal of interest in parallel architectures in the computer design community. The Thinking Machine, a massively parallel architecture designed by Danny Hillis at MIT, generated excitement in the research and the industrial communities including our own. A company (Thinking Machine Inc.) was created with the task of disseminating this approach to computing that attracted some of the very best minds in the field. I was fortunate to be called to participate to the company as Fellow since its inception and during my sabbatical year at MIT in 1987, another defining moment of my career. The atmosphere of the company had some of the characteristics of the renaissance period in arts and literature. For example, the routing algorithms for the communication among processors were designed (and implemented) by Dick Feynman, the Nobel Prize winner in physics who spent a great deal of time on Thinking Machines premises. Impromptu debates about algorithms and applications were common; seminars by the most prominent scientists of the time were organized every day. Andrea Casotto, Roberto Guerrieri, and Don Webber developed algorithms for circuit and logic simulation as well as for placement and routing customized to the Thinking Machine. However, the lack of understanding that complete solutions rather than hardware was what mattered to final users, limited the industrial use of the machines to research laboratories and eventually led to its death, much of our chagrin since it was a wonderful attempt at marrying scientific excellence to business. Other parallel architectures such as the N-cube, the Sequent, and the Intel hypercube, were actively pursued and used in EDA during this period. However, there has been no commercial tool sold on these machines. Parallel computing with various degrees of heterogeneity is still a partially untapped source of important results for EDA and other engineering applications but the fundamental problem of software support for parallel computing has to be solved before this technology can be of widespread use. In particular, the advent of multi-core microprocessors is opening up a wealth of great opportunities for EDA algorithms. A flurry of early announcements of existing tools converted to multi-core architectures has hit the press release treadmill.

High-Level Design

High-level or system-level design is a bridge to the future. We all agree that raising the level of abstraction is essential to increase design productivity by orders of magnitude. The foundational work started in the 1980s with high-level synthesis (e.g., Thomas, Parker and Gajski). While the work started almost in parallel with logic synthesis and several commercial tools were developed, there has not been a wide acceptance in the design community for this approach. Still much needs to be done. The basic question to ask is then what made logic synthesis successful and what made the adoption of high-level synthesis so difficult? I believe that the original work on high-level synthesis was too general; too many alternatives had to be explored and tools had a difficult time to beat humans at this game. However, when system-level design was focused on constrained architectures such as DSPs and microprocessor based architectures, there have been results with some degree of success in the industrial domain: the work at IMEC by DeMan and Rabaey on the Cathedral system and the hardware-software co-design approaches embedded in system such as Flex developed by Paulin, Cosyma at Braunschweig and POLIS at Berkeley.

During this period, Edward Lee developed Ptolemy and Harel developed State-Charts for design capture and verification at the algorithmic level; work that is affecting the present approaches to embedded system design. In the domain of software design, Berry at Ecole des Mines, Benveniste and Le Guernic at INRIA, Caspi and Halbwachs at Verimag proposed synchronous languages (Esterel, Signal, and Lustre).

Transfering Technology to Industry

Donald O. Pederson

The second and third generation EDA companies were formed during this period. The second-generation companies, Daisy, Mentor and Valid, were created in the 1980-1981 timeframe to serve the digital design market with schematic data capture and simulation on workstations. Daisy and Valid would build their own workstation in line with the traditional approach of the first generation company, while Mentor sold Apollo workstations with an OEM agreement. For all three, hardware sales were a very substantial part of their revenues.

In 1982, SDA and ECAD were founded. These companies were the first examples of "software-only companies" that did not base their business model on hardware. ECAD and SDA were actually supposed to be one company since the very beginning, but Paul Huang had already completed Dracula and wanted to go to market quickly while the SDA side of the equation was not quite ready. They merged because of the decision of SDA to go public in 1987 on the day that is still remembered as Black Monday. The stock market conditions prevented any IPO for a few quarters after Black Monday; Joe Costello, Jim Solomon, Paul Huang and Glen Antle felt that the best strategy was to merge with ECAD. The joint company was called Cadence.

The creation of Cadence was an exciting period. Don Pederson's policy was to leave all the results and the software that was developed at Berkeley in the open domain. Indeed, the birth of the Open Source movement could be traced to this decision. This idea made it easy to spread the research results and favored the adoption of the tools. On the other hand, it created support problems since several engineers were needed to customize and improve tools such as SPICE inside companies. Jim Solomon was among the most vocal people to suggest that a new company be formed. Jim Solomon volunteered to leave National Semiconductors and to start ISIS (the original name of the company that unfortunately was already taken). Solomon Design Automation was a placeholder name that was never actually replaced until Cadence was formed. Richard and I helped in writing the business plan, a team of former Berkeley students and Bell Labs researchers was assembled and the adventure began. The funding mechanism was novel as the majority of the funds came from large companies (Ericsson, GE, Harris Semiconductors, and National Semiconductors who contributed 1 Million each) while several VCs led by Don Lucas contributed about 1 Million total. The company had a difficult period until 1986 for product delays and other difficulties but in the end with his partnership model (again a new business idea) SDA was finally sustainably profitable. Joe Costello was behind the push towards this business model. The role of Jon Cornell, General Manager of Harris Semiconductors, in the early days of SDA cannot be overemphasized as he sustained the company in all aspects including a substantial new round when the company was in distress. James Spoto who was in charge of EDA at Harris Semiconductors later joined Cadence when the Analog Division was formed heading the engineering group of the Division, one of the most (if not the most) successful "start-up" inside an existing company.

During this period, Silicon Compilers and Silicon Design Labs were founded around the concepts of silicon compilation and symbolic layout. View Logic was pursuing the EDA market from the PC angle aiming at low cost solutions for digital design. Gateway was formed to commercialize Verilog and its associated simulator. In 1987, Optimal Solutions Inc. (OSI) was incorporated in North Carolina. Not many people know that this was the original name of Synopsys! OSI was born out of the activity of Aart de Geus at GE with Socrates, and out of Richard Newton's and my conviction that logic synthesis was indeed the next big step in EDA. The funding model for OSI was the same as SDA: a substantial support from GE and Harris Semiconductors accompanied by a VC funding that was raised via the auspices of Richard Newton who was starting to work in the VC community as an Associate Partner at Mayfield. The original offering that was based on an engineered version of Socrates was then enriched and substantially evolved by Rick Rudell and Albert Wang, who introduced the algorithms used in MIS to yield the Design Compiler. The partnership model experimented with SDA was also offered to a number of companies and provided a very strong commercial foundation for the company. By 1990, the company changed its name to Synopsys and had moved to California; shortly thereafter, it went public in one of the most successful IPO of the time.

By that time, it was clear that selling workstation hardware was not an appealing business model given the difference in margins even though as late as 1990, DeBruggere then Chairman of Mentor commented negatively on the business model of "software only" companies saying that he knew of no long-term successful EDA company who did not sell workstation hardware. Specialized hardware acceleration companies were also founded during that period: Quickturn, PiE Design, IKOS, among others. Mergers and acquisitions have left no independent hardware-acceleration company today.

Even though there was a strong incentive to adopt commercial solutions for IC and system companies, the strategic value of EDA kept internal investment high. In particular, Bell Labs and IBM were pulling ahead of the competition in tools and environments.

The second-generation business model (hardware + software) was proven not sustainable due to the dominance of general purpose workstations in the years to follow. Daisy and Valid died a slow death through acquisitions, while Mentor re-invented itself to sustain competition and be economically viable. Silicon Compilers and Silicon Design Labs also disappeared through a series of mergers and acquisitions. Silicon compilation in its basic form did not pay enough attention to performance and area of the final result; layout languages and symbolic layout systems were not accepted by a community that was used to using images and geometries to represent their designs.

The Maturity Age (1993-today)

1993 was the beginning of a new phase in our community. Technical innovation started slowing down, the vendor community became mature from Wall Street's point of view and was much less inclined to risk-taking. Quoting the keynote address at the 1995 DAC by Richard Newton:

"If there is a single point I wish to make here today, it is that as a discipline, both in industry and in academia, we are just not taking enough risks today."

This period coincides with the explosion of the Web and its applications. We may draw the conclusion that the best energies and minds in electrical engineering and computer science were attracted by this emerging field and that the funding from Venture Capital was targeting internet enterprises. This situation would then reflect in a lower rate of innovation in EDA. At the same time, the semiconductor sector continued to drive technology along the lines of Moore's law increasing the technical challenges to EDA. System-on-Chip (SOC) started attracting attention and by 2007 gained the center of attention.

System-on-Chip means many different things to different people. I found that in Japan and Korea SOC meant the integration of memory and microprocessors, elsewhere anything that would use a very large number of transistors would qualify as an SOC. In my opinion (now shared by most companies), SoC is about integrating different design styles in a coherent whole. Interdisciplinary approaches are necessary to solve the problem of designing complex problems posed by the advances in electronics.

Relevant Contributions

Physical verification attracted a great deal of attention as the geometries marched down towards the nano range. Self-test emerged as the only solution for the raising cost and requirements for test equipment. Asynchronous design methods and the associated synthesis problem had been studied as potential solutions to performance problems due to the unpredictability of wire delays and to power consumption. Designers who are interested in pushing the envelope are still grappling with this paradigm change to see what the limits that can be achieved are and whether the gains are worth the risks of the approach. As chips incorporate an increasing number of functionalities, **analog design** became the bottleneck in design. In SOCs, the name of the game is to find the best matching between analog and digital components not to optimize "to death" the performance of the analog part. Because of the dependency of analog circuits on all kinds of second order effects, the analog design activity has been more a craft than a science. When the wiring delays became relevant in chip design, the separation of concerns originated from the layering of logic synthesis and layout started showing problems in achieving "design closure", i.e., circuits that are designed at the logic layer to satisfy timing constraints, still satisfy them after final layout. Obviously, not achieving design closure would generate unacceptable time-to-market delays and costs. The proposal put forward in this period was to merge **layout and logic synthesis** in a giant optimization loop.

As embedded system design moved towards increasingly software-rich solutions, the issue of achieving the same design productivity gains that logic synthesis brought to hardware, created a strong interest in **hardware-software co-design** where from a high-level functional model of the design, the detailed and optimized software and hardware implementation could be derived. In this perspective, the issue of **software synthesis** became relevant. Note the difference between synthesis and compilation here. I talk about synthesis when I use a mathematical representation of the original design that is not biased towards a particular implementation style. Compilation implies the translation from a programming language to assembly or machine code. In this case, the mathematical abstraction is the same. Opening a parenthesis, I was surprised to see that as early as 1967 papers on software design were presented at the Design Automation Conference (DAC). Back to the future then? The basic difference is in the type of software that was of interest. At the beginning of DAC, papers presented at the Conference spanned all activities from buildings to structures, from electronic circuits to "standard" software, i.e., database and airline reservation software.

Technology Evolution in Industry

I already alluded to the stress on the EDA industry created by the Internet and high-tech financial frenzy. Established vendors had unwanted terminations in the range of 20% of their work force losing them to start-ups in the Internet domain and in EDA. In 1999, there were of the order of 80 startups in EDA! During this period, companies such as Avanti, Ambit, Magma, Monterey, Get-to-Chip, Verisity, and Verplex were started with the idea of challenging the dominant players. All of these companies were later acquired by Cadence and Synopsys. However, if we look at the post Internet bubble years, we see a different landscape: very few IPOs if any, acquisitions as the only exit strategy.

This period had been a roller coaster with adjustments in business model going to increasingly higher retable revenues to make EDA more stable and predictable but creating steep transients.

The Future of EDA

EDA must adapt to changed business conditions and structures. We are witnessing a substantial change in the client-vendor relationship. Partnerships are increasingly important as semiconductor and system companies rationalize their investments in EDA technology; all of this in view of a fundamental technology change in the semiconductor technology that requires investments of a size never seen before and that exposes the limitations of the present design methodology and tools. The cost of ownership of ASIC design is increasing rapidly due to NREs and mask costs. The cost of designing a new chip has reached the 150M$ mark thus restricting the number of companies who can afford this endeavor. Consequently, there is a constant reduction of design starts in favor of standard solutions and of customization by software. If we extrapolate the data, there is only one message: the traditional EDA market centered on ASICs has largely evaporated. There is no other choice for the EDA community than looking at other areas of applications for growth. While the semiconductor industry, the main customer of EDA, is looking for the next killer applications for its products after PCs and smart phones, the neighboring areas for EDA, mechanical CAD and CAE and embedded software design, have grown at a much faster rate. It makes sense for the EDA industry to look at opportunities in these areas by leveraging the algorithmic and methodological knowledge accumulated in its history. Mergers and acquisitions in this area accelerated dramatically in the last five years as the design problems faced by the system industry grow in complexity while time-to-market requirements are becoming increasingly tougher in face of declining prices.

The Move to System Design

Emerging Applications: Societal-Information Technology-Systems (SiS)

The next drivers for the high tech industry will refer to the global interests of society. There is a consensus forming that electronics still has a long way to go to penetrate application domains of great interest. Potential applications are pursued in the Center for Information Technology Research in the Interest of Society at Berkeley, a very broad program sponsored by the State of California and industry (see Fig. 1) and invented by Richard Newton.

The center role in this research is going to be played by devices such as the Smart Dust, developed by Kris Pister and colleagues at Berkeley, a wireless communication sensor and information elaboration node of a very rich network. Jan Rabaey's leadership in research programs such as the Berkeley Wireless Research Center, the Gigascale System Research Center and the Multiscale System Research Center as well as in ... producing excellent wines cannot be overemphasized.

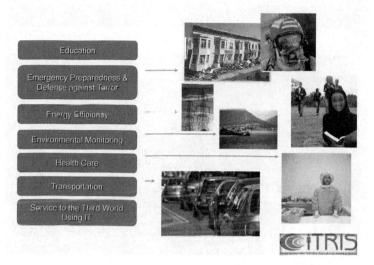

Fig. 1 - Applications in the Interest of Society

IC Design Styles

Given for granted that these applications are going to dominate the future landscape of electronics, what do we need to do to support them? As already mentioned, the design style of choice should favor re-use in all its forms and, given the constant increase in NRE and mask costs, make software even more pervasive than today. Ad hoc communication protocols will play a substantial role in the design process. If we look at the history of design methods, we see that the changes in design productivity were always associated with a raise in the level of abstraction of design capture. In the 1970s, the highest level of abstraction for integrated circuits was the schematic of a transistor; ten years later it was the "gate", by 1990, the use of HDL was pervasive and design capture was done at the Register Transfer Level.

We need to work with blocks of much coarser granularity than what we used to do to be able to cope with the increase of productivity that we are asked to provide. We need to bring system-level issues into chip design. The recent emphasis on Systems-on-Chip (SoC) is a witness to this trend. All companies involved in IC design are looking at this approach to shorten design time and cost. The issue here is one of integration of Intellectual Property (IP) blocks. It is essential that the blocks be designed so that their properties are maintained when connected together (compositionality) to allow re-use without the need for extensive checking. In the past few months, there has been a flurry of acquisitions that are shaping the battle for the SOC market and the virtual prototyping market (Vast and Co-ware have been acquired by Synopsys). The role of interconnect cannot be overemphasized here. Indeed the characteristics of the communication infrastructure determine the amount of verification that

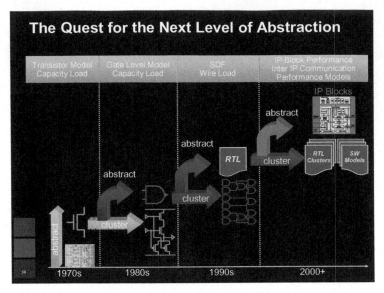

Fig. 2 - Raising the Level of Abstraction

must be carried out. Standard interconnects such as the ones proposed by ARM, Sonics and Arteris, aim at reducing the cost of assembly of IPs in an SOC design methodology. The standards, which dictate rigorous rules on coordination and separation of concerns, are offering a shortened path to correct implementation at the potential cost of reduced performance. In particular, Time-Triggered Architectures as proposed by Herman Kopetz at Vienna that spawned the FlexRay standard in the automotive domain, have definite advantages with respect to event driven architectures in terms of predictability, orthogonalization of timing and functionality, and fault isolation giving up some performance. The concept of Network-on-Chip (NOC) was introduced in the late 1990s to offer an alternative to busses and point-to-point interconnections that were running out of steam in high performance, complex, SOCs. The seminal papers by DeMicheli and students analyzed the advantages of these interconnect architectures. My research group and most notably Luca Carloni and Alessandro Pinto looked at automatic synthesis procedures for the selection of protocols and topologies that could provide optimal trade-offs between performance and robustness. In particular, Carloni developed a family of latency insensitive protocols that could reduce the overhead of using the synchronous paradigm in chips where the relative lengths of the interconnects were making the reduction in clock speed unacceptable. Gadi Singer was heading an internal Intel effort, called 6+6, trying to reduce design cycles to 6 months for design and 6 months to productization. This effort was later extended and strengthened to include microprocessor design after the Atom design started.

The System Industry and the Design Chain

The trend has been rather clear over this period: the electronic industrial sector has been segmenting at a rapid pace. System companies have retrenched in their core competence of product specification and market analysis, while shifting to others the task of delivering the engineering and the components of the system. For example, companies like Ericsson are doing increasingly less in the design of chips. Semiconductor companies are asked to do more for their strategic customers. Some of the engineering responsibilities have transferred over. At the same time, semiconductor companies have been increasingly relying on IPs provided by specialized companies, for example, ARM for processor cores and Artist for libraries. Manufacturing has been successfully transferred for some of the semiconductor companies to the likes of Global Foundries, SMIC, TSMC and UMC who are now at the leading edge (if we exclude Intel) in bringing new technology in the market place. In the area of printed circuit boards, specialized manufacturing companies such as Flextronics are taking the lion's share of manufacturing. Supporting this movement requires a view of design as a highly integrated activity across company boundaries; a task that is far from easy. Competences in engineering disciplines as diverse as mechanical engineering and electronics, RF and MEMS, have to become part of an integrated environment. Economic analysis of alternative solutions has to be provided together with a set of tools capable of exposing the trade-offs of available designs. Design representations with rigorous semantics have to be supported for clean hand-offs between design teams and to favor more robust design verification methods. Databases capable of handling design data together with manufacturing data and company databases will have to emerge. We call this emerging field *design-chain support*, which I believe is a great opportunity for EDA at large to increase its relevance both in terms of value-added and of economic opportunities.

In the last few years, we have witnessed a resurgence of interest in chip design (not manufacturing) from a number of very successful system companies such as Apple, Google and Amazon. Apple has designed a new generation of processors, the A6 and A7, that epitomizes the SoC design style and the advantages that can be reaped with a customized design if the financial structure of a company allows developing chips of this complexity. This interest may open up a new era in design approaches and opportunities.

Parallel to these developments in traditional sectors, the interest in the Internet of Things (IoT) has grown exponentially. IoT refers to the capability of objects of interacting directly without human intervention. The market estimations are wildly optimistic indicating a market size of over a trillion dollars in few years. IoT can become a reality when the cost, power and size of devices that allow the interconnection among "things" reach a new plateau. All semiconductor industries are positioning in this market trying to lower the cost, size and power consumption of communication devices, system industries are investigating the potential disruption created by IoT in their businesses, manufacturing companies are developing new plants based on IoT (the industrial aspects yielded the much discussed German initiative called Industrie 4.0). Designing the devices, the systems and their application will be a major challenge considering safety and security as primary concerns. The state-of-the-art allows some preliminary developments but much research needs to be done here. Given the immense amount of data that trillions of devices will be capable of gathering, data science will be needed to understand what these data mean to us. Design methods and tools will be likely radically different from what we have now, with off-line design merging with on-line operation due to the necessity of continuously adapting IoT systems to a changing environment and mortality of the components.

Conclusions

We have outlined the grand challenges for EDA. I believe the EDA industry needs to invest the right amount of attention to enlarge the business boundaries where it operates in these spaces. However, it needs to do so when the overall business situation is difficult. Innovation is an expensive and risky proposition that sometimes clashes with the status of a public company of medium size. We have seen over the years a wave of mergers and acquisitions carried out by the major EDA companies. While "Buying companies is a legitimate way of doing research and development in this industry," as Robert Stern, a Smith-Barney analyst said a few years ago, I believe it is important to find additional mechanisms to innovate. The innovation from start-up in EDA has been incremental except for a few albeit important cases, i.e., finding better ways to do things we already knew how to do. In addition, in this economic climate, which we hope is going to change soon, larger companies tend to outperform the market in terms of both earnings per-share and capitalization. If you look at the VC investment, it is in a decline. The IPO window closed at the end of 2001. In 2002 and 2003 up to today, not a single EDA company went public.

Thus, while the start-up concept is still a valid one and needed to foster an important part of the innovation landscape, we need to think of new ways to foster large-scale innovation as necessary to cope with the challenges that the electronic industry is facing in the wake of the era of the IoT, of synthetic biology and of precision medicine. However, there is less money to spend on innovation just at the moment when we need it the most.

Design Science as embodied today in EDA (but in the future I am sure extensions in fields such as biology and medicine will be very relevant!) is a unique, wonderful field, where research, innovation, and business have come together for many years as demonstrated by its history over the past 47 years. Can the EDA community continue its quest for better ways of doing design increasing the productivity of electronic system designers by orders of magnitude while also increasing quality? It is not going to be an easy path. Many difficulty lay in front of us.

Alberto Sangiovanni–Vincentelli

The Edgar L. and Harold H. Buttner Chair of Electrical Engineering
Department of EECS, University of California at Berkeley

The Business of Design Automation

This section is an essay on the business of Design Automation. It is not an exhaustive history, but a set of postcards that depict the evolution of this small, but meaningful segment of the Semiconductor Industry and some thoughts on where it may go next.

Moore's Law and Design Automation

The Semiconductor Industry is aware of and has admired Gordon Moore for 50 years for identifying such an incredible trend and opportunity. Dr. Moore noted that chip complexity was growing exponentially and expressed his vision that this is what would characterize the growth of the industry for the foreseeable future.

A consequence of Gordon Moore's vision becoming a reality is that other areas of the industry had to change and adapt as rapidly. That was the case for Design Automation.

In a sense, we could say that there's a corollary to Moore's Law. Design Automation has to provide to the Semiconductor Industry the methodology and tools to design the next-generation chip. That chip will be twice the complexity and will be designed in the roughly the same time at the roughly the same cost. The Design Automation industry evolved just to address and meet the challenge of the technology and advanced it to become real and effective. Technology without designers doesn't have the impact. Technology has to be brought to fruition by designers. This is Design Automation.

Aryeh Finegold, Daisy Systems

The complexity of one SoC designed today would have been impossible in the early 1980s. It would have been impossible then to conceive of billions of silicon transistors designed in months. An industry pioneer contributed significantly in making the use of the technology possible Aryeh Finegold – Daisy Systems.

The Emergence of Powerful IC Design Tools for the Design of ASICs

In the early days, Design Automation was known as CAE or computer aided engineering. CAE systems were specialized computers with both hardware and software to support IC design. Daisy Systems and Valid Logic, two of the *"Big Three"* vendors from those days, sold proprietary hardware/software systems. Mentor Graphics, the third, developed software that worked only on Apollo Computer workstations.

All three contributed significantly to the creation of this innovative segment of the industry. A more complete history of EDA would be needed to fully express how meaningful the contributions were in the first stage of EDA and it was not just software. It also was a new generation of specialized graphics systems that was a complex mix of hardware and software and that helped

Daisy Systems workstation (left). Harvey Jones (center). Tony Zingale (right).

Design Automation take off. It happened in a way that's often common with new technology and ideas.

Designing ICs was done within powerful and successful Semiconductor companies with computer-aided design (CAD) groups full of highly skilled designers accustomed to the complexity of logic and physical design, library and process development, packaging and several other specialties. New tools to support the growing design challenges were developed in-house at the time as well. Replacing internal tools with external tools has never been an easy task, especially when so much company-specific expertise is available internally.

That scenario changed when application specific ICs (ASICs) were introduced. ASICs weren't all that different from ICs, but their designers were. This new community of designers didn't need to understand the physical layout, process technology or indeed any non-digital aspects of an IC. They were designers working on the next-generation product in the system design environment who wanted to ride the same wave of Moore's Law. The way to get a system to ride Moore's Law was to make it possible to access new technology without a major change to their know-how.

The Design Automation industry realized this and created semi-custom and custom methodologies so the system designer didn't need to have the same level of understanding as a CAD engineer to write silicon. Moore's Law became meaningful to a much larger community.

CAE workstations and EDA systems rapidly expanded by supporting ASIC design because the volume was far higher and demands easier to satisfy than those of internal CAD teams of silicon companies. System designers were far more open to this than an internal CAD group.

Many notables in Design Automation were at the time working for Daisy Systems, including Harvey Jones, who became CEO of Synopsys, and Tony Zingale, former CEO of Clarify and Mercury Interactive and, most recently, Jive Software, where he serves as Executive. While at Daisy, they recognized the power of the system and ASIC design community and rolled this into what became tremendous market growth.

Increasing Levels of Abstraction Enable the Continuation of IC Design

The challenge of new products increased the level of abstraction, and higher levels of abstraction became a way for designers to cope with the exponential growth of technology complexities. Early IC design was labor intensive - physical designers dealt with each and every transistor in the design, even transistors that comprised logic gates, such as NAND, NOR and other logic functions. IC area was sacred and *"hand packing"* a design at the basic polygon level was "the way to do it" in the early days.

As semiconductor processes improved and allowed for larger die that had more capacity, it became impractical for designers to deal with every transistor. Working at that level required too many engineers making too many low-level decisions, which was too error prone to bring a product to market on schedule.

The industry turned to abstraction - the concept of designing at a higher level and relegating lower-level details to libraries and CAE tools, a similar model to what the software industry has done. Cell libraries, libraries of pre-designed logic gates with characterized properties, were born and designers could design larger circuits in less time and with less effort. Credit for leading this charge goes to Mark Templeton, president of VLSI Libraries Inc. that later became intellectual property (IP) success known as Artisan Components.

Semiconductor processes continued to improve with die capacity following Moore's Law. Cell libraries had many desirable properties. They allowed designs to be more easily ported from one fab to another and larger designs could be brought to market, giving the industry some independence from the challenge of having to cope with the increasing complexity of process technology details.

Even so, designing large circuits continued to be a daunting task. Another level of design abstraction was needed to get to the next level of productivity. This one, however, was much more difficult to accomplish because it affected design style. At that time, many designers were printed circuit board designers using transistor-transistor logic (TTL). To minimize gate count, they often used JK flip-flops and RS flip-flops in asynchronous and self-timed circuits. As designs grew, this asynchronous design style became error prone because it was difficult to ensure that a circuit would function over the entire process range.

Mark Templeton.

Once the design community made the transition to synchronous design using finite state machines (FSMs) to implement the logic in a circuit. Designers had not only cell libraries but synchronous FSMs with which to continue filling the rapidly expanding die that newer processes were able to manufacture.

When the design community turned to synchronous design and FSMs, the CAE community was able to further improve productivity with two key computer-based tools - logic minimization and optimization, and logic synthesis from a higher level representation of a design.

General Electric and GTE were pioneering companies in logic minimization, optimization and synthesis. Silc spun out from GTE Labs in 1987 and was acquired in 1990 by Racal Redac.

Dr. Aart de Geus and a team of engineers from General Electric's Microelectronics Center started Synopsys in 1986 with the synthesis technology they developed at GE. Synopsys moved up in unbelievable fashion, took the market by storm and a new level of abstraction, introduced by the design community but made productive by CAE, became standard in the industry. This became known as the Register-Transfer Level (RTL) of abstraction. Existing Design Automation companies realized they needed to move to RTL and were dedicated to the movement to make designers more productive. The move to RTL had an enormous impact on the quality of design and allowed new contributors and technologists to be effective chip designers without the need to be experts in semiconductor technology. RTL further expanded the chip design community, much as system design tools expanded the ASIC design community.

Quickturn   emulation box   (Source: Mike D'Amour).

Synopsys made a tremendous contribution to the industry by pushing the frontier of abstraction forward. Its leadership role is obvious today, but the creativity of this company strategy and execution cannot be overstated.

While the need for design to move to a higher level of abstraction has been widely recognized, a single approach has yet to be adopted. High Level Synthesis is one strong contender, and design using IP building blocks is another. Object-oriented constructs have been added to RTL languages such as SystemVerilog but these have had more impact on IC verification. At this writing, while progress has been made in behavioral synthesis as applied to specific domains, RTL continues to be the main abstraction level workhorse.

Verification - Simulation and Emulation

With increasing complexity, all the previous methods of prototyping a chip became completely inadequate. The Semiconductor Industry had to find a way to describe a chip and simulate a description of it. And, it did. The tool was called a simulator. This was a new world for electronics and the challenge was existential.

The innovators in Design Automation worked hard to develop verification tools. The demand was growing exponentially. A new company, Gateway Design Automation, came out moving faster and more deliberately, introducing the simulator known as Verilog. Gateway was acquired in 1989 by Cadence whose chief executive Joe Costello had an unstoppable vision. He devised a way to build the company by recognizing external contributions, merging competent people and their companies into one successful Design Automation powerhouse.

The Verilog simulator became part of the Cadence ecosystem and the success continued. The contributions of Prabhu Goel and Phil Moorby at Gateway

From left to right: Joe Costello, Michael D'Amour, Prabhu Goel, Phil Moorby.

cannot be overstated.

At the same time, one of the challenges for simulation was that it ran at an intrinsically slower speed than hardware. Software simulation might run 10,000 or more times slower than hardware, making simulation of large designs impractical. In order to get closer to the speed of hardware, the application needed to run on exiting silicon.

Design Automation had to come up with something other than simulation and it was hardware emulation. Emulation appeared in two guises - field programmable gate array (FPGA) emulation where the circuit was synthesized or compiled onto a network of FPGAs, and simulation acceleration where the circuit was compiled to run on a hardware-based processor. In the former category, leading companies were Quickturn Design Systems, now Cadence and IKOS Systems with VStation, now Mentor Graphics Veloce Emulation Platform. The latter were IBM with its proprietary EVE hardware accelerator, now Cadence Palladium, IKOS Systems had NSIM, Daisy Systems offered MegaLogician and Zycad offered the LE and FE machines.

Emulation supported the ability to rapidly evolve electronic systems into new generations. Without emulation, the speed of creating new electronic systems compatible with previous generations before committing to silicon would have been much slower. The speed of moving to new generations of chips was supported by the ability to do emulation in a satisfactory way.

In the early days, emulation was based on in circuit emulation (ICE) where the emulator was connected to the real world through an adapter. Today, a variety of techniques are available, including the use of high-level testbenches that run on the emulator.

And, with emulation, the industry can't forget the immeasurable contributions of Mike D'Amour and Phil Kaufman for today emulation is in every successful Design Automation ecosystem.

The Development of the IP Market

IP was a hugely successful and meaningful step forward in the design of systems in silicon. The introduction of silicon IP offered designers pre-designed blocks of functionality guaranteed to work in silicon. For example, a designer could use an arithmetic unit as a piece of IP that becomes a block of working silicon that can be reused over and over again.

IP has several basic components. Gate libraries guaranteed to work in a given process freed the designer of electronics systems from being concerned with the fundamental knowledge of the silicon process. Silicon-proven IP could be reused and would be the same in each design, enabling the design of the next-generation chip and further raising the level of abstraction.

ARM Holdings and Artisan Components, a company originally designing libraries every designer employed, independently were devising a new business and business model that became the standard IP business model, forever virtualizing components. ARM acquired Artisan Components in 2004.

Left: Phil Kaufman. Right: Sir Robin Saxby

It was innovative for a company to decide to make a processor delivered not as a physical implementation but as an IP block. The processor was the IP block becoming most important to the VLSI and SoC designers for a broad variety of applications. That's why having an IP provider take responsibility for a more and more powerful processor and the infrastructure to support it, such as compilers, runtime libraries, debuggers, protocol stacks and bus interfaces, with increasing capabilities in silicon became a humongous contribution to the industry. Consequently, the impact of Sir Robin Saxby, former chief executive and chairman at ARM Holdings, is difficult to match. And, Mark Templeton, as well.

Of course, other IP blocks for communications, graphics and memories became part of an SoC ecosystem, which made designing any SoC more of an assembly process. The designer could now concentrate on the unique and differentiated portion of the chip.

The concept of IP was to reduce the number of things to allow Moore's Law to continue unabated by making the most of a complicated SoC not requiring redesign. From this standpoint, IP is a huge contributor to Design Automation into the Semiconductor Industry.

The Next Challenges

There is no doubt that a demand from society to the Semiconductor Industry to provide more powerful computing capabilities will be a continuing challenge, especially as new steps of Moore's Law become more and more difficult to achieve.

Many times we have heard industry watchers forecasting the slowdown and maturing of the Semiconductor Industry or of the tiny Design Automation segment. We should wonder whether this is really happening or whether we just are not creative enough to see different avenues of growth.

Today, the entire IP phenomenon can be seen as an agent of growth for the Semiconductor Industry and for the ability to design more complex devices. And, there are so many more challenges to be met! For instance:

- The move to 2.5 and 3D is already showing another dimension of growth

- Silicon will need to support the full SoC, not just hardware on chip, but software as well

- Software productivity will have to grow exponentially to keep development costs under control

- The interaction of the changing requirements, along with systems and silicon complexity, will push the industry to produce programmable versions of silicon in which a capability, such as evolving security, will be supported

- Flexible and reconfigurable systems will become a reality

- The ability of computers to control different physical variables needs to expand dramatically

- All layers of incredibly complex silicon will become more flexible, and have the ability to learn and adapt

- Systems will be supported by all levels of the technology deck, from learning machines to the cloud

- Sensors will bring in more and more phenomena for computers to manage

- The evolution of computing capabilities will allow more and more applications to improve society, the obvious being self-driving cars and personalized healthcare

And, true new computers will appear. Twenty years from now, we will look back at today's silicon and computer industry and be amazed at how primitive we were. Revolutionary learning systems will be created and capable of solving human problems in an intuitive way with one-millionth the power now required. The intelligent computer will be a reality.

I know that this excursus on Design Automation has missed people and companies, but I hope that these postcards have helped paint a picture of this great journey and a flash into the future.

Lucio Lanza
Managing Director Lanza techVentures

Circuit and System Education

Engineering in general and electrical engineering in particular are evolving disciplines, and hence their educational processes have not stagnated either over the past 90 years. Equally important is the dynamical balance between on the one hand the improved understanding of circuits and systems, and the accumulated CAS knowledge over the past century and on the other hand the different characteristics of the new generations of students with natural exposure to new ICT media, and devices and the students' limited attention span. This tension requires creative processes that incorporate advanced technologies of blended learning, MOOCs, and learning analytics as well as rethinking the educational trajectories and the choice of relevant CAS topics for the new students.

Clearly there is a process of rethinking the CAS education, where we can learn from the past 90 years, and project in the future, in order to prepare the students with CAS skills and insights for their role in society during the coming 40-50 years. This is the context of the CAS Education and Outreach TC of IEEE CAS Society that was set up in 2009.

Ernst Guillemin is generally considered [1,2] as the founding father of circuit theory education with a magisterial series of six single-authored books written at MIT. Under the leadership of Vannevar Bush a modernization of the EE curriculum at MIT was launched in the late 1930s and 1940s that is more based on sciences like physics and mathematics than on craftmanship. Guillemin had moved temporarily from MIT to the University of München, Germany, for a doctorate in mathematical physics in 1926 with Arnold Sommerfeld.

Dr. Ernst A. Guillemin teaching at MIT in 1939.

Conversely in 1930-1931, Wilhelm Cauer came from Germany to MIT to work in the team of Bush and Guillemin. In his textbook [3] of 1931 Guillemin introduced new topics like transient and steady-state response, network theory, the Heaviside approach, and Fourier analysis. In the preface he gave a message both for students and teachers that describes the dynamical relationship between research and education and that still resonates today: *"Methods are frequently designated as advanced merely because they are not in current use. To the student the entire field is new; the advanced methods are no exception. If they afford better understanding of the situation involved, then it is good pedagogy to introduce them into an elementary discussion. It is well for the teacher to bear in mind that the methods which are very familiar to him are not necessarily the easiest for the student to grasp.*" His continuing work on course revision and development led him [4] to take a particular interest in engineering mathematics with topics like determinants, matrices, vectors (including vector calculus), functions of a complex variable, Fourier and Laplace transformations, and conformal mapping. All these topics appear currently in basic linear algebra and calculus courses in EE worldwide, however with a lesser role of determinants, and more attention on eigenvalues and singular values. Moreover these methods still play a central role in courses of systems and control, circuit analysis, and filter design.

The book [5] entitled *Introductory Circuit Theory* of 1953 takes a fresh approach in the power and generality of essentially quite simple ideas. It even introduces the concepts of graphs, networks and trees, cut-sets, duality and so on before even mentioning Kirchhoff's laws. The book had a great impact, and for example the graph theory is a standard subject in EE and CS. But the book did not itself

R. Fano (left) and C. A. Desoer (right) [Wikipedia].

enjoy great popularity. In [6] he presented a more systematic and rigorous approach for the relatively new discipline of network synthesis. Several famous CAS researchers like Fano, Desoer, Newcomb, and Anderson and educators from the group of Guillemin continued to write classical textbooks on CAS. In a historical perspective Guillemin has produced a whole genealogy of generations of PhD supervision descendants in the Mathematics Genealogy Project. In fact he had 11 PhD students and up until today a total of 2906 descendants. A landmark textbook by Desoer and Kuh (1969) [7] also strongly influenced the teaching of circuits. Because of the great mathematical care it has proven attractive in classes with students having superior mathematical interest, but for many universities worldwide it turned out to be too theoretical. For a deep historical account of the different textbooks, and the respective choices of topics over the period 1934-1984 there is a nice overview of Van Valkenburg [8]. The invention of the transistor and the IC and their enormous capabilities had a major impact and lead to a growing interest in nonlinear components and circuits in basic circuit education. After Chua joined the EECS department in Berkeley the new version of the textbook [9] was consequently extended with nonlinear components and circuits.

The development of inexpensive computers and computational facilities together with strong circuit simulation programs, like SPICE, and system analysis programs, like MATLAB, offered new opportunities for designers and the education of design. This lead some universities to drop or reduce the basic circuit analysis course in favor of a SPICE initiation, so that the students could quickly move to the more exciting design. But some students started thinking that design is merely a trial and error game with SPICE simulations for some random values of components until they hit upon a useful circuit. Rohrer [10] reacted to this evolution in 1990, and stressed that the introductory circuits course should do serious circuit analysis and not just give an introduction to SPICE and a shallow memorization of methods like Thévenin, without foundation. This argument of depth was further confirmed in the overview of Diniz [11] in 1998. With the growing interest in digital signal processing there was a move [12] to reverse the traditional order in EE education of analog circuits and signal processing first and then digital signal processing DSP. There are several arguments in favor of DSP first and the approach was worked out carefully [12]. The notion of impulse signal as a discrete time signal that is 1 at t=0 and 0 elsewhere and the impulse response in discrete time are much easier to grasp by students than the Dirac impulse and the corresponding continuous time impulse response. Indeed the Dirac impulse signal as a limit of a block wave of unit area for a duration going to zero is a quite unnatural continuous time signal, and the solution of differential equations with Dirac impulses is not simple. Also block diagrams with components like delay elements, scaling element and adders are closer to the tangible world of students. Moreover discrete time systems tie in very nicely with the digital systems and computer systems and the system simulation tools on digital computers. However since our physical world is continuous time, it is not recommended to postpone the continuous time systems and circuits education. So there is a growing consensus that EE education should rather deal with discrete time and continuous time in a carefully designed interlaced way. Then the concepts can be introduced in the realm of discrete or continuous time that is most easy to grasp by the student. It is then the ambition that the students can build in their mind concepts of circuits and systems that integrate continuous time and discrete time models.

The basic circuits and systems education (see Fig. 1) has currently in most EE programs a rather central role [13] between the basic sciences like mathematics and physics and subsequent courses of signal processing, control, electrical energy, biomedical circuits and systems, microwave and telecommunication systems. An impressively comprehensive overview of the whole field is given in the handbook [14] edited by W. K. Chen. CAS education is encountering other obstacles. The IC miniaturization and the virtualization of the systems render CAS devices clearly less visible. Moreover our world is dealing with more complex electrical systems in telecommunication and power distribution. Fortunately the progress in computing facilities allow for the simulation of larger circuits and systems. Hence, in order to touch all physical senses of the students, CAS education should involve more lab oriented courses where students learn to explore, experiment and analyze basic circuits and build and measure these. This is often followed by a more mathematically oriented course on circuit and system analysis. Also over the years one can observe a growing divergence [11, 13] between the research agendas and themes of young professors and the basic circuits and systems courses in EE. This implies that many young professors of EE consider the teaching of a basic circuits and systems course as a real burden that is not enriching their research. This however does not stimulate the quality or the attractivity of basic CAS courses. Tsividis [15], counteracted this evolution at Columbia University quite convincingly by exposing the students at an early stage in their EE education to real circuits and their capabilities. It has also triggered a renewed interest in EE. Moreover methods of direct teaching of systems as advocated by Ayazifar [16] at U.C.Berkeley have lead to strong motivation of students.

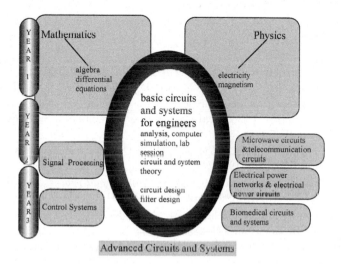

The present positioning of the basic circuits and systems education among the neighboring topics.

The evolutions in the last 30 years have not only influenced the basic CAS education but also the more advanced CAS courses with topics related to large circuits and especially VLSI circuits, sensors, computer-aided design, layout and routing problems, and graph theory and optimization strategies, analog circuit design, analog signal processing, biomedical circuits and systems, communication circuits, electrical energy distribution systems, electrical power circuits. Over these 30 years the CAS Society has bestowed excellent educators for their teaching of basic and advanced courses with the CAS education award: Eliahu I. Jury, Adel Sedra, Gabor C. Temes, Bede Liu, Sanjit K. Mitra, Yosiro Oono, Mohammed S. Ghausi, Rudolf Saal, Thomas Kailath, Adel S. Sedra, Lawrence P. Huelsman, George S. Moschytz, Wai-Kai Chen, John Choma Jr., M.N.S. Swamy, Robert W. Newcomb, Leonard T. Bruton, Magdy Bayoumi, Paulo Diniz, Christofer Toumazou, Wayne Wolf, Nirmal Kumar Bose, Josef A. Nossek, Andreas Antoniou, Yannis Tsividis, R. Jacob Baker, Sitthichai Pookaiyaudom, Luiz P. Caloba, Sung-Mo (Steve) Kang. This proves the continued importance of CAS education worldwide. Indeed it is vital that the research progress in CAS should continue to be intensely linked to the CAS educational processes.

References

[1] E. A. Guillemin, "*Teaching of circuit theory and its impact on other disciplines*," Proceedings of the IRE, May 1962, pp. 872-878.

[2] C. C. Bissell, "*He was the father of us all.: E. A. Guillemin and the teaching of modern network theory*," History of Telecommunications Conference, IEEE Histelcon 2008, pp.32-35.

[3] E. A. Guillemin, "*Communications Networks*," New York: Wiley, 1931,1935.

[4] E. A. Guillemin, "*The Mathematics of Circuit Analysis*," New York: Wiley, 1949.

[5] E. A. Guillemin, "*Introductory Circuit Theory*," New York: Wiley, 1953.

[6] E. A. Guillemin, "*Synthesis of Passive Networks*," New York: Wiley, 1957.

[7] C. A. Desoer and E. S. Kuh, "*Basic Circuit Theory*," New York, NY: McGraw-Hill, 1969.

[8] M.E. Van Valkenburg, "*Teaching Circuit Theory: 1934-1984*," IEEE Trans. Circuits and Systems, Vol. CAS-31, N0. 1, January 1984, pp.133-138.

[9] L. O. Chua, C. A. Desoer, and E. S. Kuh, "*Linear and Nonlinear Circuits*". McGraw-Hill Series in Electrical Engineering: Circuits and Systems, New York: McGraw-Hill, 1987.

[10] R. Rohrer, "*Taking circuits seriously*," IEEE Circuits and Devices Magazine, vol. 6, no. 4, pp. 2731, July 1990.

[11] P.S.R. Diniz, "*Teaching circuits, systems, and signal processing*," IEEE ISCAS Conference Proceedings, 1998, Vol. 1, pp 428-431.

[12] J. H. McClellan, R. W. Schafer, M. A. Yoder, "*DSP First: A Multimedia Approach*," Prentice Hall, 1997.

[13] J. Vandewalle, L. Trajkovic and S. Theodoridis, "*Introduction and outline of the special issue on Circuits and Systems education : Experiences, Challenges and Views*," IEEE CAS Magazine, Vol. 9. No 1, First Quarter 2009, pp. 27-33.

[14] W.-K. Chen (Ed.), "*The Circuits and Filters Handbook*," CRC Press, 2nd edition, 2008.

[15] Y. Tsividis, "*Turning students on to circuits*," IEEE CAS Magazine, Vol. 9. No 1, First Quarter 2009, pp. 58-63.

[16] B. Ayazifar, "*Can we make signals and systems intelligible, interesting and relevant?*," IEEE CAS Magazine, Vol. 9. No 1, First Quarter 2009, pp. 48-58.

Joos Vandewalle, IEEE Life Fellow

Emeritus Professor, Katholieke Universiteit Leuven, Belgium
Chairman of the CAS Education and Outreach Technical Committee CASEO

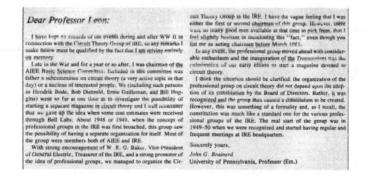

Dear Professor Leon:

I have kept no records of the events during and after WW II in connection with the Circuit Theory Group of IRE, so any remarks I make below must be qualified by the fact that I am relying entirely on memory.

Late in the War and for a year or so after, I was chairman of the AIEE Basic Science Committee. Included in this committee was either a subcommittee on circuit theory (a very active topic in that day) or a nucleus of interested people. We (including such persons as Hendrik Bode, Bob Dietzold, Ernie Guillemin, and Bill Huggins) went so far at one time as to investigate the possibility of starting a separate magazine in circuit theory and I well remember that we gave up the idea when some cost estimates were received through Bell Labs. About 1948 or 1949, when the concept of professional groups in the IRE was first broached, this group saw the possibility of having a separate organization for itself. Most of the group were members both of AIEE and IRE.

With strong encouragement of W. R. G. Baker, Vice-President of General Electric, Treasurer of the IRE, and a strong promotor of the idea of professional groups, we managed to organize the Cir-

cuit Theory Group in the IRE. I have the vague feeling that I was either the first or second chairman of this group. However, there were so many good men available at that time to pick from, that I feel slightly hesitant in mentioning this "fact," even though you list me as acting chairman before March 1951.

In any event, the professional group moved ahead with considerable enthusiasm and the inauguration of the Transactions was the culmination of our early efforts to start a magazine devoted to circuit theory.

I think the situation should be clarified: the organization of the professional group on circuit theory did not depend upon the adoption of its constitution by the Board of Directors. Rather, it was recognized and the group thus caused a constitution to be created. However, this was something of a formality and, as I recall, the constitution was much like a standard one for the various professional groups of the IRE. The real start of the group was in 1949–50 when we were recognized and started having regular and frequent meetings at IRE headquarters.

Sincerely yours,

John G. Brainerd
University of Pennsylvania, Professor (Em.)

Letter from John Brainerd.

- On 5 April 1949 the IRE Board approved a petition signed by 26 IRE members to form the IRE Circuit Theory Group.

- 20 March 1951 - First meeting of the IRE Professional Group on Circuit Theory. The first issue of the Transactions, PGCT-1, was dated December 1952. It refers to a Newsletter for members but no copy of the original

231

issue has been found [1]. By 1953 there was a Los Angeles and a Chicago Chapter, because they have a brief report in PGCT-2, and later Chapters in Philadelphia and Seattle report, the former stating that they have completed their first year of operation, and the latter having their first meeting in March 1954.

- 25 March 1963 - Name change to IEEE Professional Technical Group on Circuit Theory.

- 1966 - Became IEEE Group on Circuit Theory.

- In 1971 the opportunity to gain greater independence by a change from Group to Society status arose, and the IEEE Board of Directors approved the change in May 1972. Later that year (December 1972) the Board approved the change in name to Circuits and Systems Society. This data comes from a History of the Society written by Ben Leon (in the CAS Magazine Centennial Issue), and differs from another claim (below) that the name change was in November.

- 2 November 1972 - Name change to IEEE Circuits and Systems Society.

Conferences and International Outlook

The first annual symposium was held in Miami Beach, Florida, in December 1968. Prior to that the Circuit Theory Group had co-sponsored (apparently without financial involvement) several meetings in the Circuit Theory field:

the Midwest Symposium on Circuit Theory, the Allerton Conference on Circuit Theory, etc. A decision was taken to concentrate on supporting only one Symposium per year soon after the Miami Beach event, and these were the forerunners of ISCAS.

The IRE Circuit Theory Group had developed an international perspective at a very early stage, and after holding the annual symposium a couple of times in USA, by 1969 there was a wish to hold it outside the USA. A bid in May 1969 from the Circuit Theory Chapter of the IEEE United Kingdom and Republic of Ireland Section to host it in London, England in 1971, was accepted in August 1969. Following a Symposium in Toronto in 1973, a policy to hold the event outside the

---

[1]The failure to find this Newsletter might be due to a misunderstanding as to what is being looked for, because PGCT-1, pages 2 and 3 have a heading of Newsletter and so there need never have been a separate publication at all!

Left: Guest of Honour B.D.H. Tellegen speaking at the 1971 Symposium banquet, George S. Brayshaw, Symposium Chairman, seated.
Right: Guest of Honour B.D.H. Tellegen, Tony Davies (giving 'welcome speech'), George S. Brayshaw, Symposium Chairman, Arthur Stern (Magnavox Co.), a past Chairman of the Circuit Theory Group and future IEEE President.

USA every third year was adopted, starting with Munich in 1976. Following this pattern, the Symposium returned to London, England in 1994. In the 13 years from 2000 to 2015, it was in the USA only three times! Since it will be in Montreal in 2016, the next time in USA will be 2017 in Baltimore after a gap of 9 years since Seattle in 2008.

Although the Society has sponsored and co-sponsored many different series of conferences and continues to do so, its primary annual conference is still the International Symposium on Circuits and Systems (ISCAS). While the Presidents were all from the USA until 1999, a look at a chronological list shows that since then, Presidents have been increasingly drawn from a worldwide base.

Publications of the Society

The publication in the Electrical Engineering field has a long and rich history. One of the first publications dates back in 1909; it is the Proceedings of the Wireless Institute. That Institute become the Institute of Radio Engineers (IRE). The Proceedings of the IRE, the official publication of the Institute, published outstanding contribution by Armstrong, de Forest, Hopper, Marconi, Mauchly, and Zworykin.

From the very first issue of the IRE Circuit Theory Group, PGCT-1, the Transactions have been the source of many of the fundamentals which have led to the spectacular achievements of modern electronic engineering.

The reputation and achievements of the CAS Society are substantially dependent upon the Transactions, and while the content is created by the authors

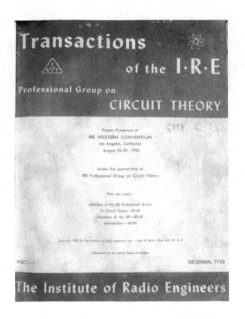

Cover of first publication, PGCT-1.

(with not insignificant contributions from reviewers), the editors are responsible for the overall quality and timeliness of what is published. Increasingly, content came from authors outside the USA and a more recent trend has been many more editors from outside USA.

Although what started out as the Transactions on Circuit Theory (which has undergone various changes and bifurcations into Part I and Part II, etc) is the central publication of the Society for leading research, it has moved to cover various other areas, sometimes by joint publications with other IEEE Societies for example the Transactions on Computer Aided Design and Transactions on Video Technology.

Additionally it has produced various styles of Newsletter over the years, and now has a Magazine, containing high quality tutorial papers.

The Core Subjects in the Early Years

Many of the early members and contributors were passive-filter designers. The design of high-performance passive filters was a very specialized topic, understood by only a few experts, but it was crucial to the implementation of line-based telecommunications systems and quite important in radio communications.

Filter designers were often regarded as 'a race apart' - engaged in using abstract theories in an almost 'black art' of which most engineers had no understanding. Even when digital computers became available to assist in the design calculations, it was at first necessary to use triple-length arithmetic (or more) in order to obtain sufficient prec

The beginnings of the Circuit Theory Group were also involved in educational aspects of Circuit Theory with a strong mission to teach fundamentals of the subject, as opposed to the ad hoc approaches which characterized the circuit-teaching in much of the university electrical engineering curriculum. The subject offered some key advantages: the possibility of an axiomatic approach, rigorous develop-

Steven Butterworth [Wikipedia].

ment of a theory uncontaminated by the imperfections of practical components and experiments, and the prospect of formal synthesis - being given a 'requirement' and producing, by a step-by-step procedure guaranteed to succeed, a circuit implementation. Although the implementations were 'theoretical' ones, requiring idealized linear, time-invariant and often lossless components, there was a real sense in which this represented an alternative to much of engineering practice in other subjects. It also laid a pedagogical foundation, widely believed (at least by the Circuit Theory people) to be a strong contender to be the basis for an engineering education in all disciplines. It also links strongly with fundamental ideas in Computer Science, where an assurance that an algorithm will terminate in a finite number of steps with a successful outcome has analogies with the requirements of formal circuit synthesis methods. For example, the proof that every positive real rational function can be implemented as an electrical network of linear lumped passive (e.g R,L,C) components, following a synthesis procedure guaranteed to terminate in a predictable number of steps.

An additional feature considered important in teaching engineering was normalisation: filter design simplified the process by designing a low pass filter with a cut-off frequency of 1 radian/sec, and terminated in a 1Ω resistor which could then

Post stamps honoring Pafnuty Chebyshev.

be easily scaled in frequency and impedance level to the actual requirement, including transformation from low pass to high pass, band pass or band stop. Students could also be easily introduced to approximations, becoming familiar with the Butterworth, Chebychev and Elliptic approximations in an easily understood context.

There was much hope from 'analogies'. The realization that the successful field of electrical circuit theory could apparently be applied just as well to mechanical, thermal, acoustical and other dynamical systems seemed to suggest that electrical circuit theory could become the foundation for many branches of engineering and not only electrical engineering - unfortunately, the lack of good practical implementations of the 'ideal linear lumped time-invariant' circuit elements in non-electrical systems severely restricted the extent to which this hope could be realized.

The linear, lumped, passive, finite, bilateral, time-invariant assumptions limited the scope of much Circuit Theory in these early years but nevertheless provided a broad field in which significant fundamental research could be done, and provided the 'training ground' for many graduate students and professors. There were failures to adequately deal with electronic components - the practice of actual electronic circuit design (at that time involving thermionic valves/tubes) did not conform well to the formalized design processes advocated by many of the leaders in circuit theory, in many cases, it involved non-linearity in an essential way (so that a linear time-invariant assumption was simply not useful) and there was a lack of clear and agreed ideas about how to extend the set of ideal linear passive elements to take into account 'activity' in a suitably idealized way to make a formal extension of passive circuit theory. Active circuits were often simply defined as those circuits which were 'not-passive', and little more was said about them.

Active filters were occasionally suggested, but never used in practice except for very low frequency (sub-audio) applications, such as mechanical servomechanisms - the only available active element was the thermionic valve (tube) which was expensive, unreliable, and required high voltage power supplies.

The Minutes of a 22 March 1956 meeting of the AdCom of the IRE Circuit Theory Group includes the statement " ... *only by taking in more fringe areas (e.g. Transistor Circuits) can we really obtain more members..*" "... *consensus ... not to have a membership drive ...*". This extract appears to indicate that the AdCom members did not consider that the theory or design of circuits containing transistors as either important or within the real scope of Circuit Theory. In view of subsequent developments in electronics, the description of Transistor Circuits as a 'fringe area' of Circuit Theory seems rather quaint, and certainly an indication of not foreseeing the future.

What can be claimed with confidence is that many of the early members of the Society who helped to define it were pioneers who had immense influence.

The often quoted 1676 statement of Isaac Newton "*If I have seen further, it is by standing on the shoulders of giants*" surely applies to all those of us who likewise were influenced by the early 'giants' of the Circuit Theory field, many of whom went on to influential positions of significance (IEEE Presidents, Deans of major engineering schools, founders of innovative companies. etc.).

Topics initiated by the Society often 'spun off' into the formation of other IEEE Societies, or changed the scope of some existing Societies. While this is not always recognised or acknowledged, it will be understood easily by those who have had a long involvement in the CAS Society and regarded it as their professional home or birthplace.

Examples include

- digital filters, which enabled a small IEEE Society called 'Audio and Electroacoustics' to develop into the IEEE Signal Processing Society of today.

- the use of computers for on-line simulation and design of electronic circuits which has become the norm and necessity in the design of modern electronic and electromechanical systems. By 1968, the Circuit Theory Group had set up the 'CANDE (Computer Aided Network Design)' committee to promote the use of computers in the analysis and design of circuits. The December 1970 Circuit Theory Group has an announcement (page 9), written by CANDE chairman, Ronald Rohrer, of a plan for a 'computer program information pool' to document available programs. CAS Newsletter about the first CANDE workshop held in Quebec, 28-28th September 1972. CANDE could be regarded as a foundation for the much later formation of the IEEE Council on Electronic Design Automation, comprising six member IEEE Societies.

- the applications of graph theory initially to electrical network analysis and now to communications systems, transportation, chemistry and much more.

A notable feature is the often long gestation time for fundamental ideas to move into useful applications - an example is the Memristor, initially proposed by Leon Chua in 1971 as a 'missing' non-linear circuit component linking charge and magnetic flux, which did not find practical implementations and applications for nearly 40 years.

I believe that all members of the IEEE CAS Society can be proud of the great heritage that its formation and development comprises.

The Real Fundamentals of the Circuit Theory

The driving point impedance of any linear time-invariant passive system / circuit / network is a positive-real function of complex frequency. Further, if a circuit is constructed from a finite number of linear lumped passive time invariant components, e.g. from the familiar ideal {R, L, C, M, ideal transformer, gyrator} set, then this driving point impedance is a positive real rational function, for which a formal synthesis procedure is available.

Brune [O. Brune 'Synthesis of a finite two-terminal network whose driving point impedance is a prescribed function of frequency', J. Math. Phys. 10, 191, 1931] showed that every such rational function could be systematically implemented by a systematic construction (though requiring, in most cases, inconvenient mutually coupled inductive elements or ideal transformers). Finally, Bott and Duffin [R. Bott and R.J. Duffin 'Impedance synthesis without the use of transformers', J. Appl. Phys. 20, 816, 1949] were able to use the Richards transformation [P.I. Richards 'A special class of functions with positive real

part in a half-plane', Duke Math. J., 14, 777, 1947] to provide a transformerless
(e.g. R,L,C) synthesis procedure.

SYNTHESIS OF REACTANCE 4-POLES WHICH PRODUCE
PRESCRIBED INSERTION LOSS CHARACTERISTICS

INCLUDING SPECIAL APPLICATIONS TO FILTER DESIGN

BY S. DARLINGTON*

Outline

INTRODUCTION

Of the various types of electrical networks which are frequently found
useful, one of the commonest is the 4-terminal transducer of reactances,
more briefly referred to as the reactance 4-pole.[1] In particular, the
selective networks or filters which are commonly used for transmitting
certain frequencies while blocking others are almost always reactance
4-poles, and these filters form essential parts of most communication
systems.

Detailed methods of designing filters and related reactance 4-poles
are well known and have been in general use for a considerable period.
For the most part these fit into one general filter design scheme which
may be referred to as the image parameter theory, since it is based upon

*Bell Telephone Laboratories, Inc. This paper has been accepted as a Doc-
tor's thesis by the Faculty of Pure Science of Columbia University. The manuscript was received by the Editors May 18, 1938.

[1] Throughout this paper, the term 4-pole will be used to indicate a 4-terminal
transducer—i.e. a network with two pairs of accessible terminals subject to the
restriction that no external connections can be made between terminals of
different pairs. The term has been widely used in this sense and also to indicate
a network with four terminals to which external apparatus can be connected in
any desired manner.

257

First page of a Darlington publication.

These results appeared to be of great
significance at the time - especially given
the analogies with non-electrical systems
- and the lack of practical utility of many
of the synthesized circuits was more-or-
less overlooked. However, it represented
the achievement of an ideal missing in
much of engineering then and today.

Starting with a precise, formal
(mathematical) statement of the problem to be solved and achieving a realization by a systematic process guaranteed in advance by theory to succeed
(in a finite number of steps) represented
a major achievement, the importance
of which can hardly be overstated, and
it was also a philosophy which seemed
ideal for the educational foundation of
electrical engineers. (Would it not be
nice if today's office-PC software could
be designed by such procedures).

Darlington made an outstanding contribution, which must have appeared
to many at the time to be of no practical significance whatsoever. He was able
to extend the synthesis methods for driving-point impedances to show that any
positive real rational function could be implemented as a structure of lossless
(e.g. L, C) elements and exactly one positive resistor.

Despite the apparent practical uselessness of this theoretical result, relationships between the magnitude of scattering parameters of lossless two-ports
enabled this to be related to the implementation of a prescribed magnitude-
frequency response as the insertion loss of a lossless two port with resistive
terminations. This led directly to a solution to the problem of designing the
high-performance frequency selective filters upon which the whole of the analog
frequency-division multiplex based line and radio communications industry
depended at least until digital technology increasingly replaced them.

Note the absence of any consideration of non-reciprocity in the foregoing -
Tellegen invented the Gyrator - a 'missing' element was needed to complete the
theory for the 'passive' domain, in order to have non-reciprocity without activity.
The ideal gyrator provided just such a passive device [B.D.H. Tellegen,'The
gyrator; a new network element', Philips Research Report, 3, 81, 1948].

Subsequently, it became fashionable to try to devise 'new circuit elements',
many of which did not survive or achieve importance. Among the more abstract
and at first apparently useless concepts are a two-terminal element for which
both the voltage and current are always zero and a two-terminal element for
which both the voltage and the current are undefined. The description of such

Raoul Bott(left). Recommendation letter signed by Richard Duffin for John Nash, Nobel Price in Economic Sciences. Boff and Duffin implemented a transformerless synthesis procedure (right).

elements might cause the practically-minded engineer to suspect an 'April Fool Joke' and to check if the cover date of the publication he/she was reading was 1st April, yet these two elements, combined together as a 'nullor', represent a practically useful model of an ideal transistor and of an operational amplifier, and have found a permanent place in Circuit Theory.

The concept of an electrical circuit as a linear graph formed the foundation for much of the theory and a basis for systematic methods of analyzing complicated networks, and as a result, Graph Theory laid the basis for the computer based analysis methods and simulators (such as SPICE) which we now take for granted, and which provide one of the foundations upon which the suc-

Larry Nagel (left) developed SPICE under the supervision of Prof. Donald Pederson (right) [Wikipedia].

cess of modern integrated circuit technology stands. Kirchhoff's first and second laws were 'graph theory based' but what Weinberg called Kirchhoff's third and fourth laws were almost unknown, yet provide a foundation for much of the network theory developed (using concepts of trees and cutsets, etc.) which is essential for the systematic analysis of large networks. For example, the determinant of the nodal admittance matrix is the sum of all the tree-admittance products - so enabling all numerical processing to be side-stepped and demon-

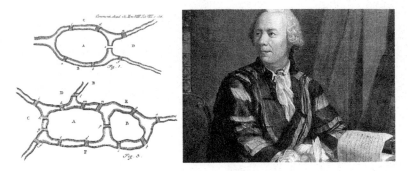

Left: Königsberg seven bridges and fifteen bridges problems. Right: Leonhard Euler.

strating immediately and dramatically the link between circuit topology and circuit transfer-functions. A book by S. Seshu and M. B. Reed (Linear Graphs and Electrical Networks, Addison Wesley, 1961) had a very strong influence on its readership. It must be remembered that Euler created Graph Theory when providing a formal solution to the unimportant question of whether the citizens of Königsberg, Prussia on the banks of the Pregel River, could take a walk returning to their starting point and crossing each of the bridge exactly once.

Controlled sources were a natural way of introducing activity into otherwise passive circuits. However, since they were also used in the representation of passive circuits (for example in modelling mutual inductance, ideal transformers and gyrators) they did not offer the convenience of being a distinctive element to be added to the passive set to introduce the concept of activity. For a while it seemed that the unlikely candidate of the negative impedance convertor (NIC) was going to fill this role. Although the concept of a 'negative resistance' was easily understood at least at a superficial level by most practicing electronics engineers, the notion of a negative capacitance or inductor was not easily accepted by such people.

The motivation for developing active filters was mainly the elimination of inductors (on grounds of their size, weight, and non-ideal properties).

Linvill [J.C. Linvill,'RC Active Filters', Proc. IRE, 42, 55, 1954 'Synthesis of Active Filters', Poly. Institute Brooklyn, MRI Symposia series, 5, 453, 1955] showed that by using just a single voltage-inversion negative impedance convertor (NIC) as the active element, any rational function could be realized as the transfer function of an RC-active circuit, and shortly afterwards, Yanigisawa [T. Yanigisawa,'RC active networks using current inversion negative impedance convertors', Trans IRE, CT-4, 140, 1957] provided a simplified synthesis procedure, using a current-inversion NIC. This stimulated intensive research into Active-RC synthesis using the NIC, and many ideas for implementing such an idealized component using transistors. However, the enthusiasm was soon dampened by the realization that the sensitivity of the circuits was so high that they were almost useless in practice. In the following decade of work with the

NIC the principal value seems, in retrospect, to have been in the production of doctoral theses and the launching of academic careers. Very few systems went into actual production as a result of this work!

What was apparently not realized by many from a passive filter background was well-known to most practicing electronics engineers: in order to get a low sensitivity with only one amplifier, a very high loop gain is needed - and the many inventive schemes to make a highly accurate NIC with two or three low gain transistors were doomed to failure. It was not until the invention of the integrated circuit OP-Amp by Bob Widlar at Fairchild that a cheap high-gain component became available to implement active-RC filters. It was then mainly the much older Sallen and Key structures [R.P Sallen and E.L. Key, 'A practical method of designing active RC filters', Trans. IRE, CT-2, 74, 1955], that survived the transition to engineering practice.

An important breakthrough came with the observation by Orchard [H.J. Orchard, 'Inductorless Filters', Electronics Letters, 2, 224, 1966] that the sensitivity to component tolerances in the classical doubly-terminated passive, lossless, LC ladder filters is exceptionally good, especially in the passband, because of the non-negative property of the insertion loss of such a passive structure, and this led to the understanding that imitating this behavior in active and digital filters was a route to getting the low sensitivity needed in practical circuits.

The large silicon area required for accurate resistors prevented successful single chip implementations of Active RC filters, and Switched-capacitor filters were the first practically successful approach to widespread implementation of high-performance filters in silicon monolithic form. Despite the need to distribute high-frequency clocks around the chip, they found their way into many real systems.

Digital filters were an inevitable development, although for a long time, their practical implementation was severely limited for real-time signal processing over the frequency ranges needed by communications systems. Although there were many attempts to implement digital filters in integrated circuit form, the development of the TMS 320 series of DSP chips by Texas Instruments was a major stimulus to converting these ideas to widespread use.

The usefulness of wave digital filters may sometimes be questioned, but the theory developed by Fettweis [A. Fettweis, 'Wave Digital Filters', Proc. IEEE, 74, 270, 1986] showed the unanticipated result that concepts from classical network theory (including passivity) could be transferred into the field of purely numerical processing of data, and that classical network theory did, after all, have something important to contribute to the emerging field of digital signal processing. In June 2001, the company Infineon Technologies AG celebrated the delivery of 50 million units of a subscriber-line filter product, each one of which contained several wave digital filters Non-linearity has not been mentioned so far - it was often considered unwelcome, to be avoided, either by pretending it was not there or by modifying designs so as to minimize its effects. A perfect world was often assumed to 'linear, lumped, finite, time-invariant, passive, bilateral', and anything falling outside this was regarded as unwelcome and harmful. It took the modern developments in dynamics and the discovery of chaos and fractals to demonstrate more widely that reality is non-linearity and that the real world is non-linear in a way that engineers need to understand and to exploit.

Consequently, non-linearity was at first considered a minor and mostly unwanted feature in Circuit Theory; only such aspects as harmonic distortion and intermodulation effects. Even though a proper explanation of oscillator behaviour required non-linear dynamics, this was generally avoided. Not until the late 1980s did non-linear dynamics begin to take a significant place in the interests of the Society, concurrently with discoveries about and applications of chaos in engineering, and the development of over-sampled systems based on sigma-delta modulators and other inherently non-linear systems.

The rest of the story is not history, it is going on around us! What looks like useless theory today can often turn out to be an essential foundation for tomorrow's technology, a message that is frequently not understood by many of today's senior and usually well-paid administrators who want to oversee immediately useful and profitable results from the people they control.

Acknowledgements

Any mention of the basis of Circuit Theory as a discipline in its own right must mention Ernst Guillemin: his impact through a few outstanding textbooks and the notable careers of his many students laid a sound pedagogic foundation of the subject, replacing a previous ad-hoc approach to electrical circuit education. (C.C. Bissell, *"He was the father of us all."* Ernie Guillemin and the teaching of modern network theory. History of Telecommunications Conference (HISTELCON '08), 11-12 Sept 2008, Paris.)

Ben Leon collected much useful material for his *"History of the Circuits and Systems Society"* published in the Centennial Issue (December 1983) of the IEEE Circuits and Systems Magazine (vol. 5, No. 4). George Moschytz persuaded me, rather against my wishes, to create the first CAS Society Directory in time with the Society's 50th Anniversary in 1999. The CAS Executive Committee at that time gave me a free hand to define the concept, structure and content of the first Directory, in which I decided to include some historical data about the subject and the Society, from which much of this Chapter has been based.

Anthony C Davies, Life Fellow IEEE,
Emeritus Professor, King's College London, England

CIRCUITS AND SYSTEMS SOCIETY

1951 - First meeting of the IRE Professional Group on Circuit Theory.
1963 - Name change to IEEE Prof. Technical Group on Circuit Theory.
1966 - Became Group on Circuit Theory.
1972 - Name change to IEEE Circuitscuits and Systems Society.

CAS SOCIETY OFFICERS (first ten years)

1949-51
Chairman   Justohn G. Brainerd (Acting)

1951-52 (Officers to June 1952)
Chairman   John G. Brainerd
Secretary-Treasurer   Herbert J. Carlin

1952-53 (Officers to June 1953)
Chairman   R. L. Dietzold
Vice Chairman   Chester H. Page
Secretary-Treasurer   William H. Huggins

1953-54 (Officers to June 1954)
Chairman   Chester II. Page
Secretary-Treasurer   Milton Dishal

1954-55 (Officers to June 1955)
Chairman   Chester H. Page
Vice Chairman   Herbert J. Carlin
Secretary-Treasurer   Milton Dishal

1955-56 (Officers to June 1956)
Chairman   Herbert J. Carlin
Vice Chairman   William H. Huggins
Secretary-Treasurer   Milton Dishal

1956-57 (Officers to June 1957)
Chairman   Herbert J. Carlin
Vice Chairman   William H. Huggins
Secretary-Treasurer   James H. Mulligan, Jr.

1957-58 (Officers to June 1958)
Chairman   William H. Huggins
Vice Chairman   Sidney Darlington
Secretary-Treasurer   James H. Mulligan, Jr.

1958-59 (Officers to June 1959)
Chairman   William H. Huggins
Vice Chairman   Sidney Darlington
Secretary-Treasurer   James H. Mulligan, Jr.

CIRCUITS AND SYSTEMS SOCIETY PRESIDENTS

Date	Name	Affiliation
1949-1952	John G. Brainerd	University of Pensylvania
1952-1953	Robert L. Dietzold	Bell Labs
1953-1955	Chester H. Page	National Bureau of Standard
1955-1957	Herbert J. Carlin	Polytech. Inst. of Brooklyn, NY
1957-1959	William H. Huggins	Westinghouse Electric Corp.
1959-1961	Sidney Darlington	Bell Labs
1961-1963	James H. Mulligan, Jr.	New York University
1963-1965	Ralph J. Schwarz	Columbia University
1965-1966	John G. Linvill	Stanford University
1966-1967	Mac E. Van Valkenburg	University of Illinois
1967	Franklin H. Blecher	National Bureau of Standard
1968-1969	Arthur P. Stern	Bell Labs
1970-1971	Benjamin J. Leon	The Magnavox Company
1972	Ernest S. Kuh	University of California, Berkeley
1973	M. Robert Aaron	Bell Labs
1974	Sydney R. Parker	Naval Postgraduate School
1975	Belle A. Shenoi	Wright State University
1976	Mohammed Ghausi	Wayne State University
1977	Leon Chua	University of California, Berkeley
1978	Omar Wing	Columbia University
1979	Thimothy N. Trick	University of Illinois, Urbana
1980	Carl F. Kurth	Bell Labs
1981	Stephen W. Director	Carnegie Mellon University
1982	Bede Liu	Princeton University
1983	Kenneth R. Laker	University of Pennsylvania
1984	Alan N. Willson Jr.	University of California, LA
1985	W. Kenneth Jenkins	University of Illinois
1986	Sanjit K. Mitra	Univ. of California, S. Barbara
1987	Ronald A. Rohrer	Carnegie Mellon University
1988	Ming Liou	Bell Labs
1989	Anthony N. Michel	Notre Dame University
1990	Rolf Schaumann	University of Minnesota
1991	Sung Mo (Steve) Kang	University of Illinois
1992	Randall L. Geiger	Texas A & M University
1993	Philip V. Lopresti	AT&T
1994	Wai-Kai Chen	University of Illinois,
1995	Ruey-Wen Liu	Notre Dame University
1996	Michael R. Lightner	University of Colorado
1997	John Choma Jr.	Univ. of Southern California
1998	Rui J.P. de Figureido	University of California, Irvine
1999	George S. Moschytz	Swiss Federal Inst. of Tech.
2000	Bing J. Sheu	Nassda Corp., Santa Clara,
2001	Hari C. Reddy	California State University
2002	Josef A. Nossek	Munich University of Technology

2003	Giovanni De Micheli	Stanford University
2004	M.N.S. Swamy	Concordia University
2005	Georges Gielen	Katholieke Univ. Leuven
2006	Ellen Yoffa	IBM Corporation
2007	Ljiljana Trajkovic	Simon Fraser Univ.
2008	Maciej Ogorzałek	Jagiellonian Univ. Krakow
2009	David J. Allstot	University of Washington
2010	Gianluca Setti	University of Ferrara, Italy
2011	Mani Soma	University of Washington
2012-2013	Athanasios G. Stouraitis	Univ. of Patras, Greece
2014-2015	Vojin G. Oklobdzija	University of California, Davis
2016-2017	Franco Maloberti	University of Pavia, Italy

Annual Symposium of the CAS Society

The first annual symposium was held in Miami Beach, Florida, in December 1968. Prior to that the Circuit Theory Group had co-sponsored (apparently without financial involvement) several meetings in the Circuit Theory field: the Midwest Symposium on Circuit Theory, the Allerton Conference on Circuit Theory, etc. A decision was made to concentrate on supporting only one event per year soon after the Miami Beach event.

The international perspective which characterized the Circuit Theory Group can be judged from the decision to hold this event in London, England in 1971, a decision in principle to do so must have been made not later than 1969. Although the Society has sponsored and co-sponsored many different series of conferences, and continues to do so, the primary annual conference is the International Symposium on Circuits and Systems (ISCAS). The following list gives the date, location and general chairman of ISCAS since its inception.

Date	Venue	General Chairman
1968	Miami Beach, Florida	Omar Wing
1969	San Francisco, CA	unavailable
1970	Atlanta, Georgia	H.E. Meadows
1971	London, England	George S. Brayshaw
1972	Los Angeles, CA	Sydney R. Parker
1973	Toronto, Canada	Kenneth C. Smith
1974	San Francisco, CA	Sanjit K. Mitra
1975	Boston, MA	John Logan
1976	Munich, Germany	Rudolf Saal
1977	Phoenix, Arizona	William Howard
1978	New York City	H.E. Meadows
1979	Tokyo, Japan	Yosiro Oono
1980	Houston, Texas	Rui J.P. de Figureido
1981	Chicago, Illinois	B. J. Leon , M. Van Valkenburg

1982	Rome, Italy	Antonio Ruberti
1983	Newport Beach,	CA George Szentirmai
1984	Montreal, Canada	M.N.S. Swamy
1985	Kyoto, Japan	Toshio Fujusawa
1986	San Jose, CA	George Szentirmai
1987	Philadelphia, Penn.	Samuel Bedrosian
1988	Helsinki, Finland	Yrjo Neuvo
1989	Portland, Oregon	Tran Thong
1990	New Orleans, Louisiana	Anthony Michel, M. Sain
1991	Singapore	J.C.H. Phang
1992	San Diego, CA	Stanley A. White
1993	Chicago, Illinois	Wai-Kai Chen
1994	London, England	Robert Spence
1995	Seattle, Washington	Robert J. Marks II
1996	Atlanta, Georgia	Philip E. Allen
1997	Hong Kong	Tony T.S. Ng, Ming Liou
1998	Monterey, California	Sherif N. Michael
1999	Orlando, Florida	Wasfy B. Mikhael
2000	Geneva, Switzerland	Martin J. Hasler
2001	Sydney, Australia	G. R. Hellestrand, D. J. Skellern
2002	Phoenix, Arizona	David J. Allstot, S. Panchanathan
2003	Bangkok, Thailand	S. Pookaiyaudom, C. Toumazou
2004	Vancouver, Canada	Andreas Antoniou
2005	Kobe, Japan	Nobuo Fuji
2006	Kos, Greece	Thanos Stouraitis
2007	New Orleans	Magdy Bayoumi
2008	Seattle, Washington	David Allstot
2009	Taipei, Taiwan	Jhing Fa Wang
2010	Paris, France	Amara Amara
2011	Rio De Janeiro, Brazil	Paulo Diniz
2012	Incheon, Korea	Myung Sunwoo
2013	Beijing, China	C. W. Chen, W. Gao, J. Vandewalle
2014	Melbourne, Australia	Jugdutt Singh, David Skellern
2015	Lisbon, Portugal	Jorge Fernandes, Wouter Serdijn
2016	Montreal, Canada	Mohamad Sawan

Editors of CAS Society Publications

The reputation and achievements of the CAS Society are substantially dependent upon the Transactions, and while the content is created by the authors (with not insignificant contributions from reviewers), the Editors are responsible for the overall quality and timeliness of what is published. From the very first issue of the IRE Circuit Theory Group, the Transactions have been the source of many of the fundamentals which have led to the spectacular achievements of modern electronic engineering. Because the names of the Editors are often recorded only on the cover pages of the Transactions, and these pages are frequently discarded by libraries as part of the process of binding the annual volumes, the identities of the Editors are easily lost, especially for early issues. The list below is intended to help to preserve this heritage. For many years, the change in editor took place in mid-year. More recently, the changes have been on a Calendar year basis (normally a two year term starting in January). Notice that initially and for several years, all editors were from USA, though increasingly, content came from authors outside the USA. A more recent trend has been many more editors from outside USA.

Editors of the Transactions on Circuit Theory
The first issue was December 1952

W.H. Huggins	John Hopkins Univ., Baltimore	1954-1957
W.R. Bennett	Bell Telephone Labs	1958-1960
M.E. Van Valkenberg	Univ. of Illinois	1961-1963
Norman Balabanian	Syracuse University	1963-1965
Dante Youla	Polytechnic Inst. of Brooklyn	1965-1967
Benjamin Leon	Purdue University	1967-1969
Gabor Temes	UCLA, CA, USA	1969-1971
George Szentirmai	Cornell University	1971-1973

At the end of 1973, the title *"Circuit Theory"* was dropped and replaced by *"Circuits and Systems,"* used from January 1974.

Editors of the Transactions on Circuits and Systems

Leon Chua	Univ. of California, Berkeley	1973-1975
Omar Wing	Columbia University	1975-1977
Alan N. Willson	UCLA, CA	1977-1979
Ming L. Liou	Bell Telephone Labs	1979-1981
Anthony Michel	Iowa State University	1981-1983
Rolf Schaumann	Univ. of Minnesota	1983-1985
Andreas Antoniou	Univ. of Victoria, Canada	1985-1987
Yen-Long Kuo	Bell Labs, North Andover	1987-1989
Ruey-Wen Liu	Notre Dame University	1989-1991

In January 1992, the Transactions were split into two parts, TCAS-I and TCAS-II, subsequently each with its own editorial team. TCAS-I was allocated to *"Fundamental Theory and Applications"* and TCAS-II to *"Analog and Digital Signal Processing"*. After a while this distinction was not easy to sustain, and the two parts were later discontinued and replaced by a new format with Part 1 being for 'regular papers' and Part 2 being for 'express briefs' (short papers).

Editors of the Transactions on Circuits and Systems, Part I

Wai-Kai Chen	Univ of Illinois, Chicago	1991-1993
Martin Hasler	EPFL, Switzerland	1993-1995
Josef Nossek	TU München, Germany	1995-1997
Pier Paolo Civalleri	Politecnico di Torino, Italy	1997-1999
M.N.S. Swamy	Concordia University, Canada	1999-2001
Tamás Roska	Sztaki, Hungary	2001-2003
Keshab Parhi	Univ of Minnesota	2004-2005
Sankar Basu	National Science Foundation	2006-2007
Gianluca Setti	University of Ferrara, Italy	2008-2009
Wouter Serdijn	Technical University Delft	2010-2011
Gabriele Manganaro	Analog Devices	2012-2013
Shanthi Pavan	ITT Madras	2014-2015

Editors of the Transactions on Circuits and Systems, Part II

Wai-Kai Chen	Univ of Illinois, Chicago	1991-1993
Dave J. Allstot	Univ. of Washington, Seattle	1993-1995
John Choma	USC, Los Angeles	1995-1997
Edgar Sanchez-Sinencio	Texas A&M Univ.	1997-1999
Chris Toumazou	Imperial College London	1999-2001
Ian Galton	UC, San Diego	2001-2003
Sankar Basu	National Science Foundation	2004-2005
Gianluca Setti	Univ of Ferrara, Italy	2006-2007
U.-K. Moon	Oregon State Univ.	2008
Tony C. Carusone	University of Toronto	2009
Yong Liang	National University of Singapore	2010-2013
Jose Silva-Martinez	Texas A&M Univ.	2014-2015

CAS Society Awards

The IEEE Circuits and Systems Society's annual awards program recognizes member achievement in education, industry, technological innovation and service. The purpose of the program is to illuminate the accomplishments of CAS Society members and celebrate their dedication and contributions both within the field and to the CAS Society. Award recipients are nominated by their CASS peers in order to honour the service and contributions that further strengthen the CAS Society. There are two general categories of awards, the IEEE Technical Field Awards and the Society and Achievement Awards.

The IEEE Technical Field Awards are:

> Gustav Robert Kirchhoff Award
> IEEE Biomedical Engineering
> IEEE Fourier Award for Signal Processing

The Society and Achievement Awards are:

> Mac Van Valkenburg Award
> Charles A. Desoer Technical Achievement Award
> John Choma Education Award
> Vitold Belevitch Circuits and Systems Award
> Industrial Pioneer Award
> Meritorious Service Award

The **IEEE Gustav Robert Kirchhoff Award** was established in 2003. This award is named for Gustav Robert Kirchhoff, the physicist who made important contributions to the theory of circuits using topology and to elasticity. Kirchhoff's laws allow calculation of currents, voltages and resistances of electrical circuits extending the work of Ohm. His work on black body radiation was fundamental in the development of quantum theory. The recipients are:

2005	Leon Chua
2006	Gabor Temes
2007	Yannis P. Tsividis
2008	Alfred Fettweis
2009	Ertnest S. Kuh
2010	Hitoshi Watanabe
2011	Charles A. Desoer
2012	Ronald A. Rohrer
2013	Sanjit K. Mitra
2014	Chung Laung Liu
2015	Yosiro Oono

The **IEEE Biomedical Engineering Award** was established in November 2010. The first presentation was scheduled for 2014. The recipients are:

| 2014 | Lihong Wang |
| 2015 | Christofer Toumazou |

The **IEEE Fourier Award** for Signal Processing was established in November 2012. The first presentation was scheduled for 2015. The recipients are:

| 2015 | Georgios G. Giannakis |
| 2016 | Bede Liu |

The **Mac Van Valkenburg Award** honors the individual for whose outstanding technical contributions and distinguishable leadership in a field within the scope of CAS Society are consistently evident. The award is based on the quality and significance of contribution, and continuity of technical leadership. The recipients are:

1985	Sidney B. Darlington
1987	Mac Elwyn Van Valkenburg
1988	Ernest S. Kuh
1989	Omar Wing
1990	Ronald A. Rohrer
1991	Hitoshi Watanabe
1992	Stephen W. Director
1993	Vitold Belevitch
1994	Timothy N. Trick
1995	Leon O. Chua
1996	Charles A. Desoer
1997	Bede Liu
1998	Leon O. Chua
1999	Sanjit K. Mitra
2000	Ming-Lei Liou

2001	Alfred Fettweis
2002	Rui J. P. de Figueiredo
2003	Alan N. Willson, Jr.
2004	Gary Hachtel
2005	Sung-Mo (Steve) Kang
2006	George S. Moschytz
2007	Ruey-Wen Liu
2008	Jan M. Rabaey
2009	Gabor C. Temes
2010	Lawrence Pileggi
2011	David J. Allstot
2012	Giovanni de Micheli
2013	Franco Maloberti and Tamás Roska
2015	Georges G.E. Gielen

The **Charles A. Desoer Technical Achievement Award** honors the individual whose exceptional technical contributions to a field within the scope of the CAS Society have been consistently evident over a period of years. Contributions are documented by publications (including but not limited to patents) and based on originality and continuity of effort. The recipients are:

1985	Irwin W. Sandberg
1987	David A. Hodges, Paul R. Gray, Robert R. Broderson
1988	Alfred Fettweis
1989	Gabor C. Temes
1990	Hermann K. Gummel
1991	Dante Youla
1992	Robert K. Brayton
1993	Leon Chua
1994	Rui J. P. De Figueiredo
1995	Anthony N. Michel
1996	Stanley A. White
1997	Sung-Mo Kang
1998	Chung Laung (Dave) Liu
1999	Alfred E. Dunlop
2000	Alan N. Willson, Jr.
2001	Ruey-wen Liu
2002	Yrjo Neuvo
2003	H. John Orchard
2004	David Allstot
2005	Andreas Antoniou
2006	"KT" Krishnaiyan Thulasiraman
2008	Edgar Sanchez-Sinencio
2009	Yoji Kajitani
2010	Jason (Jingsheng) Cong
2011	Kaushik Roy
2012	Keshab K. Parhi

2013   Eby G. Friedman
2014   Paulo Sergio Ramirez Diniz
2015   Massoud Pedram

The **IEEE CAS John Choma Education Award** honors the individual with exceptional contributions to education in a field within the scope of the CAS Society. Contributions are quantifiable by publication of textbooks, research supervision of both graduate and undergraduate students, short course development and personal participation in continual education within the field. The award is based on quality, continuity and originality of contribution. The recipients are:

1985   Eliahu I. Jury
1986   Adel Sedra
1987   Gabor C. Temes
1988   Bede Liu
1989   Sanjit K. Mitra
1990   Yosiro Oono
1991   Mohammed S. Ghausi
1992   Rudolf Saal
1993   Thomas Kailath
1994   Adel S. Sedra
1995   Lawrence P. Huelsman
1996   George S. Moschytz
1998   Wai-Kai Chen
1999   John Choma Jr.
2000   M.N.S. Swamy
2001   Robert W. Newcomb
2002   Leonard T. Bruton
2003   Madgy Bayoumi
2004   Paulo Diniz
2005   Christofer Toumazou
2006   Wayne Wolf
2007   Nirmal Kumar Bose
2008   Josef A. Nossek
2009   Andreas Antoniou
2010   Yannis Tsividis
2011   R. Jacob Baker
2012   Sitthichai Pookaiyaudom
2013   Luiz P. Caloba
2015   Sung-Mo (Steve) Kang

The **Vitold Belevitch Circuits and Systems Award** honors the individual with fundamental contributions in the field of circuits and systems. The recipients are:

2003   Alfred Fettweis
2005   Dante C. Youla

2007   Leon O. Chua
2009   Ronald A. Rohrer
2011   Patrick Dewilde
2013   Alan N. Willson, Jr.
2015   Yuval Bistritz

The **Industrial Pioneer Award** honors the individual(s) with exceptional and pioneering contributions in translating academic and industrial research results into improved industrial applications and/or commercial products. The recipients are:

1999   Quentin C. Cassen
2000   Henry Samueli
2001   Aart de Geus
2002   Howard C. Yang
2003   Prabhu Goel
2004   Ya-Qun Zhang
2005   Yervant Zorian
2006   John A Darringer
2007   Rob A. Rutenbar
2009   Paul E. Jacobs
2010   Jos Franca
2012   Robert Bogdan Staszewski
2013   Rajiv V. Joshi
2014   Babak Parviz

The **Meritorious Service Award** honors the individual with exceptional long term service and dedication to the interest of CAS Society. The award is based on dedication, effort and contributions. The recipients are:

1985   James H. Mulligan
1986   Benjamin J. Leon
1988   Timothy N. Trick
1989   George Szentirmai
1990   Kenneth W. Jenkins
1991   Ming L. Liou
1992   Belle A. Shenoi
1993   Rolf Schaumann
1994   Sung-Mo Kang
1995   Tatsuo Ohtsuki
1996   Randall L. Geiger
1997   Wai-Kai Chen
1998   Ruey-wen Liu
1999   Franco Maloberti
2000   John Choma, Jr.
2001   Philip V. Lopresti
2003   Hari Reddy
2004   Bing Sheu

2005	Michael K. Sain
2006	Martin Hasler
2007	Georges G. E. Gielen
2009	Ellen Yoffa, Magdy A. Bayoumi
2011	Zhen-Ming Chai
2012	Maciej J. Ogorzalek
2012	Ljiljana Trajkovic
2013	Gianluca Setti and Tor Sverre Lande
2015	Ricardo Reis

Paper Awards are also annually granted by the CAS Society to recognise outstanding contributions published in the Society publications. They are:

Guillemin-Cauer Best Paper Award
Darlington Best Paper Award
Biomedical Circuits and Systems Best Paper Award
Circuits and Systems for Video Technology Best Paper Award
Very Large Scale Integration Systems Best Paper Award
Outstanding Young Author Award
IEEE Circuits and Systems Society ISCAS Student Best Paper Award

Current and Emergent Topics

The strength of CASS is its vast and in-depth spectrum and diversity of topics. This should be taken advantage of so as to address both multidisciplinar and interdisciplinar topics. On the one hand, multidisciplinar topics should be encouraged to cover research across different technical committees addressing challenges of transversal nature, thus giving answer to the needs of emerging engineering fields, as well as from industry, and ultimately humanity grand challenges, which demand complex and heterogeneous systems. On the other hand, interdisciplinar topics should explore the fertile research areas that lie in the intersection and overlap of conventionally disjoint areas.

In this section, and in order to close this archival book mainly devoted to revisiting the history of CASS, each of the topics and related research communities within CASS elaborate a statement covering the mission and current activities of each of the technical communities structurally composing CASS as well as a discussion on emergent and future topics. Despite a prospection into the future is intrinsically challenging, CASS will undergo a strong interplay of activities among and across these topics and communities, to both spin off new enabling technologies, ideas and topics, as well as to combine them to address new future applications, system and in turn the grand challenges of engineering and in turn humankind.

Analog Signal Processing

Joseph Chang ASP TC chair, Nanyang Technological University, Singapore

Analog (and Mixed-Signal) circuits and systems, sometimes deemed as 'classical' and 'traditional' circuits and systems, in reality remain as pertinent and emerging as ever in the 21st century and beyond – congruous to the pertinence of this Technical Committee to the global scientific and engineering community-at-large. The emerging topics, certainly not exhaustive, include the following. The extension of analog circuits to power management now includes energy harvesting for the ever-greening and carbon-neutrality of electronics, including to the Internet-of-Things for extended battery life. Analog circuits can further facilitate carbon-neutrality with analog computation in conjunction with digital means such by neuromorphic computing. With increasing number of computation cores in many complex microprocessors where the wiring between said cores is increasingly a bottleneck, analog radio frequency 'wireless network on chip' could offer a viable alternative. Further on radio frequency where much of the radio frequency spectrum is already allocated and crowded, analog radio frequency designs are reaching to the tens of gigahertz range and into the terahertz, and mm-wave. The terahertz range offers novel exciting applications once envisioned only in science fiction. To realize the full frequency-spectrum range ranging from DC to the terahertz, emerging analog circuits will exploit non-traditional semiconductors/materials, including carbon nano tube/graphene and III/V semiconductors, and including new integration 2.5D, 3D and even '4D' that embodies integrated III/V-on-CMOS. Emerging analog circuits are also realized on flexible substrates such as PET plastic films in the form of printed/organic (large-area printed) electronics, either printed-only or as hybrid electronics embodying printed electronics and classical semiconductors. Such mechanical flexibility offers a flexible form factor where electronics can be molded and bent to fit on/into uneven surfaces and odd-shaped enclosures, including the human skin, clothes, etc. Beyond earth into the extra-terrestrial, the emerging 'space-tronics' where most satellites are expected to be not only subminiature pico-satellites but swarms of satellites, analog circuits will be designed by means of radiation hardening by design, ultra power-efficient such as subthreshold operation, and based on commercial nano-scaled CMOS.

Biomedical and Life Science Circuits and Systems

Sameer Sonkusale

BIOCAS TC chair, Tufts University, USA

Biomedical and lifescience system trends today are going towards increased miniaturization and integration at the micro- and nano scale and towards realization of portable and implantable devices to address the grand engineering challenges in health and medicine. Solutions to these challenges can be expected to happen at the interface of artificial and biological systems. Circuits and systems engineering approaches applied intimately to biological systems will thrive new research and development efforts. At the same time, there will be infusion of new ideas and approaches based on our evolving and deeper understanding of biology, which will spawn new bio-inspired engineering solutions for artificial man-made systems. Sensors and actuators will become increasingly autonomous and unobtrusively ubiquitous within and on the body, and in the environment, observing, preserving, repairing, aiding, and eventually enhancing natural functions. For example, the role of circuits and systems is critical in realization of the next generation of brain-machine neural interfaces, which will help unravel the complexities of how brain works and also help spur development of neuroprosthetics to solve debilitating neural and brain related disorders. The internet of things (IoT) revolution in the biomedical and lifesciences arena is in the offing where the research activities of the biomedical circuits and systems are at the core. For example, miniaturized sensor nodes on body or textiles monitor will key biometrics of your health and well-being, and communicate this information wirelessly to the subject and also include their doctor/caregiver in the loop in real-time. The biomedical circuits and systems efforts will also be at the core of any point-of-care diagnostic devices, which will receive increasing relevance with aging population and increasing health care costs. Ethical issues and the potential for miss-use and hazards will increase in pace and require constant attention and careful consideration and will also be at the forefront of all discussions in the biomedical circuits and systems community.

Cellular Nanoscale Networks and Array Computing

Mustak Yalcin
CNNAC TC chair, Istanbul Technical University, Turkey

The mission of the Cellular Nanoscale Networks and Array Computing (CNNAC) Technical Committee is to foster research, development, education and industrial dissemination of knowledge relating to the emerging field of cellular dynamic computers and models briefly called CNN computing. The interest field of the Cellular Nanoscale Networks and Array Computing is moving towards massively parallel heterogeneous architectures. These architectures are used in emerging nanoscale computing devices to mimic biological systems. In the twilight of Moore's law, all these research efforts help to progress. The Cellular Nanoscale Networks and Array Computing community will seek ways to put its knowledge into practice to meet the great engineering challenges in the next years. Reverse-engineer the brain, advance health information and engineer the tools for scientific discovery are present important new challenges for this field.

Circuits and Systems for Communication

Tokunbo Ogunfunmi
CASCOM TC chair, Santa Clara University, USA

The Circuits and Systems for Communication Technical Committee (CAS-COM TC) focuses on the circuit theory and design for wireless, internet, and wired communication systems; transmission, reception, interface and protocol. The committee deals with the issues of systems' integration and interfacing. Communication Systems will continue to become more complex in order to provide higher levels of performance. In addition, due to the increased variability inherent in nanoscale technologies, the scarcity of spectrum resources and the desire to provide additional communications services systems will become increasingly flexible, adaptable, cognitive, and tolerant of increased variability in nanoscale technologies. The ongoing purpose of the CASCOM TC is to provide a networking forum for leaders in the field and promote the exchange, advancement and dissemination of information about the research, development, applications and practice of design of circuits and systems technologies for communications systems. The fields of interest in the CASCOM TC is currently sub-divided into 12 major sub-topics for purposes of member interest as follows: 1 Wireline Communications 2 Wireless Communications 3 Optical Communications 4 UWB Systems 5 MIMO & Massive MIMO 6 Modeling and Analysis 7 Software Defined and Cognitive Radio Systems 8 Error Correcting Codes 9 Cryptography and CyberSecurity 10 Sensor Networks 11 Algorithms and Implementations 12 Applications of Nonlinear Dynamics to Communications. These divisions may change over time in response to changes in technology and emerging areas of importance. These are some of the emerging areas within the CASCOM TC: Hardware Security, Cryptography and Security for Communications Systems, Internet of Things (IoT), Fog Computing, Algorithms and Implementations, Optical Communications, Sensor Networks, Low power and Wearables Devices.

Circuits and Systems Education and Outreach

Joos Vandewalle

CASEO TC chair, Katholieke Universiteit Leuven, Belgium

The education of engineers in circuits and systems will play an important role in their professional functioning within our global world during the coming 40-50 years. In order to discuss, reflect and communicate the CAS Education and Outreach TC of IEEE CAS Society that was set up in 2009. It has been observed worldwide that the current student populations are fluent with ICT, smartphones and social media, and hence typically have a limited attention span and are used to receive instant gratification. It is the vision of the CASEO TC to motivate the public and the students by relating CAS research and education to the many societal challenges of the coming decades (climate change, energy shortage, health care, needs of an aging population, infrastructure, transportation, electronic waste) and to design new attractive didactical processes for new generations of students, and spread successful innovative didactical experiences. There are several emergent topics and interesting discussion issues on CAS education and outreach. At the level of advancing technologies and didactical methodologies there are methods of blended learning, flipped classrooms, MOOCs, and learning analytics. At the level of the educational trajectories one can of course build on the accumulated CAS knowledge over the past century and the successful didactical processes of the past (see chapter 'the history of CAS education'). However the choice of relevant CAS topics and the sequence require a rethinking for the new students. What are the relevant concepts for a concepts inventory? Should the traditional sequence of analog circuits and analog signal processing before the digital circuits and digital signal processing be reversed or interlaced? Can hands on experience, and problem-based learning stimulate interest and insight? Should experience with alternative educational frameworks like CDIO (conceive, design, implement, and operate) or community service engineering starting from global challenges be shared more? Should basic CAS courses already involve environmental, societal and ethical issues?

Digital Signal Processing

Wei-Ping Zhu
DSP TC chair, Concordia University, Canada

The Digital Signal Processing Technical Committee (DSP TC) of IEEE CAS Society leads and promotes research, development, education and industrial dissemination of knowledge in the field of digital signal processing circuits and systems and their applications, aiming to produce innovative energy-efficient, flexible and scalable solutions embedded in large and small scale devices and equipment. The goal of DSP TC is to promote research and education in digital signal processing theory, algorithms, circuits and systems and their applications in electrical engineering and in related fields. The topics of DSP as far as the DSP TC is concerned include current and emergent topics encompassing (1) Digital filters approximation, realization, and implementation (2) Discrete transforms (3) Time-frequency analysis including wavelets and filter banks (4) Adaptive systems and adaptive signal processing (5) Nonlinear and statistical signal processing (6) Digital architectures specialized for DSP implementations (7) Digital signal processors and embedded systems (8) DSP for communication systems (9) Speech, image and video processing and compression (10) Optimization of DSP algorithms for VLSI implementations (11) Circuits and systems based on sparse sampling and compressive sensing (12) DSP for biomedical engineering and life science (13) Graph and distributed signal processing for sensor networks, and (14) DSP for smart grid, big data and other applications.

Multimedia Systems & Applications

Zicheng Liu
MSA TC chair, Microsoft Research, USA

It's an exciting time for multimedia researchers. We are witnessing a rapid development of a variety of new sensors such as 3D, IR, location, and health-monitoring sensors. The machine learning tools have become much more powerful and mature thanks to the deep learning technology. The fast growth of the cloud infrastructure has dramatically tightened the connections among people, between suppliers and consumers, and between developers and users. The three new waves- sensor, machine learning, and cloud infrastructure are enabling a wide range of new applications in multimedia such as home surveillance, health monitoring and diagnosis, robot assisted living, smart appliance, virtual secretary, virtual meeting room assistant, etc. Many exciting opportunities and new technical challenges are lying ahead. Multimedia content understanding is at the core for a lot of the new applications. Many traditional signal processing problems such as denoising and enhancement could be solved by new algorithms that leverage heterogeneous sensors and higher level semantic understanding of the signals. Multimedia information retrieval will become more relevant and personalized thanks to both the understanding of the content semantics and the understanding of the user. How to leverage the cloud infrastructure for multimedia data collection, labeling, training, adaptation, and analysis is a critical problem for any practical systems that require higher level semantic understanding. Social media analytics have become an increasingly important tool for companies to understand the needs of their customers. New techniques must be developed to ensure the security and privacy of a cloud-based multimedia system. How to leverage the unprecedented machine intelligence to make the human computer interaction more natural is another interesting problem.

Nanoelectronics and Gigascale Systems

Robert Chen-Hao Chang

NG TC chair, National Chung Hsing University, Taiwan

With vast amounts of research and development from high-tech industry and strong demands from personal computers, cellular phones, and Internet, the semiconductor technology continues its marvelous progress into the nanometer regime. It is a great challenge to integrate hundreds of millions and billions of nano-devices into gigascale systems. There exist significant challenges in using nano-structures/ nano-devices to form nanoelectronic integration systems. With exciting opportunities, research on nanoelectronics and that on gigascale systems are tightly coupled together. To promote and excel in the areas of nanoelectronics and gigascale systems for circuits and systems research/design community, the Nanoelectronics and Gigascale Systems Technical Committee (NG TC) was organized and approved in May 6, 2001. The purpose of the NG TC is to provide a networking forum for leaders in the field and promote the exchange, advancement and dissemination of information about the research, development and practice of nanoelectronic technologies, systems and applications. Thus, the NG TC elevates research and education in the field of using nano-structures/ nano-devices to form nanoelectronic integration systems. Besides, the NG TC is actively involved in the cooperation with other CASS TCs and IEEE societies to bridge the heterogonous integration of nanoelectronics and gigascale systems. Nano areas include, but are not limited to, nano-devices, nano-circuits & nano-architectures; nano-sensors, nano-actuators & nano-robotics; reliability and manufacturing issues; non-silicon nanoelectronics; spintronic-based technology, circuits and systems; emerging memory and memristor based technology, circuits & systems; and circuits & systems for quantum computing. Gigascale systems design aspects include system, architecture, electronic system level, hardware/software co-design, design methodologies, and test. The applications encompass computer, communication, consumer and car electronics and the heterogeneous integration for bio and green electronics.

Neural Systems, Applications and Technologies

Elisabetta Chicca
NSAT TC chair, Bielefeld University, Germany

The Neural Systems, Applications and Technologies Technical Committee vision is to exploit the principles used by biological neural systems in developing new technologies and computing paradigms that can solve long-standing and challenging engineering problems. Such approach was originally proposed in the late '80s by Carver Mead, who coined the term "Neuromorphic Engineering". Initial efforts focused on the design and application of sub-threshold analog VLSI circuits to sensing and computing with neurally inspired systems. In the last two decades this research field has been rapidly growing and nowadays it is represented by a large community of scientists exploring biologically inspired approaches based on conventional (analog and digital) CMOS as well as emerging technologies. These recent developments offer the opportunity to tackle new challenges and investigate novel approaches. Representative examples of hot topics in the field are the implementation of dedicated hardware systems for deep networks, the design of intelligent circuits and systems for autonomous agents, the development of biologically inspired sensory-motor systems, the integration of CMOS and memristors for low-power fully-parallel non-von Neumann computational and learning systems, and the development of autonomous learning systems.

Nonlinear Circuits and Systems

Sergio Callegari
NCAS TC chair, University of Bologna, Italy

The motivation for a Technical Committee dedicated to nonlinear effects within the IEEE Circuits and Systems Society is explained by the ubiquitous nature of nonlinear phenomena in circuits and systems. As new technologies emerge, and existing technologies reach their limits of performance, nonlinear effects assume greater importance. Their deep understanding, analysis and exploitation become essential. For this reason, the field of interest of the Nonlinear Circuits and Systems Technical Committee encompasses a broad range of research areas and technical challenges. The committee strongly supports cooperation with other IEEE societies and councils to address grand challenges of our age. Since these challenges are cross-disciplinary in their nature, the committee puts significant effort in applying and promoting fundamental research. In many cases, new applications or developments have arisen from nonlinear theory. The memristor serves as a perfect example of this statement. Its existence was predicted in the 1970s while the device itself was built only in 2008. On many occasions, the Committee initiated new areas of research that have then developed into independent fields. In the next years, the

Nonlinear Circuits and Systems Technical Committee will continue to leverage the expertise, curiosity and interdisciplinary vision of its members to address the needs of our society. Areas such as nanoelectronics, system biology and bioelectronics present important new challenges due to the mixed-signal nature of the involved systems. At the same time, they are crucial to growing demands related to health, age, and wellbeing. Another field of growing interest for the committee is large-scale nonlinear networks. These are characterised by a huge number of interacting agents whose structure and topology evolve and adapt in time and space. Many real world applications in communications, social and biological sciences and energy management reflect this pattern and display emergence and chaotic behaviour. Green energy management systems also lie within the scope of the Committee due to the fact that efficient and flexible power control often involves on-off transitions in energy flows and piece-wise or time-variant modeling. In addition, the research community will focus on novel computational paradigms that mimic biological computation approaches. This is essential for both understanding the reasoning and sensory organs of living organisms and building some of their properties into artificial systems for smart environments, pervasive sensing, and data processing. Even if the range of commitments seems wide, they display a fundamental common trait in features and effects that can be handled by nonlinear theory. The Committee sees its role in facilitating the analysis, design and implementation of nonlinear systems across all areas of electrical and electronic engineering.

Power and Energy Circuits and Systems

Chia-Chi Chu

PECAS TC chair, National Tsing Hua University, Taiwan

The requirement for environmentally friendly power generation systems combined with the increased cost of fossil fuels and the growing complexity-size of power grids, has led to new and emerging concepts in the generation, transmission and distribution of power. One of the proposed solutions is the idea of a smart grid driven by recent advances in using distributed generation (DG) that employ renewable energy sources (RES), and modern communication/computation technology. Driven by these emerging needs, power grids are anticipated to be complex and smart networked platforms. Major paradigms are transforming from centralized structures to more decentralized and prosumer interactive ones. The use of such a concept has also dramatically changed the structure of modern power systems where interconnected power electronic converters are extensively used under different interconnection schemes. In order to integrate growing deployments of intermittent RES and maximize the economic benefits of DG, the concept of microgrids (MGs) are employed where various sources and loads can operate autonomously either as an isolated island or be connected to the main grid. MGs can help mitigate grid disturbances to strengthen grid resilience. Traditionally, power converters have been studied as linear systems operating in a small area around the desired equilibrium point. For smart grid operations, the converter must operate in a wider area and therefore various nonlinear phenomena can take place. Therefore it is imperative that these converters are studied in great detail and subsequently properly designed to provide the best possible performances. Furthermore, new converter topologies must be proposed that are capable of tackling the challenging tasks of a power converter operating in DG systems. For example multi-level or multi-input converters operating in interleaving operation with advanced control strategies may be often required. Another crucial issue of DG systems is to use complex network theory to achieve the following objectives: (i) enhance the linkage between physical systems and cyber systems for monitoring and managing the power grid under more variable and intermittent operation conditions, (ii) analyze collected big data from monitoring systems, and (iii) develop real-time remedial control actions for avoiding system-wide cascading blackouts.

Sensory Systems

Piotr Dudek

SS TC chair, The University of Manchester, UK

The interests of the Sensory Systems Technical Committee (SSTC) span image and vision sensing, including hyperspectral, thermal, polarisation and depth imaging, THz imaging, auditory sensing, mechanical sensors, odour sensing, chemical sensors, bio-sensors, and other sensing modalities. It is not just the sensing devices themselves, but the circuits and systems issues such as the interfaces and near-sensor processing circuits, that are of particular interest. There is an ever-increasing demand for embedding intelligence into products, from complex devices and machines such as smartphones, robots, automobiles, to everyday objects such as clothing; from far-away sensors on space probes, to the ones implanted under our skin; from nanoscale sensors to complex sensor networks spanning entire cities; from disposable single-use devices, to ones that must last for years build into infrastructure. In most of these applications, it is critical that sensing and associated processing of information occurs within a very tight power budget, and in minimum physical space, posing enormous design challenges. Efficient acquisition of sensory data is of paramount importance. In many cases, this means that data has to be processed in-situ, either for immediate interpretation, or efficient transmission to other parts of the system. New sensing paradigms, such as event-based sensors, are challenging conventional approaches, but require the development of new processing strategies. New technologies, such as plastic electronics, open up new applications in low-cost disposable devices, but pose challenges not only in the sensor design, but also the associated interface and processing circuits. The increasing sophistication expected of robots, UAVs, augmented reality devices, etc., and the proliferation of sensory devices producing large amount of data, call for effective processing and data-reduction strategies at the sensor level. From energy-scavenging sensors to efficient near-sensor processing, the field calls for collaboration with many areas of circuits and systems, to facilitate the construction of future intelligent objects and devices.

Visual Signal Processing and Communications

Gwo Giun (Chris) Lee
VSPC TC chair, National Cheng Kung University, Taiwan

In the 1960's, Marshall McLuhan published the book entitled "The Extensions of Man", focusing primarily on Television, an electronic media as being the extension of human nervous system which from contemporary interpretation marks the previous stage of Big Data! Based upon mathematical fundamentals as foundations for complexity aware analytical visual algorithms including augmented and virtual reality, intelligent, flexible, and efficient system architectures including both software and hardware will be concurrently explored and designed, whereby the Visual Signal Processing and Communication Technical Committee envisions even further extension of human perceptual experiences and exchange of visual information, together with the expediting of our field into yet another new era of the Internet-of-Things over Cloud.

VLSI Systems and Applications

Masud H Chowdhury
VSA TC chair, University of Missouri Kansas City, USA

The goal of the VSA-TC is to promote research and education in the field of digital VLSI systems and its applications in electrical engineering, computer science and engineering, and related fields. System design aspects include design of architectures, algorithms, developing theory, design methodologies, and implementation, as they relate to design of VLSI systems and systems-on-chips. Application areas include many exciting and cutting edge fronts, but are not limited to, computing, signal processing, communication, mobile and wireless devices, networking systems, video, multimedia, optical and photonic devices, nanotechnologies, biotechnology, ultra-low-power devices, supercomputers, extremely high efficiency circuits and systems for space, avionics and defense applications, embedded systems, game consoles, smart phones, environment and ecological monitoring through distributed sensor network, and green energy devices. The remarkable growth of semiconductor industry and the unparalleled improvements in terms of functionality and integration density of micro- & nano-electronic circuits and systems over the last several decades have been enabled due to the extension of Moore's scaling curve down to deep nanoscale dimensions. Existing micro and nanoscale IC designs mainly focus on balancing power and performance that kept conventional silicon technologies confined in the medium-performance and medium-power range. However, there have been two new trends to push the design spectrum to the two extreme ends. On one end, many sub-nanometer applications will require ultra-low-power (ULP) operation with acceptable speed. On the other end, extremely high performance (RF to GHz/THz frequencies) circuits and systems are sought for some applications without considering power. Even in the current medium range of power and performance, conventional field effect transistors (FETs) and metal interconnects are approaching their physical and material limits. Further improvement of energy efficiency and performance is not possible beyond these limits if radical technological and design changes cannot be introduced in the next generation digital applications. Many silicon and post-silicon technologies are being investigated for those two extreme ends. The industry would ultimate want to explore a third direction and be able design circuits and systems for extremely high frequency ranges at lower power. Some two-dimensional nanomaterials have the potential to fulfil that need. With the progression of the existing silicon technology towards the end-of-the-roadmap and the emergence of new technologies and design approaches, many new critically important research needs and directions have evolved. The vision of VSA-TC are to: (i) address interconnect, device, circuit and system level challenges of current technologies to extend the life-cycle of silicon based CMOS platforms and (ii) find new solutions for power, performance and reliability problems in conventional and emerging digital circuits and systems.

The above are rational and competent projections. They are valuable perspective that can benefit scientists in defining their research strategies. However, we have always to keep in mind the following aphorism of Claude Shannon:

"we know the past but cannot control it,
we control the future but cannot know it"

Eduard Alarcón
CAS Society Vice President Technical Activities
Technical University Catalunya, UPC BarcelonaTech, Spain

Franco Maloberti
CAS Society President
University of Pavia, Italy

9

In Memoriam

Claude E. Shannon

Claude Elwood Shannon was born in Petoskey, Michigan on 30th April 1916 and spent his childhood in Gaylord, Michigan. He received the B.S. degree in Electrical Engineering and Mathematics from the University of Michigan in 1936. From 1936 to 1940, he was at M.I.T., combining graduate studies with professional experience. For two years he was a research assistant in the Electrical Engineering Department, where he operated the Bush mechanical differential analyzer. He was an Assistant in the Mathematics Department from 1938 to 1940, and during 1939-1940 was a Bolles Fellow. He received the S.M. degree in Electrical Engineering and the Ph.D. degree in Mathematics from M.I.T. in 1940.

Shannon was associated with the Institute for Advanced Study at Princeton University for one year through a 1940-1941 National Research Fellowship. Beginning in 1941, he served as a research mathematician for Bell Telephone Labs in Murray Hill, N. J. Shannon also served as consultant to the National Defense Research Committee.

Dr. Shannon's work included the following fields: the use of Boolean Algebra in relay and switching circuits, theory of communication, mathematics of cryptography, theory of differential analyzer, and the use of computing machines for numerical operations. He also has studied chess-playing and maze-solving machines, the theory of Turing machines, design of reliable machines from unreliable components, stochastic processes, the Algebra of genetics, and graph theory.

An early paper (1941) was about the number of two-terminal series-parallel networks. He introduced Boolean Algebra into engineering with his MSc thesis, working under Vanevar Bush, and showed how it could be used in the analysis and design of relay switching circuits, and it was from this start that the use of Boolean Algebra became universally used in subsequent digital logic circuit design. Best known for his invention of Information Theory, and introduction of Entropy into communications theory, his most important publication was 'A mathematical theory of communication' (BSTJ, 1948).

He was noted for riding a unicycle along the corridors and for skill in juggling, and is also remembered for many quotations, often reused by others. For example "... we know the past but cannot control it, we control the future but cannot know it ..."

In 1940, Dr. Shannon was the recipient of the Alfred Nobel Prize of the American Institute of Electrical Engineers for his work in switching theory. He received the Morris Liebmann award of the Institute of Radio Engineers in 1949 for his communication theory work. Yale University awarded him an honorary Master of Science degree in 1954, and in 1955, Dr. Shannon received the Stuart Ballantine medal of the Franklin Institute for work in communication theory. He received the National Medal of Science in 1966 and the Kyoto Prize in 1985. He is the author of approximately thirty-five technical papers, and holds several patents. He is co-author, with Warren Weaver, of The Mathematical Theory of Communication, and co-editor, with John McCarthy, of Automata Studies. Dr. Shannon was a Fellow of the Institute of Radio Engineers. He died 24th February 2001 in Medford, Massachusetts.

http://ethw.org/Oral-History:Claude_E._Shannon and additional material

Balthasar van der Pol

Dr. Balthasar van der Pol was born in Utrecht, the Netherlands, on January 27, 1889. He graduated cum laude in physics from the University of Utrecht in 1916. The same year he went to England, where he continued his studies first at the Pender Electrical Laboratory, University of London, under Prof. J. A. Fleming until 1917 and later at the Cavendish Laboratory, University of Cambridge under Sir J. J. Thompson until 1919. He returned to the Netherlands to receive cum laude the Doctor of Science degree in 1920 from the University of Utrecht with a thesis entitled "The Effect of an Ionised Gas on Electromagnetic Wave Propagation and its Flow Discharge Measurements."

From 1919 to 1922 he was theoretical assistant to Prof. H. A. Lorentz at Teyler's Institute in Haarlem, the Netherlands. In 1922 he became Head of the Philips Research Laboratories at Eindhoven and later Director of Radio Scientific Research, a position he occupied until 1949. From 1949 to 1956 he was Director of the Comité Cunsultatif International des Radiocommunications (CCIR) in Geneva, Switzerland, and technical advisor to the International Telecommunications Union (ITU) on the planning and development of radio communications.

Concurrently with his scientific activities, Dr. van der Pol was keenly interested in education. From 1938 to 1949 he was professor of theoretical electricity at the Technical University, Delft, the Netherlands, and from 1945 to 1946 he was in addition President of the Temporary University which was founded in Eindhoven to replace the other Netherlands universities in occupied territories. He was Visiting Professor at the University of California, Berkeley, in 1957 and Victor Emanuel Professor at Cornell University in 1958.

Dr. van der Pol took an active part in various societies in which he was an outstanding figure. A Fellow (1920) and Life Member of the Institute of Radio Engineers (U.S.A.), he was Vice President in 1934; Founder-member, honorary member, and onetime president of the Netherlands Radio Society (1920-1952); Vice President, Union Radio Scientifique International (URSI) (1934-1952). Honorary life member, Institute of Radio Engineers (Australia) (1938-1959); Member of the 'Nederlandsche Koninklijke Akademie van Wetenschappen" (Netherlands Royal Society) (1946-1959); President, Organisation Internationale de Radiodiffusion (O.I.R.) (1946- 1949); Member, "Radio-Raad" (Radio Advisory Board to the Netherlands Government) (1948-1949); Member, American Mathematical Society (1947- 1959); Member, Board of Governors, "Mathematisch Centrum," Amsterdam, the Netherlands (1947-1959); Member for Radio Science (URSI), Executive Board of the International Council of Scientific Unions (1957-1959).

In recognition of his numerous contributions Dr. van der Pol received many honors and distinctions. In 1927 he became Knight of the Order of Oranje Nassau for establishing the first radio-telephonic communication between the Netherlands and the Dutch East Indies. The Institute of Radio Engineers (U.S.A.) awarded him in 1935 the Medal of Honor for his contributions to Circuit Theory. He became Knight of the Order of the Netherlands Lion in 1946 for accomplishments as President of the Temporary University, Eindhoven, the Netherlands. He received the Valdemar Poulsen Gold Medal from the Danish Academy of Technical Sciences in 1953 for outstanding contributions to radiotechnics and particularly for research in the field of electromagnetic oscillations and wave propagation and for international scientific cooperation and organization of technical questions related to radio communication. He was Honorary President of URSI, and a Corresponding Member of the French Academy of Science, Paris (1957-1959). He received doctors' degrees honoris causa from the Technical University, Warsaw, Poland (1956) and the University of Geneva, Switzerland (1959).

The book "Operational Calculus Based on the Two-Sided Laplace Integral" by Dr. van der Pol and H. Bremmer, a unique treatment with applications to mathematics, physics as well as circuit theory, is well known. Most of the other of his scientific works, many of which are classics for pioneering and setting the foundations of present day "modern theories" in several areas, have been collected and are published in two volumes *Selected Scientific Papers*, Vols. 1 and 2.

Dr. van der Pol was an extraordinary individual indeed. He combined several rare talents with an inexhaustible amount of energy and dedication to science and mankind. A famous scholar, a famous scientist, a famous administrator at the international level, he was equally well known for the clarity of his lectures (in several languages), his knowledge of the classics, his warm personality and his talents for friendship, and his love for music (he played several instruments, he even composed).

On October 6, 1959, Dr. Balthasar van der Pol passed away peacefully in his home at Wassenaar, the Netherlands.

Taken from: N. DeClaris, "*Balthasar van der Pol – In Memoriam*," IEEE Transactions on Circuit Theory, 1960, pp. 360-361.

Supplementary comment by co-editor Anthony Davies:

The van der Pol ordinary differential equation has become well known in electronic engineering education because it arises in the study of oscillators when realistic non-linearities are taken into account, and represents the way in which self-sustaining oscillations arise in which energy is fed into small oscillations and removed from large oscillations, leading to variety of forms of behaviour, including near-sinusoidal oscillations and very complex ones. Van der Pol introduced it in his studies of oscillators made from circuits containing thermionic triode vacuum tubes. The equation can model the build-up to a nearly sinusoidal stable oscillation but can also exhibit complex and non-periodic behaviour, and is still a subject of research studies in non-linear dynamics.

Karl Küpfmüller

Karl Küpfmüller was born the son of a Nürnberg locomotive driver in 1897. He attended the normal secondary school (*Realschule*) rather than the more academic Gymnasium, and followed this with technical college (*Polytechnikum*) and an apprenticeship at Siemens-Schuckert. He joined the German telegraph authority in 1919, and moved to Siemens & Halske in Berlin in 1921. During his seven years with the latter company Küpfmüller published regularly on various aspects of telecommunications, network theory and closed-loop control. Three of these papers were highly original and of the greatest theoretical and practical significance. Like his U.S. contemporary Harry Nyquist, Küpfmüller derived fundamental results in information transmission and closed-loop modelling, including a stability criterion.

Briefly, Küpfmüller's major achievements of the 1920s were as follows. In 1924 he published "*Transient behaviour of wave filters*", in which he applied Fourier analysis to signal waveforms, came up with a relationship between bandwidth and rise time, and – perhaps most importantly – introduced a novel 'systems approach' in which he modelled a filter by a (non-realisable) brick-wall amplitude response and linear phase; others working on similar problems at the time considered only realizable devices consisting of discrete or distributed elements. Two other important publications followed in 1928: "*The relationship between frequency response and transient behaviour in linear systems*" and "*On the dynamics of automatic gain controllers*". Both papers applied the principle of convolution to time-domain analysis; the second modelled a closed-loop system, from which Küpfmüller's stability criterion emerged. In effect, the general closed-loop control problem was presented in terms of generalized systems and signal flow, and the common topology of a variety of control systems was illustrated.

Like the Nyquist criterion, the Küpfmüller criterion offers advantages over Routh-Hurwitz. For example, in contrast to Routh Hurwitz, both the Nyquist and the Küpfmüller stability test can be applied to higher-order systems without excessive calculation. Furthermore, both tests indicate how far a closed-loop system is from the stability boundary. Finally, both tests are easy to apply based on empirical engineering data without an explicit analytic model. The Nyquist criterion, of course, is much more general and –particularly with the introduction of the Nichols chart in the late 1940s – gives a far superior indication of the distance from instability. Nevertheless, the Küpfmüller criterion remained in German and Russian control engineering texts until the 1950s. Despite the superiority of the Nyquist criterion, Küpfmüller's work should not be underestimated. His generic systems approach was novel and informed much of the later work in this area. An obituary in 1977 puts it as follows: "*With the death of Karl Küpfmüller we have lost one of the fathers of modern communication theory ... If, today, we recognize information along with energy and matter as a third fundamental building block of the world, then Karl Küpfmüller has been a major contributor to the recognition of this fact.*"

As a direct result of his achievements of the 1920s Küpfmüller was appointed

Professor of *Elektrotechnik* at Danzig University of Technology in 1928, a highly unusual appointment as he had not completed a formal university degree, nor did he possess the PhD and Habilitation (a higher-level degree) normally required. He held various academic posts in electrical engineering, culminating as the chair of telecommunications in Darmstadt until 1963, was president of the German Electrical Engineers Association (VDE) from 1955–1957, and a prime mover in the founding of the German Society for Cybernetics. He wrote two classic undergraduate textbooks, one of which remained in print well after his death, posthumously revised by additional co-authors.

Küpfmüller was a member of the Nazi party and the SS from 1937, rising to the rank of *Obersturmbannführer* in 1944, and becoming scientific advisor to Admiral Doenitz. He was awarded two wartime decorations: *Ritterkreuz des Kriegsverdienstkreuzes mit Schwerter* (awarded to members of the Wehrmacht or to Wehrmacht employees) and the *Dr. Fritz Todt Preis in Gold* (awarded for those furthering the German war effort by, for example, inventions). His wartime activities have been little reported, however, even in the German literature. In 1946-7 he was interned for denazification, when he met Hermann Druckrey, a cancer researcher. A joint publication *Dosis und Wirkung* [Dose and Effect] appeared in 1949, in which Küpfmüller applied electrical analogue modelling to the action of carcinogens, an approach later successfully verified at Rhode & Schwarz.

Küpfmüller's name is not well known in the English-speaking world. Indeed, little has appeared in English about him or his work, although his pioneering results in systems theory informed later American work, particularly through the contributions of Ernst Guillemin, a prolific writer of influential student texts and a renowned engineering educator at MIT, and who was well acquainted with the ideas of Küpfmüller and other German electrical engineers. Küpfmüller was the recipient of numerous honorary degrees and other awards. Two current prizes bear his name: The *Karl-Küpfmüller* Ring of Darmstadt University and the *Karl-Küpfmüller-Preis* awarded by the VDE (the German electrical engineers' professional body).

Christopher C. Bissell
The Open University, England

Ernst Adolph Guillemin

Ernst (Ernie) Adolph Guillemin was born on May 8, 1898 in Milwaukee, Wisconsin. He received a BS in electrical engineering from Wisconsin in 1922 and moved to do postgraduate work at MIT the same year. In spring 1924, the renowned physicist Arnold Sommerfeld contacted MIT from Munich, suggesting an exchange program for graduate assistants, and Guillemin was offered a place in Germany. He gained a PhD from Munich University's *Institut für Theoretische Physik* in 1926. He returned to MIT and spent the majority of his career there. He became full professor in 1944, concentrating increasingly on network theory in his teaching and research.

From the mid 1930s Guillemin devoted himself almost entirely to aspects of circuit theory. During the Second World War he acted as a consultant to various groups in the MIT Radiation Laboratory, where he invented the Guillemin line, of crucial importance for generating the rectangular pulses required for radar systems. (He was granted a patent in 1949.)

Having developed a new course at MIT in communication networks, Guillemin wrote up his approach in his first textbook, the two-volume *Communication Networks* that appeared in 1931 and 1935. Volume 1 was considered by many at the time to be rather 'advanced' for a supposedly introductory text, introducing transient and steady-state response; network theory; the Heaviside approach; and Fourier analysis. Volume 2 is even more novel, and in many ways established the general approach to linear systems in both electronics and elsewhere, leading quite naturally into the field of filter theory and its related problems.

The period 1949 to 1963 saw an outpouring of the results of his teaching and research, with the publication of another four, sole-authored volumes, each consisting of several hundred pages. The 1949 publication *Mathematics of Circuit Analysis* is extraordinary for its time. Covering the now classic areas of determinants, matrices, vectors (including vector calculus), functions of a complex variable, Fourier and Laplace transformations, conformal mapping, and so on, it broke new ground in the mathematical education of what we might now call information engineers. All these topics still lie at the heart of control engineering, telecommunications, filter design, network analysis, and form the basis of advanced mathematics courses even today. The only serious omission by modern standards is perhaps discrete techniques - which, of course, were in their infancy in the late 1940s.

Guillemin's often radical approach to the teaching of his subject extended into the next text: *Introductory Circuit Theory* of 1953. Remarkably, the book begins with the concepts of graphs, networks and trees, cut-sets, duality and so on before even mentioning Kirchhoff, loops and nodes. Later in the text, of course, we have the less surprising topics of Thévenin and Norton theorems, impulse and step response, sinusoidal steady-state response, and so on – but all remarkably fresh over half a century later.

The final two of this magisterial set of books are *The Synthesis of Passive Networks* (1957) and *The Theory of Linear Physical Systems* (1963). The first of these represents in many ways the first coherent presentation of the comparatively new discipline of network synthesis. Much of the book is concerned with realization theory and methods, with particular attention paid to the non-uniqueness of realization. The book concludes with a presentation of Butterworth, Chebyshev and elliptic approximation techniques for filter design. The book again offers us insights into Guillemin's teaching style: his are some of the most self-aware textbooks in the field. Guillemin's final book is rather different: more the product of an older man (he retired the year it was published) reflecting on a beloved discipline. It partly fills in some of the gaps not covered by the earlier books, and partly offers alternative approaches and theoretical consolidation. But it is still highly original, dealing in more detail than earlier works with Fourier techniques and including chapters on convolution, sampling, numerical methods and the Hilbert transform.

Guillemin received many honors during his academic life, including:

1948 President's Certificate of Merit for outstanding wartime contributions

1960 First permanent holder of Edwin Sibley Webster Chair

1961 IRE Medal of Honor

1962 AIEE Medal in Electrical Engineering Education

Guillemin was driven to enable his students to achieve a deep understanding. He was unafraid to introduce them to topics like Fourier analysis and pole-zero plots at an early stage, and to spend whatever time necessary discussing problems and 'talking round the subject'. His books are remarkable for the general discussion, inclusion of examples and analogies, and so on. Yet at the same time he was deeply committed to the importance of a full grasp of the theoretical basis of network analysis and synthesis, and a firm believer that an understanding of this would enable his students to successfully deal with novel problems in their future careers.

Guillemin died on 6 April 1970. A eulogy written by Prof. Louis Weinberg, one of his many students includes the words: *Though grown men are not supposed to cry, many will. "After all," I was told emotionally by a circuit theorist who was born and educated in Europe and who had never met Professor Guillemin "he was the father of us all."*

Christopher C. Bissell
The Open University, England

This article has been adapted from Bissell, Christopher C. (2008). *"He was the father of us all." Ernie Guillemin and the teaching of modern network theory.* In: History of Telecommunications Conference (HISTELCON '08), 11-12 Sept 2008, Paris.

Sidney Darlington

Sidney Darlington, one of the world's most creative and influential circuit theorists, died at his home in Exeter, NH, on October 31, 1997, at the age of 91. He was a man of uncommon depth and breadth whose first love was circuit theory. He made important widely known contributions in several areas including network synthesis, radar systems, rocket guidance, and transistor networks.

Sid was born in Pittsburgh, PA. He received the B.S. degree in physics (magna cum laude) from Harvard College in 1926, the B.S. in electrical communication from MIT in 1929, and the Ph.D. degree in physics from Columbia University in 1940. In 1929 he became a Member of Technical Staff at Bell Laboratories where he remained until he retired, as Head of the Circuits and Control Department, at the then-mandatory retirement age of 65. He was a member of both the National Academy of Engineering and the National Academy of Science. In 1945 he was awarded the Presidential Medal of Freedom, the highest civilian honor of the United States, for his contributions during World War II. The award was established in that year by President H. S. Truman to reward notable service during the war. In 1975 he received the IEEE Edison Medal and in 1981 he received the Medal of Honor.

During Darlington's early days at Bell Laboratories there was much interest in electrical filter theory, mainly in connection with the exacting needs of systems using frequency-division multiplexing. At that time filter theory was very different than it is today in that it was marked by ad hoc techniques in which complex filters were designed by cascading less complex filter sections whose attenuation characteristics were specified in graphical form. This was often unsatisfactory for several reasons. For example, the theory available did not adequately take into account the loading of the various sections on their predecessors. Sid's brilliant contribution was to recast the filter design problem as two problems: approximation and network synthesis – and to give a solution to each problem. The approximation problem he addressed was to suitably approximate the desired typically idealized filter characteristic using a real rational function of a complex variable, and in this area Darlington made significant pioneering contributions involving the use of Chebychev polynomials. His main contribution, which concerned the exact synthesis of a two-port network that realized (i.e., implemented) the rational function, was the introduction of his well-known insertion-loss synthesis method. This work led to his beautiful structural result that no more than one resistor is needed to synthesize any impedance. It is interesting that his results were not widely used until many years after they were obtained. This occurred partially because more exacting computations were required than for the earlier "image-parameter" filter designs; also, due to its novelty, it was not easy for filter designers of the time to fully appreciate Darlington's contributions. This is easier to understand in the context of the history of the development of lumped-constant filter theory which originally was an extension of the theory of transmission lines, and in which originally the concepts of a propagation constant, characteristic impedance, reflection factor, etc., played a prominent role. Sid's work also profoundly influenced electrical engineering education. After World War II, the Darlington synthesis of reactance two-ports was taught to a generation of graduate students who learned that linear circuit design could be formulated precisely in terms of specifications and tolerances, and that the problems formulated could be solved systematically. With concurrent advances in communication and control theory, electrical engineers began to appreciate that higher mathematics was a powerful tool for advanced study and research. This helped pave the way for the introduction of system theory and system analysis, and thus further broadened the scope of electrical engineering education.

In addition to never losing interest in circuit theory, Sid retained an interest in military systems – and related systems – throughout his tenure at Bell Laboratories. One of his most important contributions is the invention of what is called "Chirp Radar." The Chirp idea is a way to form a pulsed radar's transmitted signal so that relatively high peak power is not needed to achieve long range and high resolution. This involves transmitting long frequency-modulated pulses. The corresponding reflected and received ("chirped") pulses are "collapsed" into relatively short pulses using a network that introduces a time delay that is frequency dependent. The idea has been widely used, and there has been much interest in the design of the needed delay networks – not only at Bell Laboratories, but at many other companies and also at universities. Darlington's IEEE Medal of Honor citation reads:

for fundamental contributions to filtering and signal processing leading to chirp radar.

Sid also did very influential work concerning rocket guidance. In 1954 he ingeniously combined radar-tracking techniques with principles of inertial guidance to develop the highly effective Bell Laboratories Command Guidance System which has launched many of the U.S. space vehicles including NASA's Thor Delta booster and the Air Force's Titan I missile. The system has proved to be remarkably reliable and has played a central role in placing into orbit many satellites including the Echo I communications satellite, Syncom, and Intelsat.

Darlington is best known for an idea that he probably developed very quickly – the Darlington transistor – a simple circuit comprised of two or more transistors which behaves as a much improved single transistor. As is well known to the circuits and systems community, this idea is widely used and has had a great impact on the design of integrated circuits.

Sid was a Visiting Professor for periods of time of from one to six weeks at the University of California at Berkeley during 1960–1972, and a Visiting Professor at the University of California at Los Angeles in 1978 for a month. He gave many lectures and enjoyed these visits very much. Colleagues and students often remarked among themselves about how impressed they were with his keen physical insights, sophisticated mathematical talent, and pursuit of definitive results. After Sid retired from Bell Laboratories, he became an Adjunct Professor at the University of New Hampshire where he received an honorary doctorate in 1982. He was a consultant to Bell Laboratories during 1971–1974. Darlington held more than 40 patents, and was active in professional society activities. During 1959–1960 he was the Chairman of the IEEE Professional Group on Circuit Theory, and in 1986 he received the IEEE Circuits and Systems Society's first Society Award.

Sid was a man of great personal and professional integrity. He was an intense but gentle man who was surprisingly modest. He was also a gregarious person who was informed about many things and had much to say. A colleague once commented that "asking Sid Darlington a question was like trying to take a drink from a fire hose."

Largely taken from: E. E. Kuh, I. W. Sandberg, *"In Memoriam–Sidney Darlington,"* IEEE Transactions on Circuits and Sysrtems, 1998, pp. 1-2.

Bernard Tellegen

Bernard Tellegen was born 24 June 1900 in the Netherlands. He attended Delft University, where he obtained his degree in electrical engineering in 1923. In 1924 he entered into the service of the Philips Research Laboratories. These laboratories had been founded in 1914 by Holst and Oosterhuis. Tellegen belonged to a fairly small nucleus (van der Pol, de Groot, Penning, Druyvestein, Bouwers) around which one of the largest research centers in the world would grow. As a scientific adviser, he later became a member of its board of directors.

Tellegen's first studies concerned vacuum tubes. He became interested in electron motions in triodes and multigrid tubes. In 1926 he invented the pentode, which was patented in a number of countries. It was the first in a series of about 57 patents, which he received either alone or in cooperation with others. In the following years Tellegen became interested in electrical circuits on which he published in 1928, 1933 and 1934.

In 1932 it was noticed, that the programs of some transmitters from Beromünster, Switzerland, when received in the Netherlands, seemed to carry also the program of Radio Luxemburg and crossmodulation in the receiver tubes was suspected. Tellegen showed that this was really a nonlinear effect in the ionosphere, caused by the powerful Luxemburg transmitter.

During his further studies in electrical networks, he became more interested in fundamental problems such as duality and geometric configurations and network synthesis in particular of resistanceless four poles. During his basic study of the classical passive network elements, Tellegen arrived at the conclusion that a further element "the gyrator" could complete the series in an elegant way. This new element does not comply with the reciprocity relations, and is antisymmetrical. Tellegen studied the properties of circuits with the "gyrator". Its first realization came in the microwave field with the use of premagnetized ferrites. The circulator was a further result of this idea. When the miniaturization of electronic circuits led to new possibilities, the gyrator soon became an important building block for selective circuits at low frequencies. In 1952 Dr. Tellegen published an important paper on a general network theorem with applications. Fundamentally "Tellegen's theorem" gives a simple relation between magnitudes that satisfy the Kirchhoff laws. Many treatises and a book on the application of this theorem were published. A paper on "Synthesis of 2n-poles by networks consisting of a minimum number of elements" proved his interest in economy. He also sought ideal non-linear circuit elements, which led to a form of idealised amplifier, corresponding to what was much later introduced as the "nullor" (presented at a conference in 23 April 1954: "La recherché pour una série complte d'éléments de circuit ideaux non-lineaires"). When he was the guest of honor at the International Symposium on Electrical Network Theory, London, 1971, he gave a paper on circuits with negative resistance elements.

In the period 1946-1966 Tellegen was professor extraordinary of circuit theory at the University of Delft. Adams, Bordewijk and Duinker were among those who received their doctoral degree working with him.

From 1942 to 1952 he was president of the Dutch Electronics and Radio Society, which made him a honorary member at the end of this period. From 1948 to 1960 he was chairman of the Dutch Committee of the International Scientific Radio Union (U.R.S.I.) He was vice-president of U.R.S.I. from 1952 to 1957. From 1957 to 1960 he was vice-chairman of its commission VI, especially charged with circuit theory. From 1946 to 1966, he was professor extraordinary of circuit theory at the University of Delft.

The Australian Institute of Radio Engineers made Tellegen a honorary life member in 1953. He received the Research Prize of the Royal Dutch Institute of Engineers in 1954, the Fellow Award of the IEEE in 1955, and the IEEE Edison Medal in 1973 *"For a creative career of significant achievement in electrical circuit theory, including the gyrator"*. Tellegen was elected a member of the Royal Academy of Sciences of the Netherlands in 1960. In 1970 the University of Delft conferred on him the degree of doctor honoris causa in technical sciences.

Tellegen and his wife, the former Gertrud J. van der Lee, had two sons and a daughter, as well as many grandchildren. The couple lived amidst the woods and fenns of the beautiful North Brabant countryside. In long walks they learned to know every part of it. In connection with his international conferences they visited many countries and made a world trip in 1952. Dr. Tellegen passed away on 30 August 1990.

from: Engineering and Technology History Wiki with corrections

Wolja Saraga

I first heard of Wolja Saraga in late 1962 when I was in the Filter Design Group at GEC (Telecommunications) Ltd. in Coventry. Some time prior to that he had given some lectures there on the Darlington method of Insertion Loss Design of filters and he was held in high esteem by the staff. He was also known to have solved the difficult problem of the design of wideband phase shift networks, which were very important for multichannel line communications systems. They provided an economical way of achieving the high performance single sideband modulators in such systems.

Most filters were then designed by the Image Parameter method, known to require many adjustments to compensate for its limitations, while the Darlington or Cauer method of 'exact design' of insertion loss filters required extensive complex calculations and was difficult to understand by most of the filter designers of that time. Someone who understood and could explain this was naturally esteemed. Wolja Saraga was also highly regarded for being a senior member of the IRE, at a time when very few engineers in the UK had the opportunity of IRE membership.

Later, as an academic in London, where my research interests included Active-RC filters, I had the opportunity, with a small group of colleagues including graduate students, to visit the research group of Wolja Saraga at an industry site at Blackheath, (a research laboratory of ATE, the Automatic Telephone and Electric Co.Ltd.) near Greenwich in South East London. ATE was a subsidiary of Associated Electrical Industries (AEI).

My memories of that first visit were how this distinguished person immediately treated us all so informally and as his equals, and how we felt we had really become his friends. I found that this to be the case with all people that he met, over the many years that I came to know him well. He acted as if we junior people were his equals. I never heard him say a bad word about anyone (perhaps even some who actually deserved it).

His group at Blackheath were a varied and talented crowd, working on various aspects of Active Filter design and Circuit Theory, at a time when there was much doubt among industry engineers that Active-RC filters would ever reach a stage of being useful in practice. They were given the freedom to follow their interests in a style more usual in a university research group. It was during this time at Blackheath (in 1962) that he was elevated to Fellowship of the IRE – primarily for his success in solving the design problem for wide-band phase-shift networks, just before the merger of IRE and AIEE to form IEEE [1-2].

The outputs from his group led to many papers at Colloquia, Seminars, Conferences, and those who knew Wolja and his co-workers in various locations where he worked after Blackheath will have noted the way in which last-minute adjustments were always still being made to papers right up to the deadline when they had to be submitted.

In those days, of course, there were no word-processors, equations had to be written in neatly using a drawing pen, and Figures also all had to be hand-drawn and hand labelled. Typically Wolja's papers used all available space on the special paper provided for the conference submission, so as to include as much as could be squeezed in, including any last-minute discoveries.

With various company restructurings and mergers, the Blackheath group did not survive and its members dispersed to other locations and organisations. Wolja later had to re-build his research group at GEC Research Labs at Wembley, and continued the same style of research with new colleagues. He worked extensively on the N-path Filter and various other filter structures, including the invention of quadrature modulation filters. He continued to present papers at Conferences and Colloquia, where his presence was both expected and welcomed. Unfortunately the GEC top management appeared to be little interested in this sort of thing, and later he and some of his key colleagues transferred in 1972 to Imperial College, where he remained for the rest of his life.

For at least several years before moving to Imperial College, he and members of his research group were regular attendees at the weekly Network Theory Seminars arranged in the Department of Electrical Engineering there by Bob Spence. These seminars were attended by a wide range of people and the presence of Wolja Saraga was almost taken for granted. Such a situation would now be anathema to many modern top managers. The idea of their research leaders and their teams going off to a university every Friday afternoon to hear presentations by famous and not-so-famous people from many organisations and many countries would be incomprehensible to them. At that time, we simply regarded it as the normal way of doing things.

While working at Imperial College, particularly with David G. Haigh, he designed active filters in collaboration with the Admiralty Underwater Weapons Establishment at Portland, evidence that active-RC filters really did have a future of real applications. A paper written with D. Haigh and Robert G. Barker received the Darlington award of the IEEE CAS Society for 1979, announced only after Wolja Saraga's death.

With David Haigh a patent was obtained via NRDC for this work (US patent number 4260968), filed 14 March 1979 and granted 7 April 1981, entitled 'Active filter network utilizing positive impedance converters'. It showed particularly how a ladder network of resistively coupled capacitor-terminated positive impedance converters could be use to make a band-pass filter with all capacitors having the same value. It seemed that Imperial College, while providing him with a place for his research to continue, did not give him as much recognition as he deserved, and it is easy to assume that his modest manner meant that he did not demand it.

He died suddenly in February 1980. This was completely unexpected by his friends, and he continued to work on the things which interested him right up to this time. Following his death, the Electronics Division of IEE founded a series of one-day events titled 'Saraga Colloquium on electronic filters' which continued annually for a number of years. A special issue on filters of the IEE Proceedings Part G had been planned with Wolja as its Guest Editor: instead it was dedicated to his memory and was the location of an obituary.

So what were the origins of this great man who approached life with such unique kindness and humility, making so many close friends and contributing so much to the Filter Design world?

Born in Berlin, with a Romanian father and a Russian mother, on 3rd September 1908, he studied science and electrical engineering, being awarded a doctorate in physics 'valde laudabile' in 1936 from the Humbolt University in Berlin. He had studied telecommunications at the Heinrich-Hertz Institut für Schwingungsforschung (oscillation-research) of the Berlin Technical University, where he became a part-time teacher and research worker. While there he developed the 'Saraga-generator' around

1931, and this was patented and shown at the 1932 Berlin Radio Exhibition. It uses a photo-electric cell, and generates sounds synchronised to the movements of performers on a stage, by responding to light from a distant neon lamp, this light being influenced by the performers. The name is still well-known in the electronic music field. He was planning to get a doctorate from the Heinrich-Hertz Institute but was expelled from there because of being Jewish.

While in Berlin, he met Lotte Isenburg, who was to become his future wife. She followed him to England where they married in 1939. She survived him by nearly 5 years. As a Jew in those politically turbulent times which led to Nazi rule and World War Two, it was not the place for the young Wolja to thrive.

He had the offer of a research position in Switzerland at what is now ETH Zürich but was not allowed to travel there because by then he had been categorised as stateless. In 1938, with the assistance of an organisation set up by the Royal Society, he received permission to travel to Britain and in February 1939 obtained a position at the Telephone Manufacturing Company at St. Mary Cray (in Kent, south east of London). At the outbreak of war he was briefly interned on the Isle of Man, but allowed to return to the Telephone Manufacturing Company, where he remained for 20 years, becoming noted for his circuit theory and filter design expertise. From 1944 he was a network development research group leader, and also did part time teaching at South East London Technical College. Publications and patent applications were evidently encouraged in those days by his employers; many of the journals then available have long since been discontinued He published regularly in Wireless Engineer.

In 1940, he met the Austrian philosopher and sociologist Otto Neurath, who was interested in the use of the Saraga-Generator, and was also interned on the Isle of Man. By that time Otto Neurath had escaped to England after the 1940 German bombing of Rotterdam, having previously taken refuge in Netherlands because of advice while visiting Moscow that it would be unsafe to return to Vienna. He was interned on the Isle of Man. Later Neurath and his wife set up the Isotype Institute in Oxford. His complex proposals for using graphical notations and symbols were directed towards children's educational material and charts for guiding housing developments, etc. and a dislike of language which he felt had been misused to support propaganda. It may be considered to have laid the foundation for the pictograms which we now take for granted in airports and road signs.

References

[1] Scanlan, J.O. *Obituary of Wolja Saraga* in International Journal Circuit Theory and Applications, p. 341 (1980)

[2] *Obituary of Wolja Saraga* in IEE Proc, 128, pt. G, No. 4, August 1981, p.145

Anthony Davies
Emeritus Professor, King's College London

Acknowledgements

The advice, support and some information from Peter and Esther Saraga are welcomed and have improved the accuracy of this account.

(photo used with permission from IEE Proc., Vol 128, Part G,No. 4, August 1981 page 145)

Wilhelm Cauer

Wilhelm Cauer was a giant in the area of synthesis of LC impedances and filters who belonged to the very first researchers who understood that filter synthesis is a proper mathematical discipline. Cauer was the first who recognized that filter synthesis can be interpreted from a mathematical point of view as an inverse problem. Following this, Cauer became a main contributor to the theory of linear AC circuits and filter synthesis developed 1925 - 1940. Between 1934 - 1936 Cauer developed the powerful concept of elliptic filters and filled it in three patents – nowadays also known as Cauer filters. At the end of this period and on the line of Cauer's ideas a general systematic theory of insertion loss filter design was developed by E. Norton, W. Cauer, G. Cocci, H. Piloty, and S. Darlington which is until now one of the most elegant and powerful theories of electrical engineering.

A first breakthrough towards a systematic synthesis of filters was achieved in 1924 by R. M. Foster in his celebrated paper *A reactance theorem*. Cauer immediately recognized the potentialities of Foster's result. In his doctorate thesis (1926) *The realization of impedances of specified frequency dependence* he presented a precise mathematical analysis of the problem and developed the first steps towards a scientific program which converted empirical approaches of the electrical filter design at that time in a mathematical concept. This program addressed three distinct classes of problems where the following questions for a network or a transfer function arise:

- Which classes of functions can be realized as frequency characteristics?

- How are the interpolation and approximation problems (which constitute the mathematical expression of the circuit problems) solved by using functions which satisfy the realization conditions above?

- Which circuits are equivalent to each other, i.e. have the same frequency characteristics?

This contribution to the systematic design of electrical filters was tantamount to the very beginning of network synthesis. Cauer's program is explicitly formulated in the lecture he held in 1928 on the occasion of his habilitation in Göttingen (academic teaching license); an extended version was published in a paper from 1930. It became a milestone in the development of network theory and in a series of papers he contributed himself to this program. For example, Cauer introduced a Poisson integral representation of impedances, together with his Ph.D. student O. Brune at MIT he obtained a complete characterization of the realization class by means of analytical function theory, he applied Pick's results to problems of complex curve fitting with impedance functions, he developed a canonical lattice realizations for symmetric two-ports and a canonical realization of an arbitrary finite passive multiport by three mesh-connected purely inductive, resistive and capacitive n-ports L, R and C, respectively. Furthermore, he developed a prototype of a electro-mechanical computer for the calculation of linear equations. In 1941 Cauer presented a monograph *Theory of Linear AC Circuits* – written in German (Theorie der linearen Wechselstromschaltungen) and since 1958 also available in English language – which contains not only his own results but also a careful preparation of the scientific work of many other researchers in the area of network theory, especially his colleagues from Bell Labs.

Wilhelm Cauer was born in Berlin-Charlottenburg on June 24, 1900, the sixth child of the family. His mother came from a family of preachers and teachers. His father, also called Wilhelm, was a Privy Councilor and professor of railway engineering at the Technical University of Berlin. In 1922 he met the famous physicist M. von Laue and began to work in the area of general relativity. As a result Cauer's first publication was a contribution to the general theory of relativity, and was published in 1923. Then he started to study problems in electrical engineering at the Technical University of Berlin. In 1924 he graduated in applied physics and entered the employ of Mix & Genest, a Berlin company working in the area of communication and telephone systems, then a branch of Bell Telephone Company. There he worked on probability theory as applied to telephone switching systems and calculations relating to time-lag relays. After completing two publications, one on telephone switching systems and the other on losses of real inductors, Cauer began to study the problem of filter design. Due to his interest in this field, he regularly corresponded with Foster, who was also working on the same problem.

At the beginning of 20th century, a few engineers began studying the design aspects of communication systems. The most prominent researchers, G. A. Campbell in the United States and K. W. Wagner in Germany, explored selective filter circuits for telephone applications in the early years of World War I, thus paving the way for the first ideas about electrical circuit synthesis.

While working as a research assistant, he presented his thesis paper in June 1926 to G. Hamel, head of the Institute of Applied Mathematics and Mechanics at the Technical University of Berlin. The second referee was K.W.Wagner. In 1927 he contacted R. Courant in Göttingen as well as V. Bush of MIT because he was interested in the construction of computing machines capable of solving systems of linear equations. Thus, he became a research assistant at Courant's Institute of Mathematics at the University of Göttingen. Subsequently, he got his habilitation and became an external university lecturer in 1928. Due to the economic crisis however, his family could not solely live from this position. Now and then some extra money from royalties meant help to the young couple who had married in 1925 and now had a child. In 1930 the Rockefeller Foundation granted Cauer a one year scholarship for studies at MIT and Harvard University, and his wife followed him to the United States. There he became acquainted with several American scholars working in network theory and mathematics. He was a member of the team around Bush, who was the developer of several electrical and mechanical machines for the solution of mathematical problems. It was also there that he completed his monograph *Siebschaltungen*. After his term at MIT, Cauer worked for three months for the Wired Radio Company in Newark, New Jersey.

Back in Göttingen, Germany, the sheer lack of funding caused by the Depression prevented him from finishing the development of an electrical calculating machine that would have been the fastest linear systems solver at the time (20 minutes for 10 unknowns with an accuracy of maximally 4 digits). Then, in early 1933, the demented apparatus of the Third Reich took control of Göttingen University. A racist, right-wing revolution swept through the entire university. Admittedly, the university received orders from the Hitler government in Berlin, but there were a large number of people who were only too ready to obey these orders, although they were not really forced to do so. The small town of Göttingen, which owed its reputation to the many famous scholars that had taught at its university, became *"cleansed of the Jews"*. Nearly 70% of the teaching staff of the world-famous Mathematics Institute there were either Jewish or of Jewish descent.

The elite, including its director, Richard Courant, had to leave. Young Nazi leaders organized student riots and summoned the remaining staff to participate in *"voluntary"* military sports camps. Cauer thought that he could come to terms with the regime by spending periods of time in sports camps of this kind. But in those days of hysterical investigations it soon became known that one of Cauer's ancestors had been a certain Daniel Itzig (1723-1799), a banker of the Prussian king Frederick II, among whose descendents were a number of well-known bankers, statesmen and composers. Although this did not mean that Cauer was going to be affected by the Nazi race laws, he was given to understand that there was no future for him at the University of Göttingen. A further problem at that time was that few people could appreciate the vast potential of Cauer's special field of work. Whenever faculties discussed making new professional appointments, they tended to look for teachers in the traditional fields of mathematics. They did not pay attention to Cauer because, for mathematicians, he seemed too involved in applied sciences, and for electrical engineers his contributions included too much mathematics. Therefore, although Cauer was nominally granted the title of professor in 1935, no chair was actually available for him. It took him quite a long time to realize that his life goal of an academic career would not work out. His small income was not sufficient to maintain his family. In 1936, after various attempts to gain a position in the industry, and a short term appointment in Kassel with the aircraft manufacturer Fieseler & Storch, be became director of the laboratory at Mix & Genest. This situation in Berlin gave him stimulation and scope, and also enabled him to give lectures on applied mathematics at the Technical University in Berlin beginning in 1939.

In 1941, the first volume of his German written monograph *Theory of linear AC Circuits* was printed, in which he provided detailed information on what he intended to publish in the second volume. In this volume also recent results of Cauer should be included which could not be published during the World War two. The manuscript to this was destroyed by Allied bombing in 1943, however, and he started anew. He actually completed this second manuscript, but it was either destroyed or taken by the Red Army from the safes at Mix & Genest in 1945. Thanks to the energy of Cauer's wife, Karoline Cauer and the editorial help of his former Ph.D. student E. Glowatzki as well as G. E. Knausenberger, W. Klein, and F. Pelz, some papers and a reconstructed version of the second volume were published posthumously. The English version *Synthesis of Linear Communication Networks* from 1958 combines the first volume with reconstructed material of the second volume.

On April 22, 1945 Wilhelm Cauer was killed when the Soviet Red Army captured the city of Berlin.

Wolfgang Mathis
Leibniz University, Hannover

Emil Cauer
Berlin

Herbert J. Carlin

Herbert J. Carlin, one of the outstanding circuit theorists of the last century, was born in New York City on May 1, 1917 and passed away at Walnut Creek, California, on February 9, 2009. He got his B.S. and his M.S. degrees from Columbia University in Electrical Engineering and a Ph.D. from the Polytechnic Institute of Brooklyn. After five years of industrial experience with Westinghouse, he rejoined this Institute where he was in the Microwave Research Institute there working under Ernst Weber and R.M. Foster, and subsequently became its Chairman and also Chair of Electrophysics Department. With Dante Youla he was associated with the Microwave Research Institute Symposia Series

in the 1950s. In 1966 he was invited to Cornell University, Ithaca, New York, as the J. Preston Levis Professor of Engineering and the Director of the School of Electrical Engineering; this last charge he kept until 1975. The years of his directorate marked an unprecedented growth and progress of the School, where the Faculty increased by more than 50 percent as did the number of undergraduate majors; the Master of Engineering program increased in size by almost 100 percent, the Doctorate expanded both in the research budget and in international weight. After retirement he was appointed Emeritus Professor on his late chair.

The fame of Professor Carlin got him worldwide invitations for lecturing and doing research. He was a Senior Research Fellow at the Physics Laboratory of the École Normale Supérieure, Paris during 1964–1965, a Visiting Scientist at the National Center for Telecommunication Research, Issy–les–Moulineaux in 1979–1980, a Visiting Professor at Massachusetts Institute of Technology, Cambridge in 1972–1973, at Tjanjin University, China, in 1983, at the University College, Dublin and the Swiss Federal Institute of Technology, Lausanne. He also was repeatedly an invited lecturer in Italy, United Kingdom, Hungary, Turkey, and Japan.

He served as Chairman of the IEEE Professional Group on Circuit Theory and received the IEEE Centennial Medal in 1984.

Among the great Masters of Circuit Theory he keeps an unique position for the broadness of his views, in many instances much ahead of time. In a memorable paper of 1959 with Dante C. Youla, he provided an axiomatic approach to Circuit Theory that went over the constraint of a finite number of lumped elements, thus opening the road to include in its field waveguides and n–port electromagnetic cavities filled with any kind of exotic linear media. The importance of such an approach deserves a particular attention in our time, in which the classical conceptual structure is reviving in Quantum Circuit Theory, that exploits the possibilities offered by technology at the nano level together with the properties of superconductive junctions. Among his important contributions must be cited the synthesis procedures at fixed frequency, the circuit models of complex physical structures and the Real Frequency Technique, a conceptually driven numerical procedure for broad–band matching and filter design that among other advantages allows to obtain better results in optimization than those offered by the classical approach with special functions; such technique is widely appreciated by theoreticians and practitioners and has been the object of intensive researc and development.

Professor Carlin was the author of numerous scientific papers and of two books "Network Theory" with Anthony Giordano (Prentice–Hall, 1964) and "Wideband Circuit Design" with Pier Paolo Civalleri (CRC Press, 1997).

He published a paper 'Singular Network Elements' in 1964 in the IEEE Transactions on Circuit Theory, in which he introduced the names 'nullator', 'norator' and 'nullor', which were to become important in active circuit theory later, with the development of the integrated circuit OpAmp, an ideal version of which corresponds to a nullor. These ideas seem to have been developed from earlier ideas of B.D.H. Tellegen and Arthur Keen. His wide and deep scientific culture was reflected in his lectures, always prepared with the greatest care, which remain a model for the clarity and the simplicity of language with which he was able to illustrate even the most difficult concepts.

He was a complete intellectual. His interests went far beyond engineering, physics and mathematics, to which he had devoted his professional life; they extended to music, literature, history, politics. His capability of extracting from any subject the essential made his conversations always highly exciting and rich of suggestions. His great humanity made him beloved by all persons that had been in contact with him; those who have had the luck of listening his lectures and enjoying his teachings, or to work with him, will never forget his great personality; their gratitude will last for life.

Pier Paolo Civalleri
Politecnico di Torino, Italy

Mac Van Valkenburg

Mac Elwyn Van Valkenburg was one of the most influential engineering educators of the Twentieth Century.   As a professor Mac was a gifted teacher, scholar, author, and mentor. He authored three textbooks, co-authored four others, and published numerous research papers and articles on engineering education. He also advised more than 50 Ph.D. students, many of whom became leaders in industry, government, and academia. He was a professor at the University of Illinois at Urbana/Champaign from 1955-1966, then joined Princeton University as professor and head of electrical engineering until 1974. He then returned to University of Illinois.
Named to the W. W. Grainger Professorship endowed chair in 1982, and became Dean of the College of Engineering in 1984.

Renowned Author

In the 1950's the core of every engineering curricula was a course in dc/ac circuit analysis. The modern concepts of time domain/frequency domain transform methodologies were revolutionary and little understood by most engineering educators of that era. Although Ernst Guillemin must be rightfully credited as the father of modern circuit and system theory in engineering education, Mac's books made the concepts much more accessible to students worldwide. Engineering curricula everywhere were changed with the publication of the first edition of Network Analysis, by M. E. Van Valkenburg, Prentice-Hall, 1955. The second and third editions of Network Analysis were published in 1964, and 1974 respectively. For over four decades hundreds of thousands of students throughout the world studied the analysis of linear circuits and systems from this book. Decades after the 1960s advent of digital integrated circuits, the topics in Mac's book remain at the core of engineering curricula. As integrated circuit active device feature sizes have shrunk their performance has become more so dominated by the linear interconnect circuitry among them.

Mac's fame as an engineering educator was elevated with the publication of his second book in 1960 entitled, Introduction to Modern Network Synthesis, John Wiley and Sons. In 1982 his final book entitled, Analog Filter Design, was published by Holt, Rinehart and Winston. His textbooks were translated into many languages and became international best sellers.

Mac was truly a gifted writer. He had a knack for explaining difficult concepts in simple terms. In 1991 Dr. Steven B. Sample, then President of the University of Southern California, echoed the sentiments of many of Mac's former students when he wrote, *"Mac Van Valkenburg was my teacher in the early 1960's when I was an undergraduate at the University of Illinois. To say that he was a brilliant teacher would be an understatement. His textbooks on linear circuit theory were at that time world famous, as they are to this day. Both through his writings and in the class room, Mac was able to explain extraordinarily complex technical topics in the most disarmingly simple and appealing manner possible. Over the years he has clearly emerged as one of the half dozen or so most distinguished teachers of engineering in the country."*

The Humble Guru

As Mac's reputation grew as one of the foremost electrical engineering educators of the latter half of the Twentieth Century, the demands on his time grew to a level that most people would find intolerable. Between his worldwide travels and his keen interest in engineering education, he became a valuable source of information. He served on endless policy and advisory committees on engineering education, wrote numerous articles on engineering education, including a column in the ASEE Prism. Engineering students, educators, and publishers from all parts of the world sought his advice and counsel. Mac treated them all equally graciously from the bewildered undergraduate student to the embattled department head, dean, or university president. Mac became the unofficial *"guru"* to electrical engineers.

Mac was quick to sense new trends in electrical engineering. In 1963 he organized the first Circuits and Systems Conference at the Allerton Center at the University of Illinois at Urbana-Champaign. Most of the notable electrical engineering educators in the field were in attendance. Many new ideas for research, new courses, and textbooks fermented from the mix of seasoned veterans and young educators and graduate students. It was the kind of intellectual stimulation that Mac enjoyed throughout his career. Later he encouraged similar conferences at Princeton University, the Asilomar Conference Center in California, and at the University of Hawaii. Approximately ten years later the IEEE Circuit Theory Society became the IEEE Circuits and Systems Society. Van Valkenburg's ability to communicate in print was matched by his ability as a classroom teacher. His famous colored-chalk lectures, delivered with infectious enthusiasm, attracted thousands of undergraduates to his courses. The *"Guru of Electrical Engineering Education"* was a master at establishing a comfortable, open atmosphere for learning. Mac relished attention, but he was a humble man who didn't like being on a pedestal. He enjoyed communicating with other people in the profession. He had a talent for bringing out a person's innermost thoughts while revealing little about himself. He liked to stimulate discussions using the shock treatment methodology, or, as one friend said, *"Mac liked to throw curve balls at you,"* that is, he would say something ridiculous or outrageous to you. Friends would recognize the sly smile on his face and enjoy fielding his *"curve balls."* Others, who didn't know him well, were often confounded. A person with a good sense of humor quickly caught on to his antics and became his lifelong friend. Mac was incredibly loyal to his friends.

The Mentor

Mac's communication skills made him a great writer and teacher, but his talents didn't end there. Mac took a genuine interest in bright and creative people and enjoyed mentoring them, regardless of their sex or ethnicity. Mac's only requirement was that they be of open mind. He was generous in his support of them, even to the point of encouraging them to write textbooks which were competitive to his books. For many young people in the profession Mac was a father figure. He mentored through example, genuine interest in people, and subtle persuasion. Because of the extensive demands on Mac's time, his typical work day began at 5:00 a.m. and ended late in the evenings, seven days a week, except for time for his wife Evelyn and church activities on weekends.

A Brief Biography

Mac Van Valkenburg was born on October 5, 1921 in Union, Utah, a son of Charles M. and Nora Louise Walker Van. The family surname had been shortened to Van. In

1928 his father was electrocuted on his dairy farm when he came in contact with a live wire that lay in his pasture. At age 6 Mac and his three younger sisters were fatherless. His mother gave birth to a fourth sister two weeks after his father's death. Nora sold her share of the dairy farm and moved nearer to her sister and brother-in-law who were childless. The children were raised by the three adults. Nora never remarried.

In grade school Mac was inspired by a neighbour boy who had done such things as amplify the sound from a hand-cranked phonograph by using a one tube, battery powered radio. Before he was a teenager Mac and a close friend, Vance Burgon, were making crystal radio receivers. Materials included copper coils wrapped around oatmeal boxes and crystals of galena found in nearby copper deposits. Mac and Vance became amateur radio operators. Their walls became plastered with QSL cards, postcards from other ham radio operators verifying that their signal had been received. The postcards from North, Central and South America were simply addressed "*Mac Van, Sandy, Utah.*" At that time the nearest post office to Union was in Sandy, Utah. Soon Mac and Vance were scripting a radio program based on their experiences and information from ham radio magazines. Their program aired late Saturday nights on radio station KSL in Salt Lake City. Mac developed a love for classical music from his interest in radio.

Mac's mother influenced him to attend college, and Mac influenced the family to restore their name to Van Valkenburg. Mac picked strawberries in the summers and saved his earnings for college. The summer before he entered college, he worked for two neighbouring farmers from Japan who had developed superior strawberry plants with an extended growing season and larger berries. Mac earned $80 that summer.

Mac graduated from Jordan High School in Sandy, Utah, and the following fall enrolled in the electrical engineering program at the University of Utah. He received financial support from the National Youth Assistance Program which allowed him to earn $20/month. At the university Mac assisted the Dean of Engineering, Dr. Taylor, for 35 cents/hour. He graduated from the University of Utah in 1943. On August 27, 1943 he married his high school sweetheart, Evelyn June Pate, in Salt Lake City. Mac's daughter JoLynne asserts that Mac was charmed by the music that Evelyn made in the high school band. Since the United States was in the midst of World War II and Mac was a top student, upon graduation he received an assignment to join the staff at MIT's Radiation Laboratory where he helped develop radar under the direction of the renowned Ernst Guillemin. Evelyn terminated her college career and joined Mac. She worked as secretary to the research group. Throughout his career Mac frequently relied on Evelyn's expert secretarial skills.

In 1946 Mac received the MS degree from MIT, and returned to the University of Utah where he taught until 1955. He took a leave of absence from 1949 to 1952 to pursue and receive a Ph.D. degree at Stanford University. Interestingly his Ph.D. thesis was on the topic of the detection of meteor trails in the ionosphere. While at Stanford Mac was given the assignment of developing a new course on servomechanisms. One can only conjecture that perhaps this daunting assignment developed his interest in circuits and systems. Perhaps the seeds had already been planted by Ernst Guillemin.

Mac joined the faculty of Electrical Engineering at the University of Illinois in Urbana-Champaign (UIUC) in 1955. There he served as Associate Director of the Coordinated Science Laboratory and, for a semester, as Acting Department Head. In 1966 he became Department Head of Electrical Engineering at Princeton University. In 1974 he returned to the University of Illinois.

In 1982 Mac was named to the College of Engineering's first endowed chair, the W. W. Grainger Professorship. In 1984 he was appointed Dean of the College of

Engineering. Upon his retirement in 1988, UIUC Chancellor Thomas Everhart stated, "*The renaissance in engineering, which has seen an explosion of new endeavors in the past three years, has been due, in no small part, to the supportive atmosphere Dean Van Valkenburg has embodied and the encouragement he has given to new initiatives.*"

During his career Mac held visiting appointments at Stanford University, University of California at Berkeley, University of Colorado, University of Hawaii-Manoa, the University of Arizona, and the Indian Institute of Technology at Kanpur. He was a delegate to the meeting of the First International Federation of Automatic Control in Moscow. He also received numerous honors and awards, including Member of the National Academy of Engineering, Fellow of IEEE, IEEE Education Medal, IEEE Centennial Medal, ASEE Hall of Fame, ASEE Lamme Medal, ASEE George Westinghouse Award, Guillemin Prize, and the Halliburton Award. In the IEEE Mac served as Vice President, Editor of the IEEE Proceedings and the IEEE Transactions on Circuit Theory, Editor-in-Chief of the IEEE Press, and he served on numerous committees in the National Academy of Engineering, the IEEE, ASEE, and the Accreditation Board for Engineering and Technology. He also served on a number of advisory committees for NSF and various universities.

To those so many he touched and enabled Mac Van Valkenburg will always live on as quite simply the greatest engineering educator ever.

Timothy N. Trick
University of Illinois
Ronald A. Rohrer
Carnegie Mellon University

Vitold Belevitch

Vitold Belevitch's parents were living in Lenin-grad, that is now again called St Petersburg, the former capital of the Russian Empire. In order to flee from the aftermaths of the communist revolution in 1917, his parents tried in 1921 to get to Helsinki in Finland. Helpers brought his mother, who was expecting her first child, across the border. But before they could also guide his father into safety, he was arrested and deported to Siberia, from where he never returned. So Vitold Belevitch was born in the small Karelian town of Terijoki on 2 March 1921. His mother did not stay there, but went as soon as possible on to Helsinki, where she registered Vitold's birth. Hence he is officially listed as been born in Helsinki. When Vitold was 4 years old, his mother moved with him to Belgium, like did a

good number of Russian refugees. There he grew up and got his education in French, even though his language at home continued to be Russian. Vitold Belevitch was a man of broad scientific and cultural interest. He studied electrical and mechanical engineering at the Université Catholique de Louvain, getting his diploma at the age of 21 in 1942. Then he joined Bell Telephone Manufacturing Company (BTMC), now Nokia in Antwerp, Belgium. There he met Cauer who was working for a sister company Mix&Genest in Berlin. Cauer introduced him to the beauty of circuit theory and its applications. Stimulated by Cauer and under the supervision of Charles Manneback, he obtained the doctoral degree in applied sciences from the Université Catholique de Louvain in 1945. In his thesis he introduced the revolutionary concept of the scattering matrix, or repartition matrix, as he called it. In fact, this concept was also independently discovered by American researchers during the Second World War in the context of distributed parameter microwave circuits. Belevitch used it in order to build a comprehensive approach to several results in circuit theory that were initially developed by Brune, Cauer, Foster, Gewertz, Darlington and others. He could extend this concept to any number of ports, and reciprocal as well as non-reciprocal, real as well as complex networks. He even made valuable contributions to the theory of nonlinear circuits with applications for rectifier circuits. The majority of his research on circuits and systems is described in his three impressive books, *"Théorie des circuits de télécommunications"* and *'Théorie des circuits non-linéaires en régime alternatif,"* both written in French and the third written in English entitled *"Classical Network Theory"*. As the title of the third suggests, this is a monumental work, that goes beyond what was customary. It has since been considered as his most important contribution and a standard work on circuit theory. His publications on filter theory and modulation techniques have since served as regular tools for many engineers in laboratories and development. His reputation in circuit theory was so strong internationally, that he was the only European researcher to be invited to write a contribution on the history of circuit theory for a special issue of the Proceedings of Institute of Radio Engineers (IRE). That special issue was set up in 1962 by IRE, one of the two predecessors of IEEE, to celebrate its 50th birthday.

In 1951 Belgian research funding agencies gave the task to design and build an electronic computer to Bell Telephone, and Belevitch became the project leader.

Nearly everything had to be designed and built by mathematicians, physicists and engineers. Upon completion of this project he left Bell Telephone and became in 1955 director of the Belgian Computing Center, the 'Comité d'étude et d'exploitation des calculateurs électroniques' in Brussels. The computer was finished in 1956. It was 13 meter long and 2,5 meter high, and contained 1000 triodes, 1200 tubes with cold cathode, 400 relays, 1000 selenium diodes, and 500 germanium diodes. All information was encoded in decimal numbers. Although the components were not very reliable, and much heat was produced by the electronic tubes, the machine was operational until the beginning of 60ies. It was free of charge to be used by academic researchers and industry was charged 5000 Belgian Francs per hour (approximately 150 US$). At the request of the director of research of Philips, Prof. Casimir, Belevitch founded in 1963 the Laboratoire de Recherche at MBLE Manufacture Belge de Lampes Electriques, later renamed as Philips Research Laboratories Belgium (PRLB). The research center had in the 80ies around 70 researchers of high caliber. He directed this very successful research center until his retirement at the end of November 1984. About 30 of his former co-workers or former students are now well-known professors at recognized universities around the world. Already in 1953 Vitold Belevitch was also appointed as a part-time professor at the UCL, where he taught subjects like circuit theory, electromagnetism, applied mathematics, information theory and coding. In 1960 he became extraordinary professor there and retired in 1985.

As head of industrial laboratories he was a full-fledged engineer, with attention to practical use, reliability and performance. But he also kept strong fundamental academic research interests, primarily linked to mathematics, with a stronger inclination toward algebra than analysis. In fact the more a problem was mathematically intriguing, the more he was interested in finding a suitable solution. This pronounced orientation towards mathematics is quite typical for research minded Belgian engineers. In Belgium, indeed, for many years a regular high-school diploma does not yet give access to university studies in engineering. One must in addition pass a rather stringent entrance examination involving nothing but mathematics at the most advanced level taught at science-oriented high-school tracks. For Belevitch the mathematical intuition often preceded and guided him to the result even before calculating it. He made major contributions to the fields of information and systems and control theory, the design of electronic computers, mathematics and linguistics. In fact, he was able to speak, and especially to read, a exceptionally large number of languages. He also applied the mathematics of information theory to obtain results on human languages. The total volume of his scientific production is estimated at 4000 pages.

Vitold Belevitch received numerous recognitions for his scientific achievements. He became Fellow of the IEEE, and was awarded the IEEE centennial medal, and in 1993 he received the Society Award of IEEE Circuits and Systems Society, now called the Mac Van Valkenburg Award. In 1975 and 1978, respectively, the Technical University of Munich, Germany, and the Ecole Polytechnique Fédérale de Lausanne, Switzerland, conferred upon him an honorary doctoral degree.

Joos Vandewalle, IEEE Life Fellow
Emeritus Professor, Katholieke Universiteit Leuven, Belgium

This article has been adapted with permission from J. Vandewalle, "In Memoriam-Vitold Belevitch", International Journal of Circuit Theory and Applications, 2000, Vol. 28, pp. 429-430. Any third party material is expressly excluded from this permission.

Radoslav Horvat

Radoslav Horvat, considered founder of circuit theory in the region of former Yugoslavia, passed away on December 21, 2004, in Belgrade, Serbia, at the age of 84. He served on the Faculty of Electrical Engineering at the University of Belgrade since 1950 and has been responsible for the education of thousands of his country's engineers in circuit theory and network synthesis. He is the author of five books and numerous research papers.

Radoslav Horvat joined the Faculty of Electrical Engineering at the University of Belgrade in 1950 and taught the course on "Theoretical Principles of Electrical Engineering." The course emanated from a course on "Fundamentals of Electrical Engineering" and dealt with topics known today as "alternating current (AC)." It was taught to the 3rd year undergraduate students in electrical engineering and covered what is today known as "circuits with lumped parameters." Professor Horvat realized early the importance of circuit theory and in 1956, shortly after returning from a sabbatical spent in UK, he established a course on "Theory of Electrical Circuits," a title that has remained on the curriculum until today. Professor Horvat introduced the most up-to-date topics at that time dealing with circuit analysis. Since he was not only an engineer but also a mathematician, he systematically organized the material used in the course offered to electrical engineering students and in 1959 published the well-know textbook "Theory of Electrical Circuits" published by Građevinska Knjiga. The book is known to thousands of electrical engineers not only at the Faculty of Electrical Engineering in Belgrade, but throughout former Yugoslavia, and it served as an example to younger faculty who later authored their own textbooks. Shortly afterwards, Professor Horvat published another well-known textbook "Special Electrical Circuits" that contained coupled circuits (transformers), two-port elements, and filters. He completed the series of textbooks with the book "Elements of Analysis of Electrical Circuits" where he used a modern approach to describe principles of analyzing circuits in the time domain.

Teaching a scientific subject calls for developing new material and forming new faculty who would continue the education process. Professor Horvat always surrounded himself with extremely capable collaborators, including the late Professor Marija Šušnjar who started as an Assistant Professor teaching Fundamentals of Electrical Engineering and then continued to teach Theory of Electrical Circuits as Associate Professor, as well as late Professor Mirko Milić, member of the Serbian Academy of Arts and Sciences, and outstanding scientist with an international reputation. After graduation, Professor Milić worked as Teaching Assistant to Professor Horvat and was promoted to Assistant Professor for Theory of Electrical Circuits, teaching in the new Department of Nuclear Engineering and later in the Department of Engineering Physics.

During his career between 1950 and 1985, Professor Horvat established programs in *Theory of Analysis and Synthesis of Electrical Circuits*, first at the University of Belgrade and then at other University centers in former Yugoslavia: Niš, Novi Sad, Podgorica, Čačak, Banja Luka, Sarajevo, Skoplje, and Priština, and directly contributed to the field in centers in Split, Zagreb, Ljubljana, and Maribor. In those centers students of Professor Horvat continued to work in the area of circuit theory. They include Professors Momčilo Bogdanov (Skoplje), Milić Đekić (Čačak), Petar Hinić (Banja Luka), Gordana Jovanović-Doleček (Sarajevo, Mexico), Dragan Kandić (Belgrade), Ljiljana Milić (Belgrade), Slobodan Milojković (Sarajevo, Priština), Ladislav Novak (Novi Sad), Radoje Ostojić (Podgorica), Branislava Peruničić (Sarajevo), Radmila Petković (Niš), Branimir Reljin (Belgarde), Dušan Starčević (Belgrade), and Dr. Borivoje Stamenković (Bern).

The School of Professor Horvat in Theory of Electrical Circuits was known and recognized worldwide, as noted in the article by Van Valkenburg in 1984 that appeared in the issue of the *IEEE Transactions on Circuit Theory* published on the occasion of the IEEE Centennial.

Besides teaching, Professor Horvat already in 1968 established a series of international symposiums on theory of electrical networks, *Int. Symp. on Network Theory (ISYNT)*, held in Yugoslavia and were attended by the best-known scientists in this field including J. Aggarwal, T. Bickart, H. Carlin, L. Chua, P. Civalleri, A. Davies, T. Deliyannis, C. Desoer, S. Dutta Roy, J. Fidler, A. Fettweis, M. Ghausi, E. Kuh, E. Laker, E. Lindberg, G. Martinelli, S. Mitra, G. Moschytz, J. Neirynck, R. Newcomb, A. Petrenko, T. Roska, R. Saal, J. Scanlan, G. Temes, Y. Tokad, M. Van Valkenburg, V. Zima, and others. The first ISYNT was held in 1968 in Belgrade and then in Herceg-Novi (1972), Split (1975), Ljubljana (1979), Sarajevo (1984), and Zagreb (1989). Professor Horvat was a program committee member of the *European Conference on Circuit Theory and Design (ECCTD)* and one of the Editors of *Int. Journal on Circuit Theory and Applications*. He was Honorary Chair of the *IEEE Conference of Artificial Neural Networks (NEUREL '2000)* and reviewer for *IEEE Trans. Circuits and Systems* and *Int. Journal on Circuit Theory and Applications*. He was a founder and active member of Society ETRAN and its Honorary President.

Based on the notice published in *IEEE Signal Processing Magazine* in September 2000, on the occasion of Professor Horvat's 80th birthday, and *In Memoriam* issued in December 2004 by the Faculty of Electrical Engineering, University of Belgrade.

Ljiljana Trajković
Simon Fraser University, British Columbia, Canada

Robert Aaron

M. Robert Aaron, a key contributor to the design of the T1 carrier system, the first practical digital system in the world designed for the exchange telephone plant, died on June 16, 2007, at the age of 84.

Bob was born on August 21, 1922, in Philadelphia, Pennsylvania. He studied electrical engineering at the University of Pennsylvania in Philadelphia, where he received his B.S. in1949 and M.S. in 1951. After graduation, he joined Bell Laboratories in New Jersey, where he worked until his retirement in 1989.

During his career at Bell labs, Bob was associated with many *"firsts"*. Initially, he worked on the design of networks, filters, and repeaters for a variety of circuits for analog transmission systems, such as the equipment for the first color transmission of the Orange Bowl football game and equalizers for the LE coaxial system. He made fundamental contributions to computer-aided design and applied these techniques to the development of the first repeatered transatlantic cable system in 1956.

Also in 1956, he began working on the development of digital transmission systems. The T1 carrier system was introduced by Bell into commercial service in 1962. One of Bob's former colleagues at Bell Labs, Dr. John Mayo, wrote: *"Six problems threatened the viability of high speed digital communications in the telephone network."* Bob Aaron made major contributions to overcoming five and had a strong supporting role in overcoming the sixth. His role was at the very heart of innovation, for he analyzed every aspect of what became the carrier system which remains an essential element in global digital communications." Dr. *Irwin Dorros, another former colleague at Bell labs, described the T1 format as an influence on "the backbone of Internet transmission."*

Bob's next focus was on the search for new techniques for high-speed digital systems. From 1968 until his retirement in 1989, he was head of the Digital Technologies Department. Initially he was involved in the development of a variety of digital terminals for transmission and switching. In later years, he was responsible for exploratory development of digital signal processing terminals and techniques.

Over the years, Bob has published more than 50 papers and was awarded many patents in circuit design, control, and communications. Several of his papers have been republished in collections of benchmark publications.

Bob was an active member of the IEEE in various capacities. He was a key player in the establishment of the IEEE Control Systems Society, was chairman of the first Papers Review Committee, and was secretary of the organization. He was an associate Editor of the IEEE Transactions on Circuits and Systems from July 1969 to June 1971, and President of the IEEE Circuits and Systems Society in 1973. He was also a member of the Publications Committee of the IEEE Technical Activities Board (TAB), a working member of other IEEE committees, a member of the TAB Finance Committee, and chairman of the Digital Systems Subcommittee of the IEEE Communications Society.

Bob received many awards and honors for his professional accomplishments. He was a Fellow of the IEEE and a Fellow of the American Association for the

Advancement of Science. He was co-recipient, with John S. Mayo and Eric E. Sumner, of the 1978 IEEE Alexander Graham Bell Medal and the 1988 NEC C&C Prize for *"pioneering contributions to the establishment of a basic technology for digital communications by development of world's first practical commercial high-speed digital communications: T1."* Bob was elected a member of the U.S. National Academy of Engineering in 1979. He received the IEEE Centennial Medal in 1984, the McClellan Award of the IEEE Communications Society, and a Lifetime Service Award from the IEEE Communications Society in 1977. In 1999, he was the recipient of the International Telecommunications 'Christoforo Columbo' Award for his contributions to the development of digital communication systems, reduced bit-rate coding, and fast packet-switching systems.

Dr. John Mayo, Bob's colleague, remembered *"Upon winning the Japanese C&C Prize, Bob decided to deliver his acceptance speech in Japanese, even though he had no knowledge of the language. When asked why, he replied 'Because they would appreciate it. That showed the sensitivity, commitment, diligence, confidence, and excellence to all his work. An when the speech was over, the Japanese said exactly about all Bob's work, Done Perfectly.*

Dr. David Messerschmitt, who worked in Bob's group in the 1970's before joining the University of California, Berkeley, as a professor, recalls *"Bob was always a friend more than a boss. He was supportive in every way imaginable and was always available, to interact with the troops. By the time I knew him, Bob took over the role of a facilitator rather than individual technical contributor. He mostly worked through us, seeding us with ideas and disabusing us of our misconceptions."*

I joined Bob's group at Bell labs in New Jersey, after a two-hour telephone interview with him, and met him for the first time in July 1965. He became my mentor and great friend, and I benefitted tremendously from my friendship and close association with him. We collaborated on several projects, which led to several papers. After about a year, I informed Bob of my intention to take an academic post on the West Coast. He was supportive of my decision as felt academe would be beneficial to Bell labs in the long run.

Professor David Messerschmitt made a similar observation. *"At one point I made it known that I was really interested in a career in academe. Although Bob was conflicted about this, he always saw academe as complementary rather than competitive."*

After being diagnosed with multiple myeloma, Bob worked with the International Myeloma Foundation to promote the education and support of people of people with this form of cancer.

I kept in touch with Bob and his wife Wilma regularly after leaving Bell labs, and visited them whenever I was near their home, otherwise by telephone. In June 2007, when I telephoned Bob and asked how he was doing, he said, in a very soft voice, *"Sanjit. I am very sick."* I did not realize at the time that would be the last I spoke to him. I and everyone else who worked with him at Bell Labs and elsewhere miss him greatly.

Sanjit K. Mitra University of California, Santa Berbara

[*] This article has been adapted with permission from Memorial Tributes, vol. 13, pp. 2-7, National Academy of Engineering, Washington, DC, 2014.

H. John Orchard

John Orchard, a pioneer of modern filter theory, passed away on June 23, 2004, at the age of 82. He received his education at the University of London and, following a five-year stint (1942-1947) as a Lecturer at the Central Training School of the British Post Office Engineering Dept., took a position in their Research Division. He moved to the U.S. in 1961 and was a Senior Staff Engineer at Lenkurt Electric Company until joining the University of California, Los Angeles (UCLA) in 1970, where he taught until his retirement in 1991. He continued to be active in research at UCLA until his death.

Prof. Orchard's early work dealt with mathematically difficult, practically important topics in the areas of passive and linear active circuit design and approximation theory. His publications solved design-oriented problems of significance to filter specialists, such as equalizer and phase-shifter design, predistortion, and efficient methods for solving the approximation problem.

As a result of this highly pragmatic, sophisticated but specialized work, John became interested in the general aspects of filters. In a key letter [1] he explained the "*secret*" behind the low passband sensitivity of doubly loaded reactance two-ports and showed how to design active two-ports that retain this key attribute. Subsequently, he used this technique in developing a useful design methodology for gyrator-based active filters [2, 3].

One of Prof. Orchard's key contributions was the development of a systematic process for the computer-aided design of filters. This was carried out in terms of a transformed frequency variable [4]. The method provided a unified framework for the solution of the approximation problem and for ladder expansion, and resulted in greatly enhanced accuracy. During the early years of computer-aided filter design, when the synthesis of a single circuit required several days of multiple precision computation, his method had an important beneficial effect.

In the 1970's John became interested in the newly introduced switched-capacitor (SC) filters. He was instrumental in introducing into SC filter design the bilinear s-z mapping, previously used solely in digital filter design, and in developing a methodology that allowed the use of arbitrary active RC models for SC filter synthesis [5].

The key attribute of Prof. Orchard's contributions was a deep understanding of the underlying physics and theory, combined with great respect for the practical aspects of circuit analysis and design. He would not publish a theoretically elegant but practically useless result, nor a useful but trivial one. He could instantly extract the essence of a problem and attack it on its most fundamental level. He sometimes exasperated his coauthors (including these writers) by spending precious days trying to simplify the proofs and derivations contained in a prospective publication to make them more basic, general and accessible to the readers. All three writers of this notice have their own unique recollections of John, some of which we share below.

I (Gabor Temes) started to interact with John in 1964, when I moved to California to join Stanford University. By that time, he was a much-admired and somewhat mysterious researcher in the area of filters. The mystery came about since John (a soft-spoken and shy man) never attended meetings and preferred to interact even with his friends by way of letters and, later, by e-mail rather than in person. Even so, for about five years, he and I met weekly for lunch to discuss circuits and, as a result, published several joint papers.

I joined UCLA in 1969 and managed to convince both John and UCLA (which was nontrivial since John didn't have a Ph.D.) that it would be in their mutual interest if he would come as well. At UCLA, after a couple of years he adapted to academic life. In spite of his quiet and introverted nature, his deep understanding of both theory and practice, and his commitment to clarity in thought and expression, made him an outstanding teacher and a wonderful research collaborator.

Although I (Alan Willson) had had a course in network synthesis in the mid-1960's, I must confess that everything I know today about filter design I learned from John. The year I joined the UCLA faculty (1973), I made it a point to audit John's graduate classes on filter design. It wasn't my field (I was very nonlinear then) but what a great joy his classes were!! Since then, we've collaborated for over 30 years.

Work began on our first joint paper when, literally a few days after I'd arrived at UCLA, John asked me if I knew whether an ideal gyrator could be implemented with just a single active element. (He'd never seen such a circuit and suspected that at least two were always required.) That question kept me intrigued for a few weeks, but the right insight finally materialized; and once we had the answer and had proved that one couldn't be built [6], we *astounded* ourselves by finding that by simply changing the requirement that a (passive) gyrator would be allowed to be "*active*," our proof broke down, indicating that it just might be possible to build such a gyrator with a single op-amp; so we designed one! [7]

More recently, I'd usually be the one to walk into John's office and say something like "*It says in this here digital filter book that... Do you think that's true? If so, I'd sure like to learn why.*" And we'd be off and running with another project. His encyclopedic knowledge of analog filter design could usually be adapted to address issues that arose in my digitally focused courses. Long past his nominal retirement, John was still devoted to the work we were doing at UCLA. For the past several years, we kept an undeviating Monday/Wednesday schedule of meetings for which he (not I) was never late.

John taught a sequence of graduate courses at UCLA in a style that achieved a perfection and flawlessness rarely found today. There was a time, before research and teaching became inextricable, when universities valued good teaching in and of itself. In the hands of a few masters, good teaching became great teaching that was inspirational to students, even life-changing. John was such a master.

John put much hard work and many hours of preparation into his lectures, and he also went to great lengths to craft fiendishly searching exam problems. He believed that an examination, instead of being a test of routine problem-solving, should be a learning experience, albeit a very different one than the lecture. The student learns under pressure; but the revelations, if they arrive, are more durable. At first sight, his problems looked very difficult, seemingly necessitating lengthy numerical calculations that could not possibly be completed in time. Yet with insight and the strategically supplied hint, the problem would dissolve into simple round number solutions readily found by manual calculation, very effectively testing the particular point being examined.

Gabor Temes, Oregon State University, Corvallis, OR
Alan Willson, *Asad Abidi* University of California Los Angeles, CA

[*] References cited herein can be found in IEEE Trans. Circuits Syst. I, vol. 51, pp. 2341-2344, Dec. 2004, from which this In Memoriam has been extracted.

Charles A. Desoer

Charles A. Desoer was a world-renowned researcher, research supervisor, and dedicated educator. His research focused on the analysis, design, and control of linear and nonlinear circuits and systems that contributed to the burgeoning growth in control applications and benefited the aerospace, transportation, process control, and other essential industry sectors.

Desoer, professor emeritus of electrical engineering and computer sciences at the University of California, Berkeley, died November 1, 2010, in Oakland, California, at the age of 84, of complications from a stroke.

Charlie, as he was known universally, was born on January 11, 1926, in Brussels. He fought with the Belgian Resistance during the German occupation in World War II and joined the Belgian Army after the liberation. He obtained a degree as a radio engineer from the University of Liége in 1949 and an ScD in electrical engineering at the Massachusetts Institute of Technology in 1953. He then went to work at Bell Laboratories in Murray Hill, New Jersey, until 1958, when he left to join UC Berkeley as a professor of electrical engineering and computer sciences. He continued to serve the campus as professor emeritus after his retirement in 1993.

He was an exceptionally gifted teacher, with a style that emphasized clarity of thought and elegance of presentation, both of which were evident in his seminal textbooks on circuit theory, linear systems theory, and feedback control. Some of his texts are still considered the most authoritative references on circuits, systems, and control, and *"widely regarded as classics in the field [that] have set a high standard for their clarity of thought and presentation, as well as a deep commitment to intellectual elegance,"* said Shankar Sastry, former PhD student of Desoer and now Dean of Engineering at UC Berkeley.

I met Charlie in the fall of 1958 when I joined the Department of Electrical Engineering at UC Berkeley as a graduate student and took his upper division course *"Linear Systems."* He was an inspiring teacher and encouraged his students to think and solve homework problems using different approaches. I had planned to study computer engineering, but after taking the course with Desoer I switched to circuits and systems and decided to work on my master's thesis under his supervision. The project involved developing a computer-based approach to design analog filters with lossy inductors and capacitors.

A much-loved colleague in the Department of Electrical Engineering and Computer Sciences, Charlie was known for his sharp repartee, yet he always had kind words for his colleagues. Former students and junior colleagues remember him for his dedicated mentoring and his strong emphasis on excellence in teaching.

Charlie graduated 42 PhD students, many of who have established themselves as leaders in their fields in academia and industry, a testimonial to his inspirational teaching and mentorship. Robert Newcomb, Charlie's first PhD student wrote: *"It was one of the great pleasures of my life to have been a doctoral student of Charles A. Desoer. Fortunately for me Charles wanted to learn about network synthesis, which was the topic I wanted to pursue for my doctorate."*

Shinzo Kodama, former PhD student and professor emeritus of Osaka University, Japan, mentions that Charlie's ever inquisitive curiosity and strong desire to pursue the essence of a subject and his engineering viewpoint with an extensive mathematical background have continued to influence him and other PhD students long after they left Berkeley. He also says that he *"was most impressed by his rare sense concerning the direction of research where significant findings lie. It makes me proud whenever I tell my students that I was one of the students of Professor Desoer."*

Professor M. Vidyasagar of the University of Texas at Dallas observes that, *"While Charlie took his own research seriously, he was just as serious encouraging the next generation of researchers. In 1970 I naïvely mailed a copy of a paper to Charlie, requesting his comments. He promptly responded and made several constructive suggestions. It is difficult to imagine, in this day and age, anyone of his stature reading a paper from an unknown person and taking the time to suggest thoughtful comments."* He adds, *"It was my privilege to have coauthored a book with Charlie. He had written the bulk of the book by mid-1973 but could not find the time to finish it. So he invited me to spend some time at Berkeley so that we could revise what he had written and write some new chapters. I was just 25 but he treated me as an equal partner in the writing enterprise."*

George Oster, professor of molecular and cell biology at UC Berkeley, remarked that, *"Although I wrote but three papers with Charles, the process of writing them taught me one of the most important lessons of my scientific career: How to think clearly and express ideas precisely. The papers were written when I was a postdoc and subsequently a new assistant professor. I had inherited from my physics and engineering training a mode of thinking that Charles considered fuzzy and impreciseand he was oh, so right! In my subsequent career, however, I fear that I have drifted a considerable way from Charles's ideal. My excuse has been that biological modeling has not yet reached the precision of physics, or even of engineering. Nevertheless, his lessons always remind me that seeking clarity in writing leads to clarity in thoughtnot necessarily the other way round."*

Sanjit K. Mitra
University of California, Santa Barbara

[*] This article has been adapted with permission from Memorial Tributes, vol. 18, pp. 62-67, National Academy of Engineering, Washington, DC, 2014.

Alfred Fettweis

Alfred Fettweis was a giant in the field of circuits and systems in a broad sense. His research was focused not only on circuit theory; it also introduced the wave digital filter concept, extended it to multiple dimensions and the numerical integration of partial differential equations and finally, addressed foundational aspects of physics from the viewpoint of circuit and signal theory.

Alfred Fettweis, professor emeritus of communication theory at the Ruhr-University Bochum, passed away on August 20, 2015 in Bochum, Germany at the age of 88. He is survived by his wife and 5 children and 16 grandchildren.

Being born in Eupen, Belgium on November 27, 1926, he lived a difficult youth in an area which first became part of Belgium in 1920. During tumultuous times, people in this region had to change their nationality back and forth. During World War II when he was a high school student, Alfred Fettweis was already being educated in high frequency electronics to repair radar equipment. He was in a group together with colleagues like Carl F. Kurth, Manfred Schroeder and Rudolf Saal who later also became well known leaders in the field of circuits and signal processing.

After the war, being Belgian again, he started his academic education at the Catholic University of Louvain and graduated as Ingénieur Civil Électricien in 1951. Subsequently, he was employed as an engineer at the International Telephone and Telegraph Cooperation (ITT) in Belgium with a 2 year stint (1954-56) in their Federal Telecommunications Laboratories in Nutley, USA. During that time he was engaged in the design of filters for carrier telephony systems and in electronic switching systems based on the resonant transfer principle. This period was important for developing practical skills and experience and led insight into the problem of sensitivity of filters and its relation to passivity and matching. This important relation, which later was exploited in the domain of digital signal processing, was independently discovered by John Orchard and is sometimes called the Fettweis-Orchard Theorem. Towards the end of this period (1963) he completed his doctoral thesis, "*Théorie des Circuits u Transfert Résonnant*," in which he laid the theoretical basis for the wave digital filter concept. The official advisor of the thesis was Vitold Belevitch, another giant in the field of circuits and systems. At that time, the concept was still an analog concept, to be realized with passive lossless components and switches. As such, it was already closely related to the so-called switched-capacitor filter concept, which became a hot topic in the 1980s.

In 1963, Alfred Fettweis became full professor at the Eindhoven University of Technology, The Netherlands. Four years later he got an offer for a full professor position in Bochum, Germany. He was a professor there from 1967 up to his retirement in 1992, but still continued to do research as an Emeritus. During the period as chaired professor in Bochum, the wave digital filter concept evolved. It is interesting to note that the first contribution submitted to the IEEE Transactions on Circuits and Systems was rejected. This, however, is not unusual if one has developed a new concept or method which is fundamentally not main stream. Later, very many follow-up contributions were published in the Circuits and Systems transactions; there was even a long invited tutorial paper (60 pages!) in the Proceedings of the IEEE in 1986. Later, this concept with applications in classical filtering was extended to multidimensional signal processing based on multidimensional Kirchhoff circuits, for which a theoretical framework did not previously exist. This turned out to pave the way to robust numerical algorithms for numerically integrating partial differential equations.

After official retirement in Bochum, Alfred Fettweis was a Distinguished Professor of Computer Science and Engineering at the University of Notre Dame, USA for two years (1994-1996). During this time, the numerical integration approach was further developed with applications in electromagnetic field problems, i.e. integrating Maxwell's equations and fluid dynamics, and integrating Navier-Stokes equations. Later, he shifted the focus of his research to basic problems of physics including considerations of special relativity and the nature of electrons and photons by rigorously assuming that Maxwell's equations are valid even at extremely small dimensions. He was convinced that losslessness and passivity with the concept of multi-dimensional Kirchhoff theory can lead to new insight into physical phenomena. It is a great loss that he did not have the resources and time to continue his research in this direction.

His outstanding contributions have been recognized by honorary doctorates from six different universities: 1986 in Linkoping, Sweden, 1988 in Mons, Belgium and in Leuven Belgium, 1995 in Budapest, Hungary, 2004 in Poznan, Poland and 2011 in Paderborn, Germany. In addition, he received numerous prestigious scientific awards: 1981 Darlington Best Paper Award of IEEE CASS, 1984 ITG/VDE Ring of Honor, 1988 Charlie Desoer Technical Achievement Award of IEEE CASS and the Technical Achievement Award of ITG/VDE, 2001 Mac Van Valkenburg Award of IEEE CASS, 2003 Vitold Belevitch Award of IEEE CASS, 2008 Gustav Robert Kirchhoff Technical Field Award of IEEE. He provided valuable service to the IEEE Circuits and System Society in quite a number of positions, most notably as Technical Program Chair of the flagship conference ISCAS in 1976 in Munich.

His outstanding scientific achievements and his highest international reputation made him a member of a number of several European academies: North Rhine-Westphalian Academy of Sciences, Humanities and the Arts, Germany, Academia Europaea, London, Academia Scientiarium et Artium Europaea, Salzburg/Vienna, and the German Academy of Technical Sciences, acatech.

Over several decades, Alfred Fettweis has made outstanding contributions to engineering and physics. He introduced robustness based on the physical principles of passivity and losslessness into digital signal processing and numerical algorithms. Therefore, he is one of the fathers who paved the way to the digital world we are living in today.

I remember him as an extraordinary human being with a great personality, and as a great scientist and engineer who deeply influenced my own professional life. I had the privilege to have contact with him very early while I was still a student. His way of clear and sharp thinking, always searching for truth, is a role model which I admire and still try to emulate. He is, however, a hard act to follow.

Josef A. Nossek
Technical University Munich

Ernest Kuh

Professor Ernest Kuh has been one of the key founding fathers of the electronic design automation (EDA). His numerous research and technology breakthroughs in the most challenging EDA domains for several decades went far beyond the scientific and technology front. Professor Kuh was instrumental in creating many EDA companies formed by his students and fellow researchers. This is not just the opinion of Chi-Ping Hsu, who studied with Kuh, but is highly believed and recognised.

Ernest Shiu-Jen Kuh was born in 1928, in Beijing, China. His father was a government official and then private businessmen. Kuh left China for the United States in 1947 to continue his education. He began studying electrical engineering at the University of Michigan in the winter of 1948 where he received his B.S. degree in 1949. He then earned a master's degree in EE from MIT in 1950, where he met circuit theorist Charles Desoer. The two would later work together and were frequent collaborators.

Kuh started a Ph.D. program at Stanford in 1950, pursuing research in network theory. He completed his degree in 1952, after just six quarters, and saw his thesis published in the Journal of Applied Physics, a rare place to find student work, especially in electrical engineering.

In 1952, Kuh went to work for Bell Laboratories in Murray Hill, New Jersey. At Bell Labs, Kuh worked in the transmission development division on issues related to telephone infrastructure, specifically transmission repeater designs and submarine cable design.

Kuh's friend from MIT, Charles Desoer, joined Bell Labs the year after Kuh was hired. For a year, the two, along with other Bell Lab colleagues, would meet after work for a study group, discussing new ways to write the differential equations that form the mathematical basis of electronic circuit function.

By 1967, Kuh completed a widely used book, Basic Circuit Theory, with Desoer (who had also joined the Berkeley faculty), and was asked to head the electrical engineering department, which had just recently integrated computer science. Kuh's appointment as chair of Berkeley's Department of Electrical Engineering & Computer Sciences began in 1968.

While department chair, Kuh also began reaching out to companies and research organizations, such as Bell Labs, GE and IBM, to develop relationships. He held industrial liaison meetings, establishing a model that would be replicated elsewhere at the university and beyond.

Kuh spent years developing relationships and raising funds for what would eventually become the Bechtel Engineering Center, complete with a new library. *"Bechtel Engineering Center was the realization of Ernie's dream, a monument to his extraordinary administrative skill and hard work,"* recalled EECS professor emeritus Edwin Lewis. *"I shall miss Ernie's intensely serious approach to all matters intellectual."*

Out of the technical and administrative skill, Prof. Kuh was a demanding but highly respected educator. Prof. Sanjit Mitra remembers him outlining his extraordinary capabilities as educator and mentor.

I joined UC Berkeley as a graduate student in September 1958. My first contact with Professor Kuh was in the Fall 1959 semester when I took a graduate course on network synthesis from him. He was a very tough professor. The homework problems and examination questions he assigned sometimes took us a long time to solve. I recall in the first take-home midterm, one of the problems took me almost a week to solve. I found out later that it was a small research problem. He was always trying to make his students expand their horizons and go beyond their comfort zones.

In spring 1960 semester, I was offered research assistantship from several professors in the department including Professor Kuh. I decided to work under him. Based on the two graduate courses I took from him, I believed he would be the best supervisor for me and guide me to develop the skills to carry out high quality research.

Professor Kuh arranged for me to work during the summer months of 1960 at the Bell Telephone Laboratories at Murray Hill, New Jersey. It was a fantastic experience. I met many giants in our field including Dr. Sidney Darlington and Dr. Harry Nyquist. I remain forever grateful for this opportunity.

After returning from Bell labs I started my thesis research. I was supported by a grant from NSF on active inductorless filters, but Professor Kuh told me that I could work on any problem that I liked; all I had to do was to meet him once a week and tell him what I had done. In the beginning it was very difficult, as I had no idea what to work on. Eventually I started my own research problems. Looking back, I realize that Professor Kuh was pushing me to become an independent researcher. I remember him, as someone who did not say much, but when he did, it was very insightful and pithy. He had these very penetrating eyes I used to feel that he could see "through me". I think in not saying much, he forced me to think and have independent thoughts. This was a real gift. It was very clear that he had incredibly high standards. I found this frustrating when I was his Ph.D. student, but today I truly appreciate. Over the years I prized his advice and wisdom. He was a highly recognized Professor around the world receiving countless awards and prizes. I feel very honored to have been his student.

Sanjit K. Mitra
University of California, Santa Barbara

[*] Largely taken from "*Ernest S. Kuh, Berkeley Engineering professor and dean emeritus*," Berkeley Engineering, July 2015.

Rui de Figueiredo

Rui José Pacheco de Figueiredo made pioneering contributions to the mathematical foundations of linear and nonlinear problems in pattern recognition, signal and image processing, machine intelligence, and neural and soft computing. In a 60-year career he played an extraordinary role as educator of thousands of undergraduate and graduate students, and leader in his field and profession, with his work internationally recognized by numerous awards and symbols of international acclaim.

Rui J. P. de Figueiredo was born in Panjim, Goa, India, in April 19, 1929, grew-up in a family as the 2nd of 4 boys and lived a long life full of personal happiness and professional achievements. From the age of 4 to 9 he was home schooled, taught in Portuguese by several tutors, in mathematics, science and music, and he entered high school when he was 9 years old. A highly gifted student, musically and academically, in 1945 he turned down admission to study piano at Trinity College of Music in London, with a fully funded scholarship, as he was only 16 years old and his parents considered him too young to live alone in London. But, his teachers in the assessment of his piano performance commented that his play of the scales was *"as graceful as the gliding of skates on virgin ice."* After graduating from the high school in India he went on a journey half-way around the world to enter the Massachusetts Institute of Technology (MIT) in 1047 as a freshman undergraduate. After distinguishing himself in electrical engineering, earning both Bachelor (1950) and Master (1952) degrees at MIT, he pursued his PhD at Harvard in 1952.

He took time off as a graduate student to move to Lisbon, Portugal, where much of his family then resided. While there, he conducted research with the Portuguese Atomic Energy Commission ('Junta de Energia Nuclear') using the experimental nuclear reactor in Sacavém, Lisbon, Portugal. He returned to Harvard to receive his PhD in Applied Mathematics in 1959, and then traveled back to Portugal, to chair the Applied Mathematics and Physics Division of the Portuguese Atomic Energy Commission, and also to represent Portugal as a diplomat and technical expert at international conferences, treaty negotiations, and foreign scientific exchanges. After marrying Isabel Colaço, also from Goa, in 1961, he returned in 1962 to the United States, accepting a tenured position as Associate Professor in the School of Electrical Engineering at Purdue University. In 1965, he moved to Houston, Texas, as a Full-Professor jointly appointed by the Departments of Electrical Engineering and Mathematical Sciences at Rice University, position that he held for 25 years. In 1990, he accepted a position as Professor of Electrical and Computer Engineering and Professor of Mathematics at the University of California, Irvine (UCI), and served there as a researcher and teacher where he was the Founder and Director of the Laboratory for Intelligent Signal Processing and Communications. At UCI, he was also Professor, Above-Scale, in the same Departments, plus, Director of the Laboratory for Machine Intelligence and Soft Computing of the California Institute for Telecommunications and Information Technology (Calit2). Finally, he was Samueli School Professor Emeritus of Electrical Engineering and Computer Science at UCI and retired in 2007, but even during retirement he continued his research activity.

Rui Figueiredo contributed in the early 1970's to the invention of generalized spline filters, and in particular, the Butterworth and Chebyshev generalized spline filters, for dynamical-source-model-based recovery of analog signals from linear observations. His work supported also a variety of NASA space exploration projects, assisted the Department of Defense in weapons detection systems, helped companies identify credit card fraud, assisted the Environmental Protection Agency in oil spill detection and source matching, developed algorithms for more efficient transmission of mobile telecommunications signals, enhanced geophysical images for well-logging, and improved the early detection of brain and neural diseases, like Alzheimer's. One of his most well-known contributions was the invention and study of the Generalized Fock space F (which he baptized as *"Neural Space"*), a Reproducing Kernel Hilbert Space of input-output maps of generic nonlinear dynamical systems, and used a *"linear"* orthogonal projection in F for optimal recovery of such *"nonlinear"* maps from the input-output data. This, extended to nonlinear systems the powerful orthogonal projection method, previously used exclusively for linear systems. The analytics behind it ultimately led to the development of the optimal interpolation and CDL neural networks. The results of his work can be found in over 400 scientific publications.

Rui Figueiredo was an IEEE Life Fellow that received the IEEE 3rd Millennium Medal in 2000. He was President of IEEE CAS Society (1998) and received CASS Golden Jubilee Medal (1999), the Mac Van Valkenburg award (2002), for seminal contributions to the field of electrical engineering, and IEEE Transactions on CAS Guillemin-Cauer Best Paper Award (2003). In 2007 was elected to the Russian Academy of Natural Sciences (RANS), as foreign member (U.S. Section) that awarded him the George V. Chilingar Medal of Honor for contributions to science and engineering. In 2009 RANS awarded him the prestigious P. L. Kapitsa Gold Medal, for pioneering contributions to the mathematical foundations and applications of signal processing, and, in 2010, the RANS Golomb/Chilingar *"Giants of Science and Engineering"* Medal of Honor, in acclaim for his sustained life-long fundamental contributions to science and engineering and extraordinary international leadership in his profession. Rui Figueiredo said, when receiving this distinction: *"This award is especially gratifying to me in view of the success of my initiative on advancing the IEEE transnational frontier in Eastern Europe, particularly in Russia."*

The department chair at UCI, Mike Green, referred about him: *"During his 17 years as a faculty member in EECS department, Rui played an important role in establishing the field of signal processing as a prominent research area at UCI. In addition to being a preeminent scholar and researcher, Rui was a genuinely kind and cordial man."* Also, G.P. Li, UCI division director mentioned: *"He understood the added value of cross-disciplinary research and believed it could really transform the learning of the next generation,"* he vividly recalls their last meeting, at a faculty workshop to discuss plug-load energy efficiency research: *"He was very ill by then, but he showed up and participated in the discussion, thinking through the problems very clearly and offering suggestions as to what direction we should take."*

Rui Figueiredo passed away on July 22, 2013, in College Station, Texas, after a long and brave fight with Parkinson's disease. He was survived by his wife of 52 years, 5 children, and 12 grandchildren who fondly remember their grandfather as a generous man with a superb sense of humor who enjoyed expanding their horizons in mathematics, music, and current events.

Rui Martins
University of Macau, Macao, China
and Instituto Superior Técnico,
Universidade de Lisboa, Portugal

Mirko Milić

An extraordinary scientist, researcher, professor, and art aficionado, Professor Mirko Milić, had two main passions in his life: passion to science and passion to travel. Last days of his life he passed enjoying his two passions: he was in Davos (Switzerland), far from his home in Belgrade (Yugoslavia), sharing experiences and ideas with his colleagues, scientists from all over the world, at the European Conference on Circuit Theory and Design (ECCTD). After the conference, he spent some time in Switzerland, and suddenly died in Bern on September 9th, 1993.

Professor Mirko Milić was born in Galace, Romania, on April 21rd, 1932 as the only child of a respectable middle- class family. His father, originally from Dubrovnik (Croatia), worked as a professional ship-pilot in Galace, a small town at the Danube-river delta. His mother was a fine lady of Italian and Austrian origin. She influenced her son Mirko to be interested in the fine arts, music, and mathematics, and provided him with a fine education in a classical lycée. From that time, Mirko Milić exhibited a strong and deep interest for art, music, languages, and philosophy.

Mirko Milić graduated from high school in Belgrade, in 1950, with the highest honors. He continued his studies at the Faculty of Electrical Engineering (EE), University of Belgrade. He graduated (B.Sc.) in 1956, with excellent grades (grade point average 9.43 out of 10.00) from the Department of Telecommunications. In 1963, he received the M.Sc. degree in EE with a diploma work entitled "*The Application of Graph Theory to the Analysis of Electrical Networks with Multi-Terminal Elements.*" In 1968, he received the Ph.D. degree in EE for his dissertation entitled "*Topological dynamic Properties of State-Space Model of Non-Reciprocal Networks.*"

As an undergraduate student, Mirko Milić worked part time in the Research Institute Vinca (Belgrade). After receiving the B.Sc. degree (1956) at the Faculty of EE, University of Belgrade, he assumed the academic positions of teaching assistant, assistant professor (1963), associate professor (1973), and full professor (1980). His main research subject was circuit theory. From 1961 to 1965, he was a researcher-consultant at the Institute "*Nikola Tesla*" and the Mathematical Institute, both in Belgrade. From 19651967, he was on a British Council Scholarship at the Imperial College of Science and Technology, London (UK). In 1977, he was invited as a visiting professor-researcher to the University of California, Berkeley (USA). In 1988, he was elected to be a corresponding member of the Serbian Academy of Sciences and Arts.

Professor Milić was an extraordinary and passionate scientist. His work, encompassing a broad spectrum of circuit theory, was undoubtedly well recognized and respected internationally. His comments, discussions, and reviews were profound, clear, and extremely valuable to his colleagues. He contributed to several areas of fundamental circuit and system theory. The main characteristic of his research was "*to be at least one step before others.*" He was one of the pioneers in the foundation of spectral graph theory, having also published a textbook (with Prof. D. Cvetković) in this field. His papers cover a variety of areas, including topologicaldynamic properties of passive and active networks, statespace descriptions of linear and nonlinear networks, qualitative analysis and bounds of the solutions of semistate models, Lagrangian descriptions of nonlinear networks, numerical analysis, modeling, and signal processing. During the last years of his life, he was interested in neural networks, particularly cellular neural networks (CNN) where he suggested a novel CNN cell having only one active element. Among others, his result concerning unique solvability of linear time-invariant RLC circuits has proved to be one of the deepest results in circuit theory. Two textbooks, two solution manuals with solved problems in circuit theory, and numerous scientific papers published in international journals and conference proceedings have marked the productive period of Professor Milić's life.

Many people knew Professor Milić as a pure and precise theoretician. It is, hence, interesting that he had a patent submission entitled "Analog nth order filter suitable for integrated technology." Furthermore, although he preferred exact solutions in closed form over the numerical solutions, he recognized an importance of computer applications and in 1970s took a course in computer program ECAP and a course in computer-aided design of electronic circuits.

Professor Milić was an active member of several international and Yugoslav societies and committees, and chairman and member of a number of conference committees (ISCAS, ISTET, ECCTD). He was a corresponding member of the Serbian Academy of Sciences and Arts, a permanent member of the Scientific Committee of the International Symposia of Theoretical Electrical Engineering (ISTET) and the Information Committee of SEFI (Société Européene pour la Formation des Ingénieurs), a senior member of the IEEE (Institute of Electrical and Electronics Engineers), and a member of the Yugoslav Society for ETRAN (Electronics, Telecommunications, Computers, Automation, and Nuclear Engineering).

With several colleagues from the Faculty of EE, University of Belgrade, he initiated the first Seminar on Neurocomputing, held from December 20-21, 1990 in Belgrade. He helped the seminar series continue, despite the disintegration of the former Yugoslavia.

During the Winter 1992-93, at the time of enormous inflation in Serbia and Montenegro, he organized the second Seminar on Neural Networks as a series of lectures held on Saturdays from November 1992 to May 1993. The seminars now continue as the biennial international conferences on Neural Network Applications in Electrical Engineering (NEUREL). In co-operation with the IEEE SP Society, NEUREL is organized by the IEEE YU Section and the CAS & SP Chapter, and it hosts authors from all over the world. The next NEUREL, scheduled for September 2004, is devoted to the memory of Professor Mirko Milić.

Professor Milić was a passionate scientist and teacher always ready to explore new research fields. When working, he never spared himself nor anyone else working with him. He loved his work, both teaching and research, and always had numerous new ideas and plans. Even though his sudden death prevented him from completing many of his projects, Professor Milić made numerous scientific contributions. He was an academic who left an exceptional mark on engineering science.

He made a clear distinction between his professional and private lives. Consequently, few of his colleagues knew him as an extraordinary expert in philosophy, arts, music, and as a jazz aficionado. His illness, which he fought over a long period of time, was perhaps the reason that he worked even harder, as though he wanted to be *"just one step ahead of the ultimate destiny that awaits us all."*

Based upon 'In Memoriam' article in IEEE CAS Magazine, 2003, by Ljiljana Trajković and Branimir Reljin.

Ljiljana Trajković
Simon Fraser University, British Columbia, Canada

Donald O. Pederson

Donald O. Pederson was born Sept. 30, 1925 in Hallock, Minnesota. He entered Iowa State College in fall 1943, but soon was drafted. He served in Germany as a private in the U.S. Army from 1943 to 1946. Post-war, he completed his undergraduate education at North Dakota Agricultural College (now North Dakota State University) where he earned his B.S. degree in electrical engineering in 1948. He earned master's and doctoral degrees in electrical engineering from Stanford University in 1949 and 1951, respectively. After receiving his Ph.D. from Stanford, Pederson stayed on for a period as a researcher in Stanford's electronics research lab. From 1953 to 1955, he worked at Bell Telephone Laboratories, in Murray Hill, New Jersey, and also taught night classes at Newark College of Engineering (now New Jersey Institute of Technology).

Soon he concluded that he enjoyed teaching better even than his work at Bell Laboratories. In 1955, he contacted acquaintances in California and subsequently was offered and accepted a position as an assistant professor at Berkeley. He was an exciting and popular teacher, well-remembered by generations of students. With colleague Ernest Kuh, he coauthored Principles of Circuit Synthesis, a leading undergraduate text of its time. Later he authored another textbook, Electronic Circuits. His tenure at Berkeley also included stints as director of the campus's Electronics Research Laboratory and as vice chair and chair of the Department of Electrical Engineering and Computer Sciences. He retired in 1991.

1959 marked the invention of the integrated circuit (IC), changing the world of electronics. Don foresaw that dramatic reductions in size and cost of electronics would become possible. He became the preeminent pioneer in university research and teaching on integrated circuits, now generally known as "microchips." Don decided that to undertake research in ICs and to teach students to design them, the university needed its own semiconductor fabrication facility. When he voiced this idea, he met a host of objections—building a fab was too complicated; his group was made up of engineers, not chemists; the university had no money for expensive fabrication equipment; and the project simply couldn't be done. Ignoring the objections, Pederson, with Professors Tom Everhart, Paul Morton, Bob Pepper, and a group of graduate students, started designing the facility. "Never wait for approval, don't tell anyone you are doing something, just do it," Pederson said later. That's my motto."

Resourcefulness trumped the many difficulties. By 1962 the new facility was operational, producing publishable research and educating a new breed of engineers. Notable leaders from industry visited and praised the facility, the first microfabrication facility at a university. Graduates of the program soon became leaders in the semiconductor industry. Microfabrication capabilities at Berkeley have advanced and grown steadily ever since. Currently (2006), several hundred students and faculty members from a wide range of academic fields make use of an extremely flexible research facility.

In the mid-1960's Don became interested in the application of computer aids to the analysis of integrated circuits. He and his students used a Bendix G15 minicomputer, (the very one now displayed in the Museum of American History at the Smithsonian) with only typewriter and paper tape input and output, to try to gain a deeper understanding of the behavior of certain circuit designs. Don became convinced that the computer would play a necessary role in the design and analysis of integrated electronics. A decade of research, involving many undergraduate and graduate students, eventually produced the integrated circuit computer simulation program called SPICE (Simulation Program with Integrated Circuit Emphasis). The program allows engineers to analyze and design complex electronic circuitry with speed and accuracy. Virtually every electronic chip, developed anywhere in the world today, uses SPICE or one of its derivatives at critical stages during its design. Don and his students have made many other contributions to electronic design automation along the way as well, in areas from device modeling, mixed-mode simulation, rule-based circuit diagnosis, to macromodels. SPICE was one of the first significant open-source computer programs. The policy established by Don was that SPICE was available free of charge to any chip designer. The only request he made was that if a bug was found, or a new feature added, a copy should be sent back to Berkeley so it could be made available to all the other users. This policy accelerated the improvement of SPICE and its enhancement with many new features.

Soon after Don retired, former students and colleagues made substantial gifts to endow a Professorship in his name, and to pay for major renovations on the 5th floor of Cory Hall in a student area now identified as the Donald O. Pederson Center for Electronic Systems Design. Don Pederson died on December 25, 2004, aged 79, of complications from Parkinson's disease.

He was elected to the membership of the National Academy of Sciences in 1982 and the National Academy of Engineering in 1974. He garnered numerous other honors and awards, including a Guggenheim Fellowship in 1968, an American Association for the Advancement of Science Fellowship in 1988, the Berkeley Citation in 1991, the Phil Kaufman Award from the Electronic Design Automation Consortium in 1995, and the Medal of Honor from the Institute of Electrical and Electronics Engineers in 1998. He also received an honorary doctorate from Katholieke Universiteit Leuven in Belgium.

Largely taken from "In Memoriam: Donald Oscar Pederson, September 30, 1925-December 25, 2004" Paul R. Gray, David A. Hodges, A. Richard Newton.

Michael K. Sain

Michael K. Sain, Frank M. Freimann Professor of Electrical Engineering at the University of Notre Dame, died Tuesday (Sept. 22 2009) morning of a heart attack at his home in South Bend. He was 72.

A native of St. Louis, Sain earned bachelor's and master's degrees from St. Louis University in 1959 and 1962, respectively, before earning his doctoral degree from the University of Illinois in 1965 and joining the Notre Dame faculty as an assistant professor that same year. He became an associate professor in 1968, a full professor in 1972, and the Frank M. Freimann Chair in Electrical Engineering in 1982. He was an active member of the University of Notre Dame faculty for 44 years, until his death. In fact, in the fall semester of 2009 he was teaching two undergraduate classes, he was working on several books and was conducting research.

Among the first of Notre Dame's chaired professors, he worked on and made significant contributions in a wide variety of areas, including statistical control theory, algebraic system theory, and also in applications, most recently in structural control for buildings, bridges and other structures subject to high winds and earthquakes. He was the author or co-author of some 400 publications. Mike was a pioneer in statistical control theory, which generalizes traditional linear-quadratic-Gaussian control by optimizing with respect to any of the cost cumulants, instead of just the mean. Mike and his students have contributed to the development of minimal cost variance control, kth cumulant control, and statistical game theory. In statistical game theory, the statistical paradigm generalized mixed H_2/H_∞ control and stochastic H_∞ control concepts.

Another major contribution of Mike's research is in the field of algebraic system theory; see for example his 1981 monograph *Introduction to Algebraic System Theory*. Mike and his collaborators (B.F. Wyman and students) expanded the algebraic system-theoretic concepts of zeros of a linear system using a module-theoretic approach, thus complementing and completing the work of R. E. Kalman for poles. Earlier, in the late 60s, motivated by the algebraic work of R. E. Kalman, Mike (with J. Massey) established the algebraic structural relationships between coding theory and control theory; they also made seminal contributions to the theory of dynamical inverses for linear dynamical systems and also presented analogous work for linear sequential circuits, work recognized as one of the key development of coding theory.

Mike was deeply committed to his Catholic faith and to Notre Dame as a Catholic University. Mike was a devout Catholic, devoted to the Virgin Mary. He attended daily Mass and visited Medjugorje in Bosnia several times. He was very much involved with Catholic causes and organizations. Mike loved his God, he loved his family, his Engineering, his Mathematics. Mike was a very deliberate, careful thinker who spent significant time pondering carefully technical, spiritual, and societal questions.

Sain began to teach an innovative, if not unprecedented, undergraduate course titled "Theology and Engineering," conducting an improbable dialogue between the two disciplines: Grace is considered alongside exogenous loop signals, the actions of the Holy Spirit alongside inner state feedback loops, the Fall alongside parameter uncertainty, virtues alongside feedback control laws, and the Beatitudes alongside coordinated loop excitations.

It was a very popular course and Sain was at least as enthusiastic about it as his students. "To integrate what I believe about God with what I think and do as an engineer is the most important thing that has happened to me here at Notre Dame," he once said. "It is a genuine joy to do this work, to give and take with the students about it, and to contribute in some small way to what makes Notre Dame more unique as a place to live and grow."

According to Sain's colleague, Thomas E. Fuja, chair of Notre Dame's Department of Electrical Engineering, the course "brought together two of his passions – system theory and the search for a better understanding of God." Mike was deeply committed to his Catholic faith and to Notre Dame as a Catholic university. This led him to find bridges between his technical discipline and Catholic moral teachings.

"He did it not because of any research glory it may have brought him –believe it or not, the research community at the intersection of engineering and theology is pretty sparse – but because he is a teacher and he wanted to share what he understands, because that's what teachers do."

Mike was a valuable contributor to the IEEE. In addition to being the Editor of the IEEE Transactions on Automatic Control, he was the founding Editor-in-Chief of the Circuits and Systems Magazine transforming the IEEE Circuits and Systems Society Newsletter. He also served on numerous award committees, including the IEEE Award Board, where he chaired the Baker Prize Committee. During his 44 years of service, he received numerous awards and honors including the IEEE Centennial Medal, IEEE Circuits and Systems Society Golden Jubilee Medal, IEEE Fellow, and University of Notre Dame President's Award.

Partially taken from: Michael O. Garvey "Michael Sain, Freimann Professor of Electrical Engineering, dies," Notre Dame News, September 23, 2009, and Panos J. Antsaklis, "In Memoriam: Michael K. Sain," IEEE Transactions on Automatic Control, November 2009, pp.2491-2492

Tamás Roska

Tamás Roska was a world-renowned researcher, research supervisor, innovator and dedicated educator. His research focused on the analysis, design, and experimental validation of nonlinear circuits and systems and in particular the cellular neural networks CNN that have been applied in aerospace, biomedical signal processing, medical diagnostics, process control, and several safety critical applications. Tamás Roska was Professor of Information Technology at the Pázmány Péter Catholic University in Budapest. Between 1964 and 1970 he worked at the Instrumentation Research Institute of the Hungarian Academy of Sciences. He passed away on June 18, 2014, in Budapest, Hungary at the age of 74.

Professor Roska is most lauded for his breakthroughs in real-time image processing as he phrased it: "*The CNN proved to be very useful in partially mimicking the visual pathway, in particular the retina and the visual cortex. Based on our insights, our team was able to develop a new kind of computer.*" The applications are numerous: real-time image processing tasks in navigation, bionic eyeglasses, collision avoidance of unmanned aerial vehicles, mission-critical operations, etc. Roska's team has also implemented the technology in tactile and auditive applications. "*The first visual microprocessors are already on the market and I predict that sensory computers will become ubiquitous. The computer architectures of the future will increasingly be cellular and - perhaps even more interesting - molecular-level computing is emerging as well.*"

Being born in Budapest in 1940 at the beginning of the world war, he had a difficult youth and adulthood due to the war and the Soviet occupation of his country until 1991. He nevertheless always had an enormously positive attitude towards mankind and nature. As a university student he was always curious and enthusiastic. He was interested in everything of intellectual beauty. He enjoyed his studies and the discovery of the world. His devotion to mathematics and his enthusiastic interest in understanding the secrets of nature were exceptional. He eagerly wanted to understand everything.

After his Diploma of electrical engineering from the Technical University of Budapest in 1964, he obtained in 1973 resp. 1982 the Ph.D., resp. the D.Sc. degree in Hungary. He subsequently held research positions and took courageous initiatives. Since 1982 he performed research at the Computer and Automation Research Institute of the Hungarian Academy of Sciences, was head of the Analogic and Neural Computing Research Laboratory and the Chairman of the Scientific Council. As a young researcher he became engaged with Circuit Theory, the "*Queen*" of Electrical Engineering, bridging physics i.e. describing the order in real nature, to mathematical models of the creative designer. Logic and mathematics take care of the correctness of the mind. The challenge of 1960's and 70's was to create better and better von Neumann type binary universes in a "*chip,*" i.e. to lay down the principles of design for very large scale circuits applied in computing, control and communication. He became one of the mentors of many talented young researchers. In the late 1980s he organized a doctoral school at SzTAKI. Soon after the fall of the iron curtain in 1992 he assisted the University of

Veszprém to launch a Department for Engineering Informatics, and later in 1998 he accepted the invitation of the Rector of the Pázmány Péter Catholic University to establish as founding Dean an Engineering Faculty (University Department) focusing on Information Technology. The Mission Statement declares that the program will be *"human-centered and nature-inspired."* It will integrate the field of electrical engineering and information technology with certain areas of life-sciences. This program has attracted many bright students, and received international attention. In 1993 he co-founded the postgraduate Centre for Neuromorphic Information Technology in Budapest.

He was exceptionally intelligent, bright and talented, he was a great scientist, but first of all, he was an extraordinary human being. He radiated optimism and kindness wherever he went. And when in the late 1980's the National Science Foundation (NSF) reopened the possibility to start joint MTANSF research projects, his joint proposal with UC Berkeley was among the first welcome and fruitful initiatives. This project and subsequent projects brought him into intense cooperations with the University of California, Berkeley and the University of Notre Dame, Indiana USA and different European universities. Young researchers from Hungary and all over the world became motivated by the new paradigm, and Tamás became the best known Hungarian scientist in the global community of the IEEE Circuits and Systems Society. He was active in setting up a sequence of CNNA conferences related to the analogic spatial-temporal supercomputing and computational complexity issues. He contributed to the IEEE CAS Society as a member of the Board of Governors and as initiator of the Cellular Nanoscale Networks and Array Computing (CNNAC) Technical Committee. In 2013 the Society recognized him by the prestigious Mac Van Valkenburg Award. His other awards and honours include 2013 Doctor Honoris Causa at KU Leuven, Fellow of the IEEE, Széchenyi Prize, Grand Prize of the *"Pro Renovanda Cultura HUngariae,"* 2002 Bolyai Prize, 1993 Hungarian Gábor Dénes Prize, Member of the Hungarian Academy of Sciences, Member of the Academia Europaea.

He has published more than hundred research papers and four books (partly as a co-author). His seminal paper on the CNN Universal Machine, co-authored with L. O. Chua, has received more than 1000 citations. Dr. Roska is a co-inventor of two US patents from UC Berkeley : the CNN Universal Machine (with Leon O. Chua) and the analogic CNN Bionic Eye (with Frank S. Werblin and Leon O.Chua). During the last 15 years he has received two NSF grants, five ONR grants, two EU Grants and several Hungarian Grants. He was a founding member of two spin-off companies, one in Berkeley and one in Budapest. For 4 years, in Hungary, he was the advisory Chair of the National R&D Program on Information and Communication Technology. In 2005-2007 he was a member of the Advisory Committee of the EU Commissioner in the Commission of Information Society and Media Technologies in Brussels. Roska has always been devoted to these emerging technologies, as well as to 'bionic systems', a new field that has emerged at the crossroads of information technology, electronics, computing and biotechnology.

Leon Chua
Department of EECS, University of California at Berkeley, CA
Joos Vandewalle
Katholieke University of Leuven, KUL, Leuven, Belgium

John Choma

John Choma, Jr., born on November 6, 1941, in Pittsburgh, Pennsylvania, was a recognised expert in the analysis, modelling, and design of high performance analog and mixed-signal integrated circuits and an outstanding teacher. He passed away August 10, 2014 recovering from a short illness. Dr. Choma was a Professor of Electrical Engineering in the Ming Hsieh Department of Electrical Engineering of the University of Southern California, where he also served as chair of Electrophysics Department. Prior to joining the USC faculty in 1980, Prof. Choma was a senior staff design engineer in the TRW Microelectronics Center in Redondo Beach, California (now Northrop-Grumman). His earlier positions include technical staff at Hewlett-Packard Company in Santa Clara,

California, Senior Lecturer in the Graduate Division of the Department of Electrical Engineering of the California Institute of Technology, electrical engineering lectureships at the University of Santa Clara and the University of California at Los Angeles, and a faculty appointment at the University of Pennsylvania.

Professor Choma authored close to 200 publications and presented more than 65 invited short courses, seminars and tutorials. He is the author of a 1985 Wiley Interscience text on electrical network theory, the co-author of a 2007 World Scientific Press text on feedback networks, and the author of a forthcoming Cambridge University Press undergraduate level text on integrated circuit analysis and design. Prof. Choma has contributed several chapters to five edited electronic circuit texts, and he was an area editor of the IEEE/CRC Press Handbook of Circuits and Filters.

Prof. Choma was a long-time member of the IEEE Circuits and Systems Society serving as a member of the IEEE CAS Board of Governors, Vice President for Administration 1995, and its President in 1997. He also served as the International Chair of the IEEE Gustav Robert Kirchhoff Awards Committee. He was an Associate Editor and Editor-In-Chief of the IEEE Transactions on Circuits and Systems, Part II, Associate Editor of the Journal of Analog Integrated Circuits and Signal Processing and a former Regional Editor of the Journal of Circuits, Systems, and Computers. His latest service to the IEEE CAS was on the Education Award committee.

Colleagues and former students were greatly saddened to have lost one of their great colleagues and a mentor. Ronald Rohner remembered John Choma with a long memorial published in the *"IEEE Circuits and Systems Magazine"*. He starts with:

> John Choma had passed away that morning, like so many who knew him, I spent the rest of the day numb with shock. That evening, thinking of my friend, I put on Frank Sinatra's "I did it my way" (rather loud) as I continued to think of John. He did indeed do it his way. John held everyone around him to the highest of standards; he held himself to even higher standards. And he lived up to them as he helped all around him do the same.
>
> But don't just take my word for it. Hear what his friends, students and colleagues have to say.

Let's read what former students wrote for Ron's paper:

Aaron Curry: "John Choma was my professor, colleague, and Ph.D. adviser; but most importantly he was my mentor and my friend. He is the reason I decided to pursue a Ph.D. and I am honored to have worked with him and spent as much time with him as I did in the last years of his life. John was one of the most outstanding men I have ever met and I would like to take this opportunity to say why. Despite the long list of achievements and accolades John had, he was very humble, down to earth, kind, outgoing, and encouraging. If you ever had the privilege of meeting him, you may not know that you were talking to such an accomplished person because he pushed formalities aside and treated you like an equal.

John was truly a man for others; he found happiness in making other people better which is best exemplified through his work at USC. John was the best teacher I have ever had. He had the unique ability to take difficult concepts and make them easily understood. On top of his ability to teach, he was personable, understanding, and always made time for his students. You felt like you could go to him for anything. If you needed a difficult concept to be explained, needed advisement on what courses to schedule, or even just talk, he was always delighted to help. He was also so kind and encouraging that you always left his office feeling confident and excited to tackle any obstacle. John was there for his students. One of the greatest joys he always expressed was to see his students go on to do great things and be "better" than he is. When I think about graduation and know that he will not be there to thank, tears swell; but then I think about all the people he has helped and lives he has impacted and I beam with happiness knowing that I knew John, and he was my friend."

Ahmet Erten: "Prof. Choma was a great mentor and it was always a pleasure to work with him. He was one of those teachers who would leave a lasting impression on students. Quite often I would run into him pacing in front of Powell Hall at USC smoking while probably solving a problem on his mind and I fondly remember our conversations there, passing on his passion for engineering to me. He was the reason for many of his former students like me to pursue a Ph.D. The generations which will not a get a chance to know him, will maybe get a glimpse of his enthusiasm for his research and teaching through his former students. However, they will never get a chance to feel the intensity, the passion he was emanating. He will truly be missed as a mentor, as a friend..."

Chris Grossman: "I first met Dr. John Choma, Jr. when I was an undergrad at USC. I had not really considered graduate school before I met him, but he inspired me to go and offered me a Research/Teaching Assistant position. His insights and great teaching skills taught me many new ways think about, understand, and analyze circuits and devices that I use to this day. He helped me get a part-time position at TRW working with GaAs HBTs which eventually led to a collaboration in which we won the IEEE Microwave prize. I owe much of my success to the engineering concepts he taught and opportunities he helped me find. Dr. Choma was my advisor, my friend, and the best teacher I have ever known."

Chris Reynolds: "I was terribly saddened to hear of Dr. John Choma's passing. He was a shining example of everything an exceptional teacher should be; gifted in his deep knowledge, in his patience and clear, thoughtful explanations, and in his ability to motivate students with gentle, consistent encouragement. I can honestly say that I owe much of my rewarding 30+ year career as an engineer to John. His help and guidance as a teacher and mentor were instrumental in helping many of us learn, and come to enjoy the challenges of engineering design.

In addition to his exceptional technical and teaching abilities, his easy manner, unshakable grace, and pursuit of knowledge were without equal. Over the years as his student and decades after, it was a genuine privilege to know John as a teacher, mentor and friend. Rest well John. Safe journey. We will miss you greatly."

Colin Daly: "I first met Professor Choma while still uncertain of what area to focus on in Electrical Engineering. His class was the perfect mix of an intellectual challenge delivered so clear and concise by someone who had an excellent grasp on the subject and a unique gift for teaching. Professor Choma's class was the first time I felt inspired and it helped me feel confident that I had indeed made the right choice of major. Professor Choma took notice of my interest and performance and offered me an internship with a company he was involved in, and I was now lucky enough to work with Dr. Choma on a professional level. It was during this year and a half of working closely together that we began to establish a personal friendship and I learned how great of a person he was, on top of a fantastic intellect. John eventually met my father when he came down from San Francisco to visit me, and John's office and position very much impressed my father a working class immigrant from Ireland. When my father passed away, John's compassionate personal side showed brightly with his great comfort. I owe a lot to John professionally, personally, and intellectually. He was a great friend, and I will truly miss him."

The Ron Rohrer Memorial continues with twenty three additional recollections of colleagues and concludes with:

> John would have been pleased but also embarrassed by these remembrances. Even with all of his accomplishments his humility always shined through. And, I know that much of the above is redundant, but that's as well it should be. John Choma was that guy: a great teacher as attested by his students and his numerous teaching awards at USC as well as the IEEE Circuits and Systems Society Education Award in 1999; a great researcher as attested by his numerous publications and, again, and most important to him, his students. Not just an academic teacher and researcher, John was an industrial-strength circuit designer as attested by his popularity as a consultant. But most of all John Choma was a great human being; a friend to his students and colleagues who is most missed by us all.

I was also a dear friend of John and his wife Lorraine and we spent rewarding hours together. When I had the chance to see John on the stage of a conference as an invited lecturer I was always impressed by his special speaking ability and capability to motivate and inspire the audience. To me John was a model but, much more, was a sincere and trusted friend; for his remembrance I have to simply reuse a large portion of what Yannis Tsividis wrote in the CAS Magazine paper: *My knowledge of John's leadership abilities began with my acquaintance with him as* President of the Circuits and Systems Society (I was VP Region 8 at that time). *Not only did he take a number of initiatives* as President of the Society, *but he also demonstrated a superb organizational ability. He did not waste time - he acted on things. His leadership, selflessness, sense of responsibility and dependability never ceased to impress me. He imparted on* the Executive Committee *a sense of unity and enthusiasm. As a result of his initiatives, several important actions were taken by the Society, notably including ones on Education.* This is the reason why the CAS Society warmly accepted the proposal to name the Educational Award the *"John Choma Educational Award"*.

Franco Maloberti
University of Pavia, Italy

[*] Largely taken from "R. Rohrer, *John Choma in Memoriam: I Did It My Way,*" IEEE Circuits and System Magazine, Fourth Quarter 2014.

Jiří (George) Vlach

Dr. Jiří (George) Vlach, Distinguished Professor Emeritus, passed away on February 11, 2015. His area of research was networks and numerical mathematics, and he published three influential textbooks, including Computer Models for Circuit Analysis and Design, which was widely used and translated into Farsi and Russian, and Basic Network Theory with Computer Applications, a fundamental text on electrical networks. His most recent book, Linear Circuit Theory–Matrices in Computer Applications, was published in February 2014.

Vlach officially retired in 1988 at age 65, but continued to play an active role at Waterloo, heading up a team that studied methods of improving electrical switching systems and in the process becoming one of the only retired professors to receive a strategic grant from the Natural Sciences and Engineering Research Council (NSERC) in 1992. Professor Vlach was also a longtime participant in Waterloo's Campus Recreation program, taking fitness classes for more than 25 years. In June 2000, he was named Distinguished Professor Emeritus.

Dipl. Ing. Jiří Vlach, was born in Prague, Czechoslovakia on October 5, 1922. His mother died while he was still a child and his father, a teacher, instilled in him the value of education and rational thought. Jiří graduated in 1947 with a Diploma in Engineering from the Czech Technical University, the first of his family to obtain a university degree. He earned his PhD in Electrical Engineering while working as an engineer in Prague. His abilities and skills not only in research but also his knowledge of many languages, including fluency in Czech, Russian, German, and English, led to a position as a visiting professor at University of Illinois, Urbana-Champaign, USA in 1966. After his stay in Illinois, he decided not to return to Czechoslovakia because of the political situation there, and as a result could not make a return visit for many years. He obtained a position at the University of Waterloo in Canada, where he became a full professor in 1969 and remained there for the rest of his career. According to then current practice, he had to retire in 1997, but remained active at the university for many years, at one point being the oldest recipient of an NSERC research grant.

Dr. Vlach received many awards during his long and distinguished career: he became an IEEE Fellow in 1982, was bestowed the IEEE CAS Golden Jubilee Medal (1999), and became Distinguished Professor Emeritus of University of Waterloo in 2000. He was a guest lecturer at universities throughout the world and a prolific publisher of academic papers.

Dr. Vlach's research helped lay the foundation of the surge in electronic innovations over the last 50 years. The many engineering students whom he advised and mentored applied their knowledge to develop the numerous electronic applications that have become essential components of our lives.

Throughout his life Jiří was an avid tennis player and a swimmer. He loved canoeing, whether on his beloved Vltava or on the numerous Ontario rivers and lakes. He traveled all around the world his whole life, and took extensive bicycling trips throughout Europe into his 90s.

Partially taken from "Remenbering Jiří Vlach," University of Waterloo, Daily Bulletin. [Online available] http://www.bulletin.uwaterloo.ca/2015/feb/25we.html

A. Richard Newton

A. Richard Newton was larger than life in the eyes of the ones who had the fortune of meeting him. His outlook on life was so different, innovative, and refreshing that one could not avoid being enthralled by his ideas long after parting ways. In this note, I remember him both as a wonderful human being and as an old and dearest friend. It is not easy for me to do justice to the great contributions that he made to the EDA community and to the world in general.

Richard was not only a wonderful engineer and a superb professor but also a great man. He would not rest on the laurels of his (many) industrial and academic successes like many of us are tempted to do; rather he pursued relentlessly noble causes that could (and did) have an impact on the human condition. Nothing was too hard for him. He would complain at times and act distressed, but would never abandon an idea he believed in. He liked BIG, audacious ideas and as soon as he saw a way of making them a reality and in good hands, he would move on to something bigger and better.

In Chancellor Birgeneau's words: "An inspired and dynamic leader, Richard understood the power of engineering and technology in entirely new ways and he connected them to addressing society's toughest problems. He had an unrelenting commitment to engineering for the betterment of society. His passing is an enormous loss to us at UC Berkeley and for engineering nationally and internationally." Richard's own words are perfect to define the kind of leadership he exercised: "A great leader is someone, first and foremost, who can build a relationship of trust with the people that he or she works with. A leader needs to create a vision that's compelling enough that everyone who is part of that vision wants to contribute to it personally, and also believes that they're part of something much greater than themselves." He was all of this and more.

His involvement with EDA started way back. A fortuitous meeting in the early 1970s with Donald Pederson, who died in 2004, was the event to kick-start Newton's lifelong interest in electronic design automation (EDA). In his own words "In 1970, I had the good fortune to bump into Professor Donald Pederson in the computer room of the University of Melbourne, Australia, and before I knew it, I was a SPICE-1 developer". He arrived in Berkeley to pursue his Doctorate in August 1975; I arrived there July of the same year. We met at the end of August when he literally ran over me turning a corner at the speed of light while I was running in the opposite direction. From that collision, where it is easy to imagine who had the worst consequences (Richard was almost 2 meters tall and I am 1.68 meters short!), a close friendship sparked. We talked and talked and talked about everything in life. We were so different in background and yet so close in our enthusiasm and passion for the intellectual endeavors that a University career would offer us.

In 1978, Newton earned his Ph.D. degree under the guidance of Don Pederson and was appointed to the engineering faculty later that year after he received several important offers in industrial research Labs and other Universities. Richard quickly scaled the academic ladder, going from assistant professor in 1978 to associate professor in 1982 and to full professor in 1985. During that period, we traveled the world together giving lectures and courses on circuit simulation, layout, and logic synthesis.

In his Ph.D. thesis, Richard saw the possibility of developing tools for mixed-mode design where analog and digital components could be developed together. He also intuited that relaxation-based techniques had great promise for MOS circuits and developed with his students what was then called Iterated Timing Analysis. This method is still the basis of the fast circuit simulators being sold today by the EDA industry. Together, we developed jointly a body of knowledge that went under the name of "Relaxation-Based Simulation." The paper that summarized our thoughts was published simultaneously by three journals, a unique case in our discipline.

In 1983, he helped found SDA, the parent company of Cadence Design Systems, Inc., a company that made the ideas of a unified framework for design its flagship. Richard had an unmatched talent in marrying technical insights with industrial needs. In these years, he also predicted the great importance of logic manipulation and optimization that led to the formation of Synopsys in 1987. After the great technical and business success of these initiatives, Richard moved his interest to the process of enterprise formation. Beginning in 1988, he advised several venture capital firms, including the Mayfield Fund, where he was a Venture Partner until 2002, and Tallwood Venture Capital, where he contributed both to the evaluation and early stage development of more than two dozen new companies including Silicon Light Machines (where he acted as president and CEO during 1994 and 1995).

During his tenure as Dean, Richard was the driving force behind the creation of the UC Berkeley-based Center for Information Technology Research in the Interest of Society (CITRIS), one of four California Institutes for Science and Innovation. CITRIS was established in 2001 to develop the next generation of technologies that will be critical to sustaining California's economic growth and global competitiveness and to solving society's most critical needs.

Richard was a strong advocate of promoting women in engineering, and while he was dean, the number of women on the faculty at the College of Engineering nearly doubled from 15 in 2000 to 27 today. Newton also served on the Board of Trustees for the Anita Borg Institute for Women and Technology, which provides resources and programs to help industry, academia and government recruit, retain, and develop women leaders in high technology careers.

Richard always acknowledged the role of his students in his career and never forgot to mention their contributions in accepting honors such as the Kaufmann Award and the DAC keynote address.

I would like to conclude with two quotes: the thoughts of fellow countryman Giovambattista Vico (1668-1744) which are a perfect rendition of what moved Richard: "The holy furor for truth lives in the eternal attempt to go beyond the limit, in the infinite possibility of self-realization and of overtaking ourselves to discover the power of the spirit and give a new push towards knowledge." And the wonderful words by Tibetan Lama Sogyal Rimpoche, chosen by Dick Blum, San Francisco financier, philanthropist and vice chair of the UC Regents, for all of us: "One way of comforting the bereaved is to encourage them to do something for their loved ones who have died, by living even more intensely on their behalf after they have gone, by practicing for them, and so giving their death a deeper meaning. Don't let us half die with our loved ones, then; let us try to live, after they are gone, with greater fervor." I certainly will do so, to carry your flag until I reach you! Goodbye Richard, my friend.

Alberto Sangiovanni–Vincentelli
Department of EECS, University of California at Berkeley, CA

Authors Biographies

Asad Abidi is Distinguished Professor of Electrical Engineering at the University of California, Los Angeles. He received the BSc (Hons) degree from Imperial College, London in 1976, and the MS and PhD degrees from the University of California, Berkeley, in 1978 and 1981, respectively, all in Electrical Engineering. He worked at Bell Laboratories, Murray Hill, NJ until 1985, when he joined the UCLA faculty. Abidi has received the IEEE Donald J. Fink Prize Paper Award, the Best Paper Award from the IEEE Journal of Solid-State Circuits, the IEEE Donald O. Pederson Award, and several awards at the ISSCC. He is Fellow of IEEE and Member of the National Academy of Engineering.

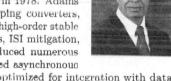

Robert Adams graduated from Tufts University in 1976 and joined Analog Devices in 1988. Robert Adams' career includes many inventions across a wide range of disciplines. One early discovery was the log-domain analog filter principle, first published by Adams in 1978. Adams was a pioneer in the area of high-resolution noise-shaping converters, introducing a series of fundamental innovations including high-order stable loops, multi-bit quantization, continuous-time loop filters, ISI mitigation, and mismatch-shaping techniques. Adams also introduced numerous advances in signal processing including the first integrated asynchronous rate-converter for audio as well as a line of DSP cores optimized for integration with data converters. Robert Adams is a Fellow of the IEEE, a Fellow of the Audio Engineering Society, and a Fellow of Analog Devices. He received the 2015 Donald O. Pederson award in Solid-State Circuits, and holds 33 patents.

Eduard Alarcón received the M. Sc. (National award) and Ph.D. degrees in EE from UPC BarcelonaTech, Spain, in 1995 and 2000, respectively, where he is professor since 1995. He was a visiting professor at CU Boulder, USA and KTH Stockholm, Sweden. He has co-authored 350 scientific publications, 6 books, 7 book chapters and 10 patents, and has been involved in different national, european and USA projects within his research interests in the areas of on-chip energy management, energy harvesting and wireless energy transfer, nanotechnology-enabled wireless communications, small satellites, and molecular communications, including projects and awards from Google, Samsung and Intel. He has been CASS DLP speaker, 2007 chair of the energy TC, BoG member (2010-2013), CASS VP Finance (2015) and VP Technical Activities (2016). He was TPC co-chair ECCTD 2007 and LASCAS 2013, AE of IEEE TCAS-I and IEEE TCAS-II and IEEE JETCAS deputy EiC (2016).

David J. Allstot received B.S., M.S., and Ph.D. degrees from U. of Portland, Oregon State U., and the UC Berkeley. He was Boeing-Egtvedt Chair at U. of Washington from 1999 to 2012 and Chair of EE from 2004-2007. In 2012 he was a Visiting Professor at Stanford and, since 2013, the MacKay Professor in EECS at Berkeley. He has advised 67 M.S. and 39 Ph.D. graduates, published more than 300 papers, and received awards for teaching, advising, and research: 1980 IEEE W.R.G. Baker Award, 1995 and 2010 IEEE CAS Society Darlington Awards, 1998 IEEE International Solid-State Circuits Conference Beatrice Winner Award, 1999 IEEE CASS Golden Jubilee Medal, 2004 IEEE CASS Charles A. Desoer Technical Achievement Award, 2005 Semiconductor Research Corp. Aristotle Award, 2008 Semiconductor Industries Assoc. University Research Award, 2011 IEEE CASS Mac Van Valkenburg Award, and 2015 IEEE Trans. on Biomedical CAS Best Paper Award. He has been active in service to the IEEE CAS and Solid-State Circuits Societies throughout his career. He is a Life Fellow of IEEE.

Andreas Antoniou (M'69–SM'79–F'82–LF'04) received the B.Sc. (Eng. Hons.) and Ph.D. degrees in Electrical Engineering from the University of London, U.K., in 1963 and 1966, respectively. He was a researcher in industry form 1966 to 1970. He taught at Concordia University, Montreal, Canada, from 1970 to 1983 and at the University of Victoria, British Columbia, Canada, from 1983 and 2003, having served as the founding Chair of the Department of Electrical Engineering during 1983 to 1990. He was awarded the CAS Golden Jubilee Medal by the IEEE Circuits and Systems Society and the B.C. Science Council Chairman's Award for Career Achievement both in 2000, the Doctor Honoris Causa degree by the National Technical University of Athens, Greece, in 2002, the IEEE CAS Society Technical Achievement Award for 2005, the IEEE Canada Outstanding Engineering Educator Silver Medal for 2008, and the IEEE CAS Society Education Award for 2009. He is a Life Fellow of the IET and IEEE.

Chris Bissell is Professor of Telematics at the British Open University, where he has contributed to undergraduate distance teaching modules in signal processing, control engineering, digital communications, and other topics. He has also taught digital signal processing at post-graduate level at Cranfield University, and has regularly assessed the quality of higher engineering education at other institutions. His research interests are in the history and social context of information engineering, and in engineering education; he has published numerous papers on all these topics during his 35-year career at the Open University. He has authored two textbooks, Control Engineering (Chapman & Hall, 1988, 1994) and Digital Signal Transmission (Cambridge University Press, 1992); and co-edited the Oxford Illustrated Encyclopedia of Invention & Technology (Oxford University Press, 1992) and Ways of Thinking, Ways of Seeing (Springer 2012).

Sergio Callegari received a Dr. Eng. degree (with honors) in electronic engineering and a Ph.D. degree from the University of Bologna, Italy, in 1996 and 2000 respectively. He was a visiting student at King's College London, UK and the University of Sussex, UK. He is currently an assistant professor at the University of Bologna. His current research interests include nonlinear signal processing, internally nonlinear, externally linear networks, chaotic maps, delta-sigma modulation, testing of analog circuits, and random number generation. Sergio Callegari has authored or co-authored more than 80 papers. In 2004 he was co-recipient of the IEEE CAS Society Darlington Award. He served as an Associate Editor for the IEEE Transactions on CAS-II in 2006–2007 and as an Associate Editor for the IEEE Transactions on CAS-I in 2008–2009. He is Chair of the Technical Committee on Nonlinear CAS and a member of the Technical Committee on Education and Outreach in the IEEE CAS Society.

Virginio Cantoni is a Full Professor in Computer Engineering at the University of Pavia. In the period 2008-2011 he has been seconded to the Centro Linceo 'Beniamino Segre' of the Italian Academy of Lincei. He has been the founder and first Director of the University of Pavia's European School of Advanced Studies in Media Science and Technology and Director of the Interdepartmental Centre for Cognitive Science. His research activity is concerned with pattern recognition, computer vision and multimedia. He is author or co-author of about 300 journal, conference papers and book chapters as well as editor or co-editor of 30 books and  co-author of three books. He has organized many International Conferences, Seminars and Workshops including a NATO Advanced Research Workshop on pyramidal systems for computer vision. An Expert and Project Reviewer for the EU Commission, he became a Fellow of the IAPR in 1994 and Fellow of the IEEE in 1997.

Sandro Carrara is an IEEE Fellow for his outstanding record of accomplishments in the field of design of nanoscale biological CMOS sensors. He is a faculty member (MER) at the EPFL in Lausanne (Switzerland). Along his carrier, he published 7 books, one as author with Springer on Bio/CMOS interfaces and, more recently, a Handbook of Bioelectronics with Cambridge University Press. He also published more than 200  scientific papers and is author of 12 patents. He is now Editor-in-Chief (Associate) of the IEEE Sensors Journal; he is also founder and Editor-in-Chief of the journal BioNanoScience by Springer, and Associate Editor of IEEE Transactions on Biomedical Circuits and Systems. His work received several international recognitions and awards. He has been the General Chairman of the Conference IEEE BioCAS 2014, the premier worldwide international conference in the area of circuits and systems for biomedical applications.

Joseph Chang received his PhD from the Dept. Otolaryngology, Melbourne Uni. He is a professor and was the Associate Dean (Research) at the Nanyang Tech Uni, and is adjunct at Texas A&M Uni. His research includes analog, mixed-signal, digital and aerospace electronics, bioengineering (audiology/acoustics and microfluidics), and printed electronics.  He has received several Best Papers awards and research grants exceeding $13M, including from DARPA, the EU, etc. He has founded two start-ups. He has 30 awarded/pending patents, and has commercialized several innovations. He has served as the Guest Editor of the Proceedings of the IEEE, IEEE JETCAS and the IEEE CAS Magazine, and as senior editor, editor and associate editor of numerous IEEE journals. He has also served as General Chair of several IEEE-NIH and international conferences/workshops, IEEE Distinguished Lecturer, keynote speaker of numerous conferences, and Chair of several IEEE Tech Committees.

Robert Chen-Hao Chang (S'91M'95SM'09) received the B.S. and M.S. degrees from National Taiwan University in 1987 and 1989, respectively, and the Ph.D. degree from the University of Southern California, LA, CA, USA, in 1995, all in electrical engineering. In 1996, he joined the faculty of the Department of Electrical Engineering, National Chung Hsing University, Taiwan, where he is currently a Distinguished Professor. Since Aug. 2014, he has become the Dean of College of Science of Technology, National Chi Nan University, Taiwan. He has published more than 100 technical journal and conference papers. His research interests include  signal processing systems and mixed-signal IC design. He is a Fellow of IET and a member of Tau Beta Pi. He is Distinguished Lecturer by the IEEE CASS for years 2013 and 2014 and an Associate Editor for the IEEE TVLSI. He served as the Chair of IEEE CASS Taipei Chapter from 2011 to 2012 and NG TC Chair from 2015.

Elisabetta Chicca obtained a Laurea degree (M.Sc.) in Physics from the Univerisity of Rome, Italy in 1999, a Ph.D. in Natural Science from the ETH Zurich and a Ph.D. in Neuroscience from the Neuroscience Center Zurich, in 2006. Dr. Chicca was Postdoctoral fellow and Group Leader at the Institute of Neuroinformatics (University of Zurich and ETH Zurich). Since 2011, she is an assistant professor at the Bielefeld University. Her current interests are in the development of VLSI models of cortical circuits for brain-inspired computation, learning in spiking VLSI neural networks, bio-inspired sensing (olfaction, active electrolocation, audition) and motor control. Dr. Chicca is Associate Editor of the TBioCAS and Senior Editor of the IEEE JETCAS.

Masud H Chowdhury is Senior Member of IEEE. He received the B.S. from Bangladesh University of Engineering and Technology in 1999 and the Ph.D. degree from Northwestern University, Evanston, Illinois in 2004. Currently, he is an Associate Professor and PhD Program Chair at University of Missouri – Kansas City. He has published more than 150 articles in his fields of research interest, which include high performance issues of deep submicron and nanoscale ICs, emerging interconnect and device technologies, on-chip voltage regulation, ultra-low-power circuit design, multi-core design issues, and the next generation post-silicon circuits and devices. Dr. Chowdhury is the Chair of IEEE VLSI Systems and Applications TC.

Chia-Chi Chu received the B.S. and M.S. degrees in electrical engineering from National Taiwan University, Taipei, Taiwan, R.O.C. and the Ph.D. from Cornell University, Ithaca, N.Y., U.S.A., in 1996. He was a member of the technical staff with Avant Corporation, Fremont, CA, USA, from 1995 to 1996. From 1996 to 2006, he was with Chang Gung University, Taoyuan, Taiwan. Since 2006, he has been a faculty member of EE with National Tsing Hua University, Hsinchu, Taiwan, and is currently a Professor. His current research interests include power system stability and microgrid control. Dr. Chu was a recipient of the Young Author Award of the IEEE Control of COC in 1997 and the 8th International Conference on PEDS in 2009. Currently, he is the Chair of Power and Energy Circuits and Systems Technical Committee (TC-PECAS), IEEE Circuits and Systems Society (CAS).

Leon Chua is widely known for his invention of the Memristor and the Chua's Circuit. His research has been recognized internationally through numerous major awards, including 16 honorary doctorates from major universities in Europe and Japan, and 7 USA patents. He was elected as Fellow of IEEE in 1974, a foreign member of the European Academy of Sciences (Academia Europea) in 1997, a foreign member of the Hungarian Academy of Sciences in 2007, and an honorary fellow of the Institute of Advanced Study at the Technical University of Munich, Germany in 2012. He was honored with many major prizes, including the Frederick Emmons Award in 1974, the IEEE Neural Networks Pioneer Award in 2000, the first IEEE Gustav Kirchhoff Award in 2005, the International Francqui Chair (Belgium) in 2006, the Guggenheim Fellow award in 2010, Leverhulme Professor Award (United Kingdom) during 2010-2011, and the EU Marie curie Fellow award, 2013.

Pier Paolo Civalleri was born in Turin, Italy, on 17 June 1934. He received the degree in Electrical Engineering from the Polytechnic of Turin in 1959. He was from 1975 to 2008 the Professor of Electrical Engineering for the EE Course of the Polytechnic of Turin, where is now Emeritus Professor. He is a national member of the Accademia delle Scienze di Torino and a IEEE Life Fellow. He has been a member of the New York Academy of Sciences from 1993 to 2002. He was a visiting Professor in Cornell University, Ithaca, NY, in 1977, 1979, 1982, 1983, 1986. His scientific interests have covered analysis and synthesis of lumped and distributed networks. Presently he works on quantum electrodynamics with focusing on quantum circuits. He is the author of over 130 scientific publications and coauthor with H.J. Carlin of the book "Wideband Network Design". He was awarded the IEEE Centennial Medal in 1984.

Fernando Corinto received the Ph.D. and the European Doctorate from the Politecnico di Torino in 2005. He is currently an Associate Professor at the Politecnico di Torino. His research activities focus on nonlinear dynamics in circuits and systems, locally coupled nonlinear/neural networks and memristor nanotechnology. He is co-author of 6 book chapters and more than 125 international journal and conference papers. He is Member of the IEEE NCAS and IEEE CNNAC Technical Committees. He is the vice–Chair of the IEEE CAS North Italy Chapter. He served as TPC in the 13th and the 14th CNNA Workshops and General co–Chair of the 5th Memristor Symposium. He was Associated Editor of the IEEE Trans. on Circuits and Systems I for 2014–2015. He is a Review Editor of the Int. J. of Circuit Theory and Applications. He is the vice–Chair of the COST Action IC1401 MemoCiS.

Paolo Di Barba DSc PhD MEng, was born in the year 1963. Since 2002 he is a full professor of electrical engineering at the Department of Electrical, Computer and Biomedical Engineering, University of Pavia, Italy. His scientific interests include the computer-aided design of electric and magnetic devices, with special emphasis on numerical methods for multiphysics analysis, field synthesis and automated optimal design in electricicty and magnetism. He has authored or co-authored more than 200 conference and journal papers, a book Field Models in Electricity and Magnetism (Springer, 2008) and a monograph Multiobjective Shape 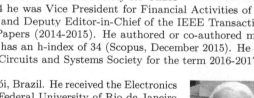 Design in Electricity and Magnetism (Springer, 2010). At the time being, he is the head of the doctoral school in Electronics, Computer Science and Electrical Engineering, University of Pavia. At the same university he chairs the Research Centre for the History of Electrical Technology. He is a visiting professor at the Lodz University of Technology, Poland. He acted as a co-editor of IEEE Transactions on Magnetics - Conferences.

Mario di Bernardo (SMIEEE '06, FIEEE 2012) is Professor of Automatic Control at the University of Naples, Italy and Professor of Nonlinear Systems and Control at the University of Bristol, U.K. On 28th February 2007 he was bestowed the title of "Cavaliere" of the Order of Merit of the Italian Republic for scientific merits from the President of Italy. In January 2012 he was elevated to the grade of Fellow of the IEEE for his contributions to the analysis, control and applications of nonlinear systems and complex networks. He is President of the Italian Society for Chaos and Complexity. From 2011 to 2014 he was Vice President for Financial Activities of the IEEE Circuits and Systems Society and Deputy Editor-in-Chief of the IEEE Transactions on Circuits and Systems: Regular Papers (2014-2015). He authored or co-authored more than 250 international scientific and has an h-index of 34 (Scopus, December 2015). He is a Distinguished Lecturer of the IEEE Circuits and Systems Society for the term 2016-2017.

Paulo S. R. Diniz was born in Niterói, Brazil. He received the Electronics Eng. degree (Cum Laude) from the Federal University of Rio de Janeiro (UFRJ) in 1978, the M.Sc. degree from COPPE/UFRJ in 1981, and the Ph.D. from Concordia University, Montreal, P.Q., Canada, in 1984, all in electrical engineering. Since 1979 he has been with the Department of Electronics and Computer Engineering (the undergraduate dept.) UFRJ.  He has also been with the Program of Electrical Engineering (the graduate studies dept.), COPPE/UFRJ, since 1984, where he is presently a Professor. He held visiting positions at University of Victoria, Canada, Helsinki University of Technology (now Aalto University), Finland, and University of Notre Dame, USA. He is a Fellow of IEEE and EURASIP, and wrote three books. He has also received the 2004 Education and the 2014 Charles Desoer Technical Achievement Award from the IEEE Circuits and Systems Society.

 **Piotr Dudek** received his mgr inż. in electronic engineering from the Technical University of Gdańsk, Poland, in 1997, and the MSc and PhD degrees from the University of Manchester Institute of Science and Technology in 1996 and 2000 respectively. He is currently a Reader in the School of Electrical and Electronic Engineering, The University of Manchester. In 2009 he was a Visiting Associate Professor in the Department of Electronic and Computer Engineering at the Hong Kong University of Science and Technology. His research interests are in the area of vision sensors, cellular processor arrays, novel computer architectures and brain-inspired systems. He is the Chair of the IEEE CASS Sensory Systems Technical Committee and Associate Editor of IEEE TCAS II. He has co-authored best student paper awards at ISCAS 2012, CNNA 2010, CNNA 2008, IJCNN 2007 and best demo awards at ISCAS 2011.

 **Maysam Ghovanloo** received the B.S. degree in electrical engineering from the University of Tehran in 1994, the M.S. degree in biomedical engineering from the Amirkabir University of Technology in 1997, and the M.S. and Ph.D. degrees in electrical engineering from the University of Michigan in 2003 and 2004, respectively. From 1994 to 1998 he worked part-time at the Industrial Development for Electronic Application Inc. From 2004 to 2007, he was an Assistant Professor at NC-State University, Raleigh, NC. He joined Georgia Institute of Technology, Atlanta, GA in 2007, where he is an Associate Professor and the Founding Director of the GT-Bionics Lab. He has authored or coauthored more than 200 peer-reviewed publications. Dr. Ghovanloo is an Associate Editor of the IEEE Transactions on Biomedical Engineering and IEEE Transactions on Biomedical Circuits and Systems. He was the chair of the 2015 IEEE Biomedical Circuits and Systems Conference in Atlanta, GA, and a member of the IEEE CAS Distinguished Lecture Program in 2015-16.

 **Adrian Ioinovici** received the degree in electrical engineering in 1974 and Doctor-Engineer from Polytechnic University, Iasi, Romania, in 1981. He has been with Holon Institute of Technology, Israel since 1982, served as a Dean and presently is a full professor. He was a professor in Hong Kong Polytechnic University (1990-1995) and the Director of the National Center of Power Electronics and Energy in Sun Yat-sen University, Guangzhou (2012-2016) within the China's "One thousand talents" program. He is the author of the books Computer-Aided Analysis of Active Circuits and Power Electronics and Energy Conversion Systems. He published over 170 papers, current recent interests being in switched-capacitor and high dc gain converters. He served as chairman of the TC on Power Systems and Power Electronics of the IEEE CAS Society and as Associate Editor for Power Electronics of the IEEE TCAS-I. He is a Fellow of IEEE.

 **Kenneth Jenkins** joined Penn State University as Professor and Head of Electrical Engineering in 1999. After receiving his B.S from Lehigh University and his M.S. and Ph.D. from Purdue University, Jenkins was a research scientist associate in the Communication Sciences Laboratory at the Lockheed Research Laboratory, Palo Alto, CA. In 1977 he joined the University of Illinois at Urbana-Champaign where he was a faculty member in Electrical and Computer Engineering from until 1999. Jenkins' current research interests include fault tolerant DSP for highly scaled VLSI systems, adaptive signal processing, multidimensional array processing, computer imaging, and bio-inspired optimization algorithms for intelligent signal processing. He received the IEEE CAS Society Distinguished Service Award, the 2000 IEEE Millennium Award, the 2000 IEEE CAS Society Golden Jubilee Medal, and the 2000 George Montfiore Foundation International Award. He is a Life Fellow of IEEE.

Marian K. Kazimierczuk received the M.S., Ph.D., and D.Sc. degrees in electronics engineering from the Warsaw University of Technology, Poland, in 1971, 1978, and 1984, respectively. He was with the Warsaw University of Technology from 1978 to 1984 with the Institute of Radio Electronics, Department of Electronics and Information Technology. In 1984, he was a Project Engineer with Design Automation, Inc., Lexington, MA. From 1984 to 1985, he was a Visiting Professor with the Department of Electrical and Computer Engineering, Virginia Polytechnic Institute and State University, Blacksburg. Since 1985, he has been with the  Wright State University (WSU), Dayton, OH, where he is currently a Distinguished Professor. His research areas include power electronics, RF transmitters, high-frequency magnetic components, applied controls, wireless power transfer, and renewable energy systems. He has written more than 400 papers, seven books, and hold seven patents. He is a Fellow of IEEE.

Michael Peter Kennedy received the Ph.D. from UC Berkeley in 1991 and the D.Eng. from the National University of Ireland in 2010. He holds the Chair in Microelectronic Engineering at University College Cork and is Scientific Director of the MCC Ireland. His research interests are in the simulation, analysis and design of nonlinear dynamical systems for applications in communications and signal processing. He was made an IEEE Fellow in 1998. He has over 350 publications and has taught courses on nonlinear circuits and systems in America, Asia, and Europe. He received the 1991 Best Paper Award from the International Journal of Circuit Theory and Applications. He served five terms as Associate and Guest Editor of the IEEE Transactions on CAS. He was awarded the IEEE Third Millennium Medal and the IEEE CAS Society Golden Jubilee Medal in 2000. He was Vice-President for Region 8 of the IEEE CAS Society 2005–07 and a CAS Society Distinguished Lecturer 2012–13.

Lucio Lanza is the Managing Director of Lanza techVentures, an early stage venture capital and investment firm, and the 2014 recipient of the Phil Kaufman Award for Distinguished Contributions to EDA. Since 2008, he has been a general partner of Radnorwood Capital, LLC, an investor in public technology companies. Previously, Dr. Lanza was the chairman of Artisan Components and became a non-executive director of ARM, the world's leading semiconductor IP company. He serves as chairman of the board of PDF Solutions, Inc., a provider of technologies to improve semiconductor manufacturing yields, and is on the board of directors of several private companies providing financial and strategic support. Dr. Lanza joined the venture capital industry in 1990 after executive positions at Olivetti, Intel, Daisy Systems. From 1990-1995, he held an executive position at Cadence Design Systems, while simultaneously serving as a venture capitalist. He holds a doctorate in EE from Politecnico in Milan, Italy.

Gwo Giun (Chris) Lee (S'91- M'97-SM'07) received his M.S. and Ph.D. degrees in Electrical Engineering from University of Massachusetts. Dr. Lee has held several technical and managerial positions in the industry including System Architect in former Philips Semiconductors, USA; DSP Architect in Micrel Semiconductors, USA; and Director of Quanta Research Institute, Taiwan before joining the faculty team of the Department of Electrical Engineering in National Cheng Kung University (NCKU) in Tainan, Taiwan. Dr. Lee has authored more than 100 technical documents. Dr. Lee also serves as the Associate Editor for IEEE Transactions on Circuits and Systems for Video Technology (TCSVT) and Journal of Signal Processing Systems. He received the Best Paper Award for the BioCAS track in IEEE ISCAS 2012. His research interests include Algorithm/ Architecture Co-design for System-on-a-Chip, Visual Signal Processing and Communication, Internet of Things, Biomedical Image Processing, Bioinformatics.

Zicheng Liu received his Ph.D. in computer science from Princeton University in 1996. He is currently a principal researcher at Microsoft Research, Redmond, Washington. Before joining Microsoft Research, he worked at Silicon Graphics Inc. as a member of technical staff. His current research interests include human activity recognition, 3D face modeling and animation, and multimedia signal processing. He was a Technical Program Co-Chair of both 2010 and 2014 IEEE International conference on Multimedia and Expo. He is a general co-chair of 2012 IEEE Visual Communication and Image Processing. He is the chair of the CASS Multimedia TC. He is a member of the IEEE Transactions on Multimedia Steering Committee. He is a Co-Editor-in-Chief of Journal of Visual Communications and Image Representation, and an associate editor of Machine Vision and Applications. He is an affiliate professor in the department of Electrical Engineering, University of Washington. He is an IEEE distinguished lecturer from 2015-2016. He is a fellow of IEEE.

Rui Martins born April 30, 1957, received the Bachelor, Masters and Ph.D. plus Habilitation for Full-Professor in Electrical and Computer Engineering (ECE) from Instituto Superior Técnico (IST), Universidade de Lisboa, Portugal, in 1980, 1985, 1992 and 2001. He has been with Department of ECE/IST since October 1980. Since 1992 is on leave from IST and is with Department of ECE, Faculty of Science and Technology, University of Macau, Macao, China, where he is Vice-Rector since 1997 and Chair-Professor since August 2013. Co-authored more than 350 papers, 6 books and holds 18 Patents. He is Founding Director of the State Key Laboratory of Analog and Mixed-Signal VLSI (Macao, China) since 2011. He is an IEEE Fellow and was VP of IEEE CASS for Region 10 (2009-2011) and for (World) Regional Activities and Membership (2012-2013). He holds 2 government decorations and is a Corresponding Member of the Portuguese Academy of Sciences (in Lisbon).

Wolfgang Mathis was born in Celle, Germany, on May 13, 1950. He received the Dipl.-Phys. degree in physics (1980), the Dr.-Ing. degree in electrical engineering (1984), and the Habilitation degree (1988). He became assistant professor (1985-1990), a professor at the Universities of Wuppertal (1990-1996) and Magdeburg (1996-2000). Since 2000 he is professor at the Leibniz University of Hanover. His research interests include nonlinear deterministic and noisy circuits and dynamical systems, computer-aided circuit analysis and design, nanoelectronics, quantum computing, analysis and numerics of electromagnetic fields, and history of electrical engineering and physics. He published 3 books and more than 400 papers. He is a member of IEEE, VDE/ITG, DPG and became IEEE Fellow in 1999. He was associate editor of the IEEE Trans. CAS (1995-1997) and is chair of the IEEE CAS, German chapter. Since 2001 he is member of the NRW Academy of Sciences and since 2008 member of the National Academy of Science and Engineering.

Sanjit K. Mitra is a Research Professor in the Department of Electrical & Computer Engineering, University of California, Santa Barbara and Professor Emeritus, Ming Hsieh Department of Electrical Engineering, University of Southern California, Los Angeles. Dr. Mitra has published over 700 papers in the areas of analog and digital signal processing, and image and video processing. He has also authored and co-authored twelve books, and holds five patents. He has presented 29 keynote and/or plenary lectures at conferences. Dr. Mitra has served IEEE in various capacities including service as the President of the IEEE CAS Society in 1986. He is a member of the U.S. National Academy of Engineering, a member of the Norwegian Academy of Technological Sciences, an Academician of the Academy of Finland, a foreign member of the Finnish Academy of Sciences and Arts, a foreign member of the Croatian Academy of Sciences and Arts, international member of the Croatian Academy of Engineering and the Academy of Engineering, Mexico, and a Foreign Fellow of the National Academy of Sciences, India and the Indian National Academy of Engineering. Dr. Mitra is a Life Fellow of the IEEE.

Josef Nossek received the Dipl.-Ing. and the Dr. techn. degrees in electrical engineering from the University of Technology in Vienna, Austria in 1974 and 1980, respectively. In 1974 he joined Siemens AG in Munich, Germany, in 1978 he became supervisor, and from 1980 on he was Head of Department. In 1987 he was promoted to be Head of all radio systems design. Since 1989 he has been Full Professor at the Technical University of Munich. He was President of the IEEE Circuits and Systems Society in 2001- 2003 respectively. He was President of VDE (Germany) 2007 and 2008. His awards include the ITG Best Paper Award 1988, the Vodafone Innovations Award 1998, the Award for Excellence in Teaching from the Bavarian Ministry for Science, Research and Art in 1998. From the IEEE CASS he received the Golden Jubilee Medal in 1999 and the Education Award in 2008. In 2008 he also received the Order of Merit of the Federal Republic of Germany and in 2009 he has been elected member of the German National Academy of Engineering Sciences (acatech). In 2013 he received a Honorary Doctorate and in 2014 the Ring of Honor from VDE. He is a Life Fellow of IEEE.

Tokunbo Ogunfunmi received the B.Sc. degree from the University of Ife, Nigeria, the M.S. and Ph.D. degrees from Stanford University. He is currently with Santa Clara University (SCU). From 2010-2014, he served as Associate Dean for Research and Faculty Development at SCU. He also was visiting professor at the University of Texas and at Stanford University. His current research interests include adaptive/nonlinear/digital signal processing, communications, multimedia (speech, video) and VLSI/DSP/FPGA implementations. He has authored 4 books, 8 book chapters, 160+ journal and conference papers in these and related areas. He served on the editorial board of IEEE TCAS-I and currently serves on the editorial boards of IEEE Signal Processing Letters, IEEE TCAS-II, IET Electronics Letters and Circuits, Systems and Signal Processing. He served as an IEEE CASS DLP (2011-2013) and currently chairs the IEEE CASS CASCOM TC.

Gaetano Palumbo joined the University of Catania in 1994; since 2000 he is full professor in the same university. His primary research interests are in analog and digital circuits. He was the co-author of four books by Kluwer and Springer, in 1999, 2001, 2005, 2014 respectively, and a textbook on electronic device in 2005. He is the author of more than 400 scientific papers on referred inter. journals (170+) and in conferences. Moreover he is co-author of several patents. He served as an Associated Editor of the IEEE Trans. on CAS part I in 1999-2001, 2004-2005 and 2008-2011, and of the IEEE Trans. on CAS part II in 2006-2007. In 2011-2013 he served as a member of the BOG of the IEEE CAS Society. In 2005 he was one of the 12 panelists in the Committee for Evaluation of the Italian Research and in 2015 he has been a panelist for the Evaluation of Italian Research Quality in 2011-2014 in the EE area. In 2003 he received the Darlington Award. Prof. Palumbo is an IEEE Fellow.

Dookun Park Dookun Park received the B.S. degree in Electrical Engineering from Seoul National University in 2007, and the M.S. and Ph.D. in Electrical Engineering from Stanford University in 2009 and 2015, respectively. His Ph.D. Thesis advisor is Bernard Widrow, and his research interests include reinforcement learning, adaptive control systems, machine learning, and signal processing. From 2011 to 2012, he was a Research Associate at Hewlett Packard Labs where he developed signal processing algorithms for HP ConnectBoard and HP SmartBed. From 2013 to 2014, he was a Teaching Fellow at Stanford University and taught a course of Digital Signal Processing. He is currently a Staff Data Scientist at Tapjoy inc. where he develops personalized AD service methodologies. He was a recipient of Samsung Scholarship and Kwanjeong Educational Foundation Scholarship, in 2007 and 2009, respectively. He lives in Palo Alto, CA, with his wife, Kiwon, and children, Katherine, Kristin, and Lucas.

Ron Rohrer is University Professor Emeritus of ECE at Carnegie Mellon. A noted innovator, entrepreneur and professor, Ron is recognized as an early developer of circuit simulation. He is credited with driving Electronic Design Automation (EDA) tools into broad industry use. In 1968, in a graduate course he taught at UC, Berkeley, he oversaw the production of an IC simulation program similar to the FairCirc program, a forerunner to SPICE. In 1971, Ron developed the foundation of the industry standard technique for simulation of analog and Radio Frequency (RF) IC noise. The founding editor of the IEEE Transactions on Computer-Aided Design of Integrated CAS, Ron also served as president of the CAS Society. He is a Fellow of the IEEE and a member of the National Academy of Engineering. He received the IEEE Education Medal and Kirchhoff Award, the NEC Computer and Communication Prize, the EDAC Kaufman Award, the IEEE CAS Van Valkenburg and Belevitch Awards and the ASEE Terman Award. Ron received the B.S. degree from MIT (1960) and the M.S. (1961) and Ph.D. (1963) degrees from UC, Berkeley.

Riccardo Rovatti received the M.S. degree in Electronic Engineering and the Ph.D. degree in Electronics, Computer Science, and Telecommunications both from the University of Bologna, Italy in 1992 and 1996, respectively. He is now a Full Professor at the University of Bologna. He is the author of approximately 300 technical contributions, and of two volumes. His research focuses on mathematical and applicative aspects of statistical signal processing and on the application of statistics to nonlinear dynamical systems. He received the 2004 IEEE CAS Society Darlington Award, the 2013 IEEE CAS Society Guillemin-Cauer Award, as well as the best paper award at ECCTD 2005, and the best student paper award at EMC Zurich 2005 and ISCAS 2011. He was elected IEEE Fellow in 2012 for contributions to nonlinear and statistical signal processing applied to electronic systems.

Alberto Sangiovanni–Vincentelli holds the Buttner Chair of Electrical Engineering and Computer Sciences, UCBerkeley. He helped founding Cadence and Synopsys, the two leading companies in EDA. He consulted for companies such as Intel, HP, Bell Labs, IBM, Samsung, UTC, Kawasaki Steel, Fujitsu, Telecom Italia, Pirelli, GM, BMW, Mercedes, Magneti Marelli, ST Microelectronics, ELT and UniCredit. He earned the IEEE/RSE Maxwell Award for "groundbreaking contributions that have had an exceptional impact on the development of electronics and electrical engineering", the Kaufmann Award for seminal contributions to EDA, the EDAA lifetime Achievement Award, the IEEE/ACM R. Newton Impact Award, the University of California Distinguished Teaching Award, and the IEEE Graduate Teaching Award for inspirational teaching of graduate students. He is an ACM fellow, a member of the National Academy of Engineering and holds two honorary Doctorates. He authored over 850 papers, 17 books and 2 patents.

Mohamad Sawan joined Polytechnique Montreal in 1991, where he is currently a Professor of microelectronics and biomedical engineering. He was Canada Research Chair from 2001 to 2015. He is leading the Microsystems Strategic Alliance of Quebec (ReSMiQ), founder of the Polystim Neurotech Laboratory, and cofounder and now Editor-in-Chief of the IEEE Transactions on Biomedical circuits and systems. He is member of the Board of Governors of IEEE CAS Society. He published more than 700 peer-reviewed journal and conference papers, two books, and awarded 15 patents. He received the Shanghai International Collaboration Award, the Queen Elizabeth II Golden Jubilee Medal, the Bombardier Award, the Jacques-Rousseau Award, the medal of merit from the President of Lebanon, and the Barbara Turnbull Award. He is Fellow of the IEEE, Fellow of the Canadian Academy of Engineering, Fellow of the Engineering Institute of Canada, and Officer of the Quebec's National Order.

Gianluca Setti received a Ph.D. degree from the University of Bologna in 1997. He is currently a Professor at the University of Ferrara, Italy. His research interests include implementation and application of chaotic circuits and systems, EMC, statistical signal processing and biomedical circuits and sytems. Dr. Setti received the 2013 IEEE CAS Society Meritorious Service Award, the 2004 IEEE CAS Society Darlington Award, the 2013 IEEE CAS Society Guillemin-Cauer Award, as well as of the best paper award at ECCTD2005, and the best student paper award at
EMCZurich2005 and at ISCAS2011. He was the EIC for the IEEE TCAS-II (2006-2007) and of the IEEE TCAS-I (2008-2009). He was also the TP Co-Chair of ISCAS2007, ISCAS2008, ICECS2012, BioCAS2013 as well as the General Co-Chair of NOLTA2006. He was member of CASS Board of Governors (2005-2008), and he served as the 2010 CASS President. He held several other volunteer positions for the IEEE and in 2013-2014 he was the first non North-American Vice President of the IEEE for Publication Services and Products.

Sameer R. Sonkusale (M'02, SM'12) is a Professor of Electrical & Computer Engineering and Biomedical Engineering (adjunct) at Tufts University, where he leads Nano Lab. He received his MS and PhD in Electrical Engineering from University of Pennsylvania in 1999 and 2003 respectively. He was an Assistant Professor at Texas A&M University, from 2002 to 2004 and at Tufts University from 2004 to 2010. Prof. Sonkusale's teaching and research interests are in the area of biomedical circuits and systems, miniaturized sensors and instrumentation, lab-on-chip devices and biomedical diagnostics. Prof. Sonkusale received the NSF CAREER
award in 2010 and has won best paper awards at FTM 2009, IEEE NANO 2008, IEEE SENSORS 2008, IEEE BIOCAS 2014 and ISDRS 2009. He is also on the editorial board of Scientific Reports (Nature Publishing Group). For 2014-16, he serves as the chair of the IEEE Biomedical and Lifesciences Circuits and Systems Technical Committee (BIOCAS TC).

Gabor C. Temes (SM'66F'73LF'98) received the undergraduate degrees from the Technical University of Budapest and Eötvös University, Budapest, Hungary, in 1952 and 1955, respectively. He received the Ph.D. degree in electrical engineering from the University of Ottawa, in 1961, and an honorary doctorate from the Technical University of Budapest in 1991. He held academic positions in Budapest, Stanford and UCLA. He worked in industry at Northern Electric R&D Laboratories and Ampex
Corp. He is now a Professor at Oregon State University. He coedited and coauthored many books. He wrote about 600 papers in journals and conference proceedings. Dr. Temes was Editor of the IEEE T. on Circuit Theory and Vice President of the IEEE CASS. He received the IEEE CAS Darlington Award, the IEEE Centennial Medal, the Andrew Chi Prize Award, the CAS Education Award, the CAS Technical Achievement Award, the IEEE Graduate Teaching Award, the IEEE Millennium Medal, the IEEE CAS Golden Jubilee Medal, the IEEE Kirchhoff Award, and the 2009 IEEE CAS Mac Valkenburg Award. He is member of the National Academy of Engineering.

Ljiljana Trajkovic Ljiljana Trajkovic received Ph.D. degree in electrical engineering from University of California at Los Angeles in 1986. She is currently a Professor in the School of Engineering at Simon Fraser University, British Columbia, Canada. From 1995 to 1997, she was a Visiting Professor in the EE and Computer Sciences Department, University of California, Berkeley. She was a Research Scientist at Bell Communications Research, NJ, from 1990 to 1997, and a Member of the Technical Staff at AT&T Bell Labs, Murray Hill, NJ, from 1988 to 1990. Her research
interests include communication networks and theory of nonlinear circuits and dynamical systems. She received the 2012 IEEE CAS Society Meritorious Service Award. She served as 2014-2015 President of the IEEE Systems, Man, and Cybernetics Society and 2007 President of the IEEE CAS Society. She is a Life Fellow of the IEEE.

Timothy Trick joined the faculty of the University of Illinois at Urbana-Champaign in 1965, where his is currently Professor Emeritus of Electrical and Computer Engineering. He served as Interim Dean of the College of Engineering (2001), Head of the ECE Department (1985-1995), and Director of the Coordinated Science Laboratory (1984-1986). He published in the areas of computational methods for integrated circuit analysis and design, analog and digital signal processing. He was elected to serve on the IEEE Board of Directors (1986-89), as IEEE Vice-President for Publications (1988-89), and as President of the IEEE CAS Society (1979). He received the Guillemin-Cauer Award, IEEE Centennial Medal, IEEE CAS Society Meritorious Service Award, IEEE CAS Society Van Valkenburg Award, IEEE Millennium Medal, IEEE CAS Society Golden Jubilee Medal, the Outstanding EE Alumni Award from Purdue University, and the Semiconductor Industry Association University Research Award (2002). Dr. Trick is a Life Fellow of the Institute of Electrical and Electronics Engineers (IEEE), and a Fellow of the American Association for the Advancement of Science (AAAS).

Joos Vandewalle received the electrical engineering degree and doctorate in applied sciences from KU Leuven, Belgium in 1971 and 1976. Until October 2013 he was a full professor at the Department Electrical Engineering (ESAT) at KU Leuven and head of a division at ESAT with more than 150 researchers. He held visiting positions at UC Berkeley and I3S CNRS Sophia Antipolis. He taught courses in linear algebra, linear and nonlinear system and circuit theory, signal processing and neural networks. His research interests are in mathematical system theory and its applications in circuit theory, control, signal processing, cryptography and neural networks. He supervised 43 PhD students, and (co-)authored more than 220 international journal papers and 5 research monographs that received over 34 000 google scholar citations. He is Life Fellow of IEEE, and Fellow of IET, and EURASIP. He is a member of the Academia Europaea and of the Belgian Academy of Sciences where he is currently chairman of the class of Technical sciences. He held several responsibilities within the IEEE CASS.

Andrei Vladimirescu received the M.S. and Ph.D. degrees in electrical engineering and computer sciences from the University of California, Berkeley, in 1980 and 1982, respectively. He received his Diploma Engineering degree in electronics from the Polytechnic Institute in Bucharest, Romania, in 1971. In 1977 he joined the IC/CAD research group of the EECS department at UC Berkeley, where he was a key contributor to the SPICE program. From 1983 to 1997 Dr. Vladimirescu was Director of Engineering at EDA companies such as Daisy Systems and Cadence. Currently he is Visiting Professor at UC Berkeley and at the Institute of Electronics of Paris, ISEP, as well as a consultant to industry in the area of Electronic Design Automation. He published "The SPICE Book" by J. Wiley and Sons in 1994; he also authored and co-authored over 100 journal and conference papers. He is a member of the IEEE CASS BoG active in organization of conference such as ISCAS, ICECS and ESSCIRC.

Bernard Widrow received the S.B., S.M., and Sc.D. degrees in Electrical Engineering from the Massachusetts Institute of Technology in 1951, 1953, and 1956, respectively. He joined the MIT faculty and taught there from 1956 to 1959. In 1959, he joined the faculty of Stanford University, where he is currently Professor Emeritus of Electrical Engineering. He began research on adaptive filters, learning processes, and artificial neural models in 1957. Dr. Widrow is a Life Fellow of the IEEE and a Fellow of AAAS. He received the IEEE Centennial Medal in 1984, the IEEE Alexander Graham Bell Medal in 1986, the IEEE Signal Processing Society Medal in 1986, the IEEE Neural Networks Pioneer Medal in 1991, the IEEE Millennium Medal in 2000, and the Benjamin Franklin Medal for Engineering from the Franklin Institute of Philadelphia in 2001. He was inducted into the National Academy of Engineering in 1995 and into the Silicon Valley Engineering Council Hall of Fame in 1999.

Alan Willson has been a Professor at UCLA since 1973. He earned his B.E.E. from Georgia Tech (1961), M.S. and Ph.D. from Syracuse Univ. (1965 and 1967). From 1961 to 1964, he was with IBM, and was a researcher at Bell Labs from 1967 to 1973. He has also served UCLA as Associate Dean of the UCLA School of Engineering (1987 to 2001). His research areas have included circuit theory and signal processing. He is the editor of Nonlinear Networks:Theory and Analysis (IEEE Press, 1974).
He is a member of the National Academy of Engineering, Eta Kappa Nu, Sigma Xi and Tau Beta Pi. From 1977 to 1979 he was Editor of the IEEE Transactions on Circuits and Systems, and was President of the CAS Society during 1984. He received the 1978 and 1994 Guillemin-Cauer Awards, the 2003 Mac Van Valkenburg Award, and the 2013 Vitold Belevitch Award (all being CAS Society awards). He was the 2010 recipient of the IEEE Leon K. Kirchmayer Graduate Teaching Award.

Mustak E. Yalcin he obtained in 1993 the degree in Electronics and Telecommunications Engineering from Istanbul Technical University (I.T.U.). In 1997, he received the MSc in Electronics and Communications Engineering from I.T.U. Institute of Science and Technology. In 2004, he recieved the Ph.D. degree in Applied Sciences from KU Leuven, Belgium. Between June 2004-December 2004, he was a postdoctoral fellow at KU Leuven, Department of Electrical Engineering (ESAT)/SCD-SISTA. He was also a Visiting Research Fellow at The Institute for Nonlinear Science, University of California San Diego (UCSD) in 2000. He is currently a
Professor with Istanbul Technical University, Turkey. His research interests are mainly in the areas of the theory and application of nonlinear circuit and systems. He is co-author of the book "Cellular Neural Networks, Multi Scroll Chaos and Synchronization" (World Scientific).

Wei-Ping Zhu (SM'97) received the B.E. and M.E. degrees from Nanjing University of Posts and Telecommunications, and the Ph.D. degree from Southeast University, Nanjing, China, in 1982, 1985, and 1991, respectively. He was an Associate Professor with Nanjing University of Posts and Telecommunications. From 1998 to 2001, he worked with hi-tech companies in Ottawa, Canada, including Nortel Networks and SR Telecom Inc. Since July 2001, he has been with Concordia's ECE Department as a full-time faculty member, where he is presently a Full Professor. Dr. Zhu was an Associate Editor of the IEEE TCAS-I from 2001 to 2003, and an Associate Editor of Circuits, Systems and Signal Processing from 2006 to 2009. Currently, he serves as an Associate Editor of Journal of The Franklin Institute. Dr. Zhu was the Secretary of Digital Signal Processing Technical Committee (DSPTC) of the IEEE CAS Society (2012-2014), and is presently the Chair of the DSPTC (2014-2016).

Domenico Zito received the Laurea and PhD degree from the University of Pisa, Italy. Since 2009 he is with the University College Cork and Tyndall National Institute, Cork, Ireland, where he established the RF SoC group and the Marconi Lab, awarded as Start-up Laboratory of the Year at the Irish Laboratory Awards 2014. His primary research interests are the design of radiofrequency system-on-chip transceivers for data communications and contactless sensing. He has published about 150 papers in international journals and conference proceedings. He had the responsibility of integrated circuits and systems in industry and academia and established pioneering contributions for the integration of radiofrequency data transceivers and contactless sensors such as radars and radiometers. In 2005 he received the Mario Boella (VP URSI) Prize for the Innovation in Wireless Technology. He is a TPC member of the ESSCIRC and associate editor of the IEEE TCAS-I. He is also TPC co-Chair of the IEEE ICECS 2016.

Author Index

CPSIA information can be obtained
at www.ICGtesting.com
Printed in the USA
LVOW02*1037031216
515615LV00002B/3/P